ADHESION MEASUREMENT OF THIN FILMS, THICK FILMS, AND BULK COATINGS

A symposium
presented at
ASTM Headquarters
AMERICAN SOCIETY FOR
TESTING AND MATERIALS
Philadelphia, Pa., 2–4 Nov. 1976

ASTM SPECIAL TECHNICAL PUBLICATION 640

K. L. Mittal
IBM Corporation
East Fishkill Facility
Hopewell Junction, N.Y.
editor

List price $39.25
04-640000-25

AMERICAN SOCIETY FOR TESTING AND MATERIALS
1916 Race Street, Philadelphia, Pa. 19103

Library of Congress Catalog Card Number: 77-84460

NOTE

The Society is not responsible, as a body,
for the statements and opinions
advanced in this publication.

Foreword

The symposium on Adhesion Measurement of Thin Films, Thick Films, and Bulk Coatings was held at the headquarters of the American Society for Testing and Materials, Philadelphia, Pa., 2–4 Nov. 1976. The ASTM Publications Committee sponsored the symposium. K. L. Mittal, IBM Corporation, presided as symposium chairman and editor of this publication.

Related ASTM Publications

Spreading Resistance, STP 572 (1975), $3.55, 04-572000-46

Surface Analysis Techniques for Metallurgical Applications, STP 596 (1976), $15.00, 04-596000-28

Composite Materials: Testing and Design (Fourth Conference), STP 617 (1974), $51.75, 04-617000-33

A Note of Appreciation to Reviewers

This publication is made possible by the authors and, also, the unheralded efforts of the reviewers. This body of technical experts whose dedication, sacrifice of time and effort, and collective wisdom in reviewing the papers must be acknowledged. The quality level of ASTM publications is a direct function of their respected opinions. On behalf of ASTM we acknowledge with appreciation their contribution.

ASTM Committee on Publications

Editorial Staff

Contents

Adhesion Measurement of Deposits and Coatings

Summary

Introduction

This volume chronicles the proceedings of the Symposium on Adhesion Measurement of Thin Films, Thick Films, and Bulk Coatings, held under the auspices of the American Society for Testing and Materials, 2–4, Nov. 1976, in Philadelphia, Pa. Thin films (<1 μm), thick films (>1 μm), and bulk coatings (>25 μm) are used for a variety of purposes in electronic, engineering, optical, biomedical, nuclear, space, and other applications. Whatever their intended use may be, the properties, structure, functional characteristics, and performance all depend, *inter alia*, on adhesion between the coating and the substrate. So the need for reliable and quantitative techniques for measuring cohesion is quite patent. Furthermore, the quantitative determination of adhesion is important to discriminate between the acceptable and nonacceptable parts or products, to optimize process variables, for example, during film deposition, and to settle disputes between the vendor and the buyer regarding adhesion-related performance of products.

In the past no common forum had been provided to discuss the many and varied so-called adhesion measurement techniques comprehensively. However, a considerable amount of effort had been and was being devoted to the development of new or the improvement of existing methods of measuring adhesion. So I approached ASTM with the proposal that a symposium on adhesion measurement was very timely and highly needed, and the response to my proposal was both warm and affirmative.

The symposium was organized to review and assess current measurement techniques, to provide a forum for the exchange of ideas, to define problem areas which needed intensified efforts, and to galvanize increased interest in developing better and more versatile techniques. The leitmotif of the symposium was Adhesion Measurement Techniques: Their Potentialities and Limitations. So the authors were prevailed upon to discuss clearly the merits and limitations of the techniques they used, and also to focus on the factors which interfere with the reliability and reproducibility of the technique; and it is a pleasure to acknowledge their cooperation in this regard.

The technical program contained 33 papers by 49 authors from seven countries, but, unfortunately, three papers were not presented. The program attracted about 80 people from Canada, Denmark, France, Germany, Japan, Switzerland, and the United States. The papers dealt with a host of adhesion measurement techniques for a variety of adherend-adherate combinations. The authors represented a broad spectrum of professional affiliations, backgrounds, and interests. The technical program consisted of both invited and contributed papers covering reviews and original research contributions.

This volume contains a total of 25 papers, divided into four sections, as eight papers are not included for a variety of reasons. The first section comprises seven papers dealing with such questions as are germane to adhesion measurement. The topics addressed include: What is meant by adhesion? What are the mechanisms of adhesion between different adherend-adherate combinations? What exactly is measured when one attempts to measure adhesion? How does the measured adhesion relate to fundamental or basic adhesion? What is the nature and the importance of locus of failure? and, What are the experimental difficulties and interpretational complexities in adhesion measurement? The second section embodies papers dealing with adhesion of thin films; adhesion of thick films constitutes the third section. The final section contains papers concerning adhesion measurement of deposits and coatings. The salient aspects or findings of all these papers are embodied in the Summary at the end of the volume. Each paper was followed by a discussion; the queries raised by the discussers and author(s)' closure are appended at the end of the respective papers.

The symposium banquet was held in the evening of November 3rd, and it was a distinct pleasure to have Dr. Harvey Alter as the banquet speaker. He spoke on "The Material Science of Waste." The technical program was followed by a panel discussion (the panelists were W. D. Bascom, J. J. Bikerman, D. M. Mattox, K. L. Mittal, and L. H. Sharpe) in which some of the relevant topics were discussed.

It is hoped that both the uninitiated and the veteran in the area of adhesion measurement will find this volume a ready source of information and guidance.

K. L. Mittal
East Fishkill Facility, IBM Corporation, Hopewell Junction, N.Y. 12533; symposium chairman and editor.

General Papers Related to Adhesion Measurement

K. L. Mittal[1]

Adhesion Measurement: Recent Progress, Unsolved Problems, and Prospects

REFERENCE: Mittal, K. L., **"Adhesion Measurement: Recent Progress, Unsolved Problems, and Prospects,"** *Adhesion Measurement of Thin Films, Thick Films, and Bulk Coatings, ASTM STP 640*, K. L. Mittal, Ed., American Society for Testing and Materials, 1978, pp. 5–17.

ABSTRACT: In this paper, the term "basic adhesion" is used to signify the summation of all interfacial intermolecular interactions, whereas the term "practical adhesion" is used to represent the forces or the work required for the disruption—either at the interface or in the interfacial region—of the adhering system. There are many factors which influence practical adhesion, and these are discussed in detail. The relationship between practical adhesion and basic adhesion is discussed. Basic adhesion cannot be determined by the commonly used techniques for measuring practical adhesion. The virtues of an ideal practical adhesion test are outlined. Some of the techniques for adhesion measurement of thin films, thick films, and coatings are discussed briefly, and references are made to the recent reviews on this topic.

For a given adhering system, different techniques yield different practical adhesion values. The difficulties involved in adhesion measurement are highlighted, and the unresolved problems are brought into sharper focus.

The failure surfaces should be examined using surface analytical tools to ascertain the locus of failure, which is important in understanding the mechanism of failure.

KEY WORDS: basic adhesion, practical adhesion, adhesion measurement, thin films, thick films, electrodeposits, coatings

Thin films (<1 μm), thick films (>1 μm), and bulk coatings (>25 μm) are used for a variety of purposes in diversified applications in many technologies. Whatever their intended purpose—functional, decorative, or protective—may be, the properties, structure, functional characteristics, and performance all depend, *inter alia,* on adhesion between the adherate and adherend. Adherend is a general term used for solid substrate to which other materials adhere; and Mittal [*1*][2] has suggested recently the

[1] Staff engineer, East Fishkill Facility, International Business Machines Corporation. Hopewell Junction, N. Y. 12533.

[2] The italic numbers in brackets refer to the list of references appended to this paper.

term adherate to represent the materiaı which adheres to an adherend. Obviously, in an adhering system of A and B, A adheres to B, and B adheres to A, but the thinner phase is called the adherate. Examples of adherates are thin films, thick films, paints, and coatings. Adhesive is a special kind of adherate in that it adheres to two adherends instead of one.

In this publication, we are concerned mainly with adherates which adhere to only one adherend. It is very important to differentiate between an adherate-adherend combination—called adhering system—and an adhesive joint (expressed in short form as adhint [*2*], because ideally in the former there is only one interface and two bulk phases, whereas in the latter there are two interfaces and three bulk phases. It should be pointed up, however, that in many real situations, there are present additional interfaces and interphases, as discussed later. The failure behavior of adhesive joints is more complicated than that of adhering systems and will not be discussed in the present paper.

The study of adhesion of adherates in the form of thin films, thick films, and coatings is vital, and some examples where adhesion plays an important role are given in the following paragraphs. The term "adhesion" simply means sticking together of two similar or dissimilar materials.

1. Thin films (usually <1 μm, and in some applications of the order of 50 nm) are so fragile that they must be supported by more substantial substrate, and the degree to which the film can share the strength of the substrate depends upon the adhesion between the two.

2. Durability, longevity, and wear of adherates depend intimately on the degree of adhesion between the adherate and adherend.

3. Whenever adherates (paints, coatings) are used as protective overcoats for environmental protection, adhesion is very important. If the adhesion is poor, the extent of deterioration of the substrates by environmental factors (humidity, corrosive gases, etc.) is greatly accelerated.

4. Polymeric substrates are metallized to reduce their permeability to gases, and one can very well appreciate the importance of adhesion between the metal and the polymer.

5. In the case of thin-films deposition (for example by evaporation), adhesion plays an important role in governing the kinetics of the growth and structure of the films, which in turn determine their functional performance. Whenever metal coatings are deposited on base metals, some of the physical properties (grain structure, integrity, uniformity, etc.) are dictated by adhesion to the base metals.

6. In the case of multilayer structures, adhesion between the individual layers is crucial.

7. The adhesion of thick-film conductors (frit bonded and reactively bonded) is important in microcircuits fabrication.

8. Artificial teeth are mostly made of acrylate, which often causes allergic changes in the mucous membranes of the mouth and hence is inconvenient

to use [*3*]. Thin films of dental gold are used as coating to decrease this unpleasant effect, and the adhesion of the coating to the tooth surface is very important.

In addition to what has just been said, basic adhesion is important in surface chemistry and physics as it depends directly on the interatomic and intermolecular forces.

There are essentially two aspects of an adhesion study program: understanding of the factors affecting adhesion and thereby improvement of adhesion strength, and the measurement of adhesion strength. The first aspect has been the subject of recent symposia [*4–6*] and in the present publication we are concerned primarily with the methodology of the measurement of adhesion in adhering systems.

The measurement of adhesion is important for a variety of reasons: (*a*) to assess the efficacy of changes in the process variables or of substrate preparations, and thereby to optimize the conditions which yield the required adhesion strength; (*b*) to discriminate parts or products which have poor adhesion strength from those which are acceptable; and (*c*) to gain fundamental insight into the mechanism of adhesion.

So the need for reliable, reproducible, and quantitative methods for measuring adhesion is quite manifest.

Can Adhesion Be Measured?

"Can adhesion be measured?" is one of the perennial questions in adhesion science and technology, and if one scans the literature, one will encounter discordant opinions on this issue. But the simple answer to the question is yes and no, depending upon what is meant by adhesion. The meaning and ramifications of the term adhesion are discussed first, followed by comments on whether adhesion can be measured.

ASTM Definition of Terms Relating to Adhesives (D 907-70) defines adhesion as "the state in which two surfaces are held together by interfacial forces which may consist of valence forces or interlocking forces or both." For a detailed discussion of the definition of adhesion, the reader should consult a recent paper by Good [*7*].

There are various theories or mechanisms of adhesion [*8–13*] but there is no single theory or mechanism which can explain all adhesion behaviors. All these mechanisms are valid to varying degrees, however, and their relative importance depends on the adhering system in question.

In an adhering system, adhesion can be expressed in terms of forces or work of *attachment,* or in terms of forces or work of *detachment.* If expressed in the former manner, then the correct appellation should be "basic adhesion," "fundamental adhesion," "true adhesion," or "interfacial adhesion;" however, the present author prefers the term "basic adhesion." Basic adhesion signifies the interfacial bond strength and should depend exclusively on the interfacial properties, without any contribution from any other sources. Basic adhesion is simply the summation of all intermolecular or interatomic interactions. These interactions could be electrostatic, chemical, or van der Waals type. Unless there is a well-defined interface between the adhering phases, the term basic adhesion does not have any significance.

On the other hand, experimentally, adhesion is measured in terms of forces or the work of detachment or separation of the adhering phases. The separation may take place at the interface, or in the interfacial region (also called interphase), or in the bulk of the weaker adhering phase. Separation in the bulk is termed cohesive failure and is related to the cohesive strength of that bulk phase. The cohesive failure of a thin coating or adhesive however, is unlikely to be the same as the cohesive failure of the same material in bulk. Mechanical constraints by the adherend(s) or differences in chemical composition or morphology due to the conditions of coating deposition or joint formation are two possible causes.

The interface is a mathematical plane and can be realized only in the case of a nonsoluble/noncompound forming adherend-adherate combination. On the other hand, an interfacial region or interphase possesses a certain thickness, and its mechanical properties are different from those of the contiguous phases. Interphases may be present on the adhering phases (for example, oxide on metal, oily layer on a surface), or they may be formed by interaction of the adhering phases (for example, diffusion type of interphase in a metal-metal system). The nature and thickness of interphases depend on the adhering phases involved [*14,15*]. The presence of these interphases would give rise to additional interfaces.

So obviously, in real adhering systems, there can be many interfaces and interphases in addition to the two bulk phases, and separation could be at any of the interfaces or in any of the interphases or in one of the two bulk phases. If the separation occurs at an interface or in an interphase, then it is suggested that the measured adhesion be labeled as "practical adhesion." It should be noted, however, that such suggestion is not entirely satisfactory as there are still many unanswered questions:

1. As the interphase is of definite thickness, then why not call the failure in an interphase the "cohesive failure" and the measured adhesion the separation strength of that particular interphase?

2. If the failure occurs in one of original adhering phases but very close to the interface, then should it be called "adhesive" or "cohesive" failure? In such a situation, should the measured stress required for such separation be a measure of practical adhesion? In the case of very thin films, failure close to the interface may be actually deep bulk failure.

3. If the failure is mixed (interfacial, plus interphasial, plus bulk) in nature, then should the measured stress be called practical adhesion?

There is no easy answer to these problems, but it is suggested that unless the failure is deep in the bulk of the adhering phases, the measured stress required to effect separation of the adhering system be labeled "practical adhesion."

The forces required to disrupt the interface or interphase can be applied in various forms (tensile, peel, shear, etc.) and the practical adhesion is expressed in terms of tensile strength, peel strength, or shear strength. Peel strength is measured in terms of force/width required to maintain the continuous detachment of a strip of adherate from an adherend at a specified detachment rate. Peel strength is also expressed in terms of work or energy per unit area.

Tensile strength is defined as the stress (force/area) required to remove a specific area of the adherate when the entire area of the adherate is pulled in a direction perpendicular to the adherend surface.

Forces of adhesion and the work or energy of adhesion can be related only if assumptions are made about the changes in force with distance of separation, so that an integration can be performed. In other words, w, the work of adhesion is

$$w = \int f(x)\, dx$$

Bikerman [*16*] has promulgated the nonexistence of interfacial failure in a "proper bond," that is, in a bond where intermolecular interaction between the adherate and the adherend has been achieved. According to him, true interfacial failure practically never occurs and what is taken for interfacial failure is actually a separation in a weak boundary layer. Cases are chronicled in the literature, however, where true interfacial failure has been observed [*17*]. On the other hand, Bikerman has recently [*18*] cited some examples of noninterfacial failure to bolster his previous views. For a critique of Bikerman's views, the reader should refer to Good [*19*]. What Bikerman calls "weak boundary layers" are actually the interphases between the adhering phases.

In any case, visual inspection is inadequate to ascertain the locus of failure in a separated system, and techniques like electron spectroscopy for chemical analysis, Auger electron spectroscopy, or secondary ion mass spectrometry should be employed for this purpose, and these have been recently used [*20,21*].

Now the question arises, What is the relationship between practical adhesion and basic adhesion? Let us look at some of the characteristics of both. Basic adhesion is strictly an interfacial property and depends exclusively on the surface characteristics of adhering phases. Basic adhesion should be independent of the thickness of adherate, thickness of adherend, specimen size, specimen geometry, temperature, measurement technique, manner of applying external forces, manner of performing the test, test rate, bulk properties of adhering phases, etc.

However, the so-called adhesion measurement techniques measure practical adhesion, which is affected by all these factors. For example, practical adhesion in terms of peel strength depends upon the rate of peel, angle of peel, etc. Tensile strength depends upon the manner of performing the test and the rate of pull.

Basic adhesion can be theoretically calculated or indirectly determined. For example, if the adherate is in liquid form, one can calculate the thermodynamic work of adhesion in terms of its wetting behavior [*11,22*]. In the case of thin-film deposition, basic adhesion between the film and the substrate can be determined by the nucleation methods, that is, by observing the kinetics of formation of these films [*23*]. In essence, basic adhesion is calculated by taking the summation of adsorption energies of individual adatoms. Of course, nucleation methods of determining basic adhesion are of limited applicability and are quite tedious.

An elegant approach based upon the molecular models was taken by Taylor and Rutzler [*24*], but it has not been extended to a stage where it could be used to quantify basic adhesion in a system of interest.

In summary, the relationship between practical adhesion and basic adhesion can be expressed as follows:

1. If there exists a sharply defined interface between the two adhering phases and the separation is clearly at this interface, then

$$\text{Practical adhesion} = f\,(\text{basic adhesion, other factors})$$

2. If the separation is in the interfacial region, then

$$\text{Practical adhesion} = f\,(\text{interatomic or intermolecular bonding within the interfacial region, other factors})$$

These other factors include intrinsic stresses (which depend, among other things, upon the adherate thickness), presence or absence of sites of easy failure, the mode of applying external stress, that is, the technique of measuring practical adhesion, the failure mode, etc.

The following conclusions are made obvious by the foregoing: (*a*) Basic adhesion can be determined only from practical adhesion provided the contribution due to "other factors" can be quantitatively accounted for in the practical adhesion. However, the existence of a clear-cut interface, and the occurrence of clear separation at that interface are not realized in most practical systems. (*b*) One can expect different practical adhesion values by different techniques. (*c*) There may not be simple and direct correlation between practical adhesion and the basic adhesion.

Methods of Measurement of Practical Adhesion

The literature is replete with techniques for measuring practical adhesion of thin films, thick films, and coatings. As these techniques have been recently reviewed in detail [*25,26*], there is no need to expatiate upon them here. So, in this paper references are made to the recent reviews, and only salient developments subsequent to these reviews are covered.

An ideal test for measuring practical adhesion should be [*25,26*]: (*a*) reproducible, (*b*) quantitative, (*c*) nondestructive, (*d*) easily adaptable to routine testing, (*e*) relatively simple to perform, (*f*) not very time-consuming, (*g*) applicable to a wide range of adherate thicknesses, (*h*) independent of operator's experience, (*i*) applicable to all combinations of adherates and adherends, (*j*) valid over a wide range of specimen sizes, (*k*) applicable to products and processes, (*l*) able to measure wide range of practical adhesion strengths, (*m*) free from interpretational complexities, and (*n*) amenable to standardization. Furthermore, machining requirements for specimen preparation should be minimum, and no specialized equipment should be necessary. But such idealization is not realized in practice, as there is no technique which fulfills even half of the foregoing attributes. All commonly used tests for measuring practical adhesion are destructive in nature.

Adhesion Measurement of Thin Films

Mittal [*25*] has recently reviewed comprehensively and critically the adhesion measurement techniques for thin films. The principle, experimental details, potentialities, and limitations of each technique are discussed, and in many cases the practical adhesion strengths obtained using these techniques are presented. Also the interpretational difficulties, factors affecting values of practical adhesion, and the precautions which should be borne in mind to obtain reproducible results are discussed.

The following techniques are discussed in this review.

1. Nucleation methods. As mentioned earlier, the information derived from the nucleation behavior of depositing thin films is used to determine the basic adhesion. These methods are not really the techniques for measuring practical adhesion.

2. Mechanical methods. Direct pull-off, moment or topple, ultracentrifugal, ultrasonic, adhesive tape, peel, tangential shear or lap shear, scratch or stylus or scribe, blister, and abrasion.

3. Miscellaneous techniques. This category includes those techniques which are either of very limited applicability or whose usefulness in routine practical adhesion measurement is not established.

Since this review was completed, a few interesting papers have appeared.

Jacobsson [*27*] has analyzed the direct pull-off method in detail and has quantified the effects of variables which affect direct pull strengths. Goldstein and Bertone [*28*] have discussed the use of the scratch test to evaluate metal film adhesion onto flexible substrates. Earlier, the scratch test had been used only for rigid substrates. Recently [*29*] scratches have been examined by very sensitive surface profilemeter and scanning electron microscopy (SEM) coupled with energy dispersive spectroscopy, and it is observed, among other findings, that the mechanism of film failure depends upon whether the film is ductible or brittle.

Stephens and Vossen [*30*a] have described a laser spallation technique for measuring adhesion. They claim that the technique can be used to measure interfacial bond strength—which is basic adhesion in terms of the terminology proposed earlier. The technique involves impinging a pulsed, high-energy, neodymium-glass laser beam (1.6 μm) onto the back surface of a substrate (made opaque in the case of transparent substrates). The explosive evaporation of absorbing material on the back side of the substrate generates a comprehensive shock wave in the substrate directed toward the film-substrate interface. The magnitude of the shock wave is varied so as to attain spallation or detachment of the film. This obviates the need to attach some sort of mechanical device to detach the film. More recent findings using this technique are discussed by Vossen in this publication [*30*b].

Krongelb [*31*] has described a technique for measuring the adhesion of thin films in which the interaction of an electric current and a magnetic field (referred to as $\vec{I} \times \vec{H}$) can be used to apply a known stress to a deposited thin film without requiring any mechanical attachment to the film. The maximum achievable stress is proportional to the available magnetic field and the current. The specimen fabrication, experimental conditions, and results obtained on specimens of copper evaporated at various temperatures on thermal silicon dioxide (SiO_2) are described.

Adhesion Measurement of Thick Films

For a detailed discussion of the practical adhesion measurement techniques for thick-film conductors, the reader should refer to Hitch and Bube [*32*], Jacobson [*33*], and Anjard [*34*]. The following tests have been discussed by Hitch and Bube: uniaxial tension, wirebond, peel, soldered wire, tensile peel, and dot.

Hitch [*35*] has most recently reviewed the practical adhesion measurement of thick-film conductors. The techniques for the measurement of practical adhesion of electrodeposits have been recently reviewed in detail by Mittal [*26*]. The principle, experimental details, potentialities, limitations, interpretational difficulties, factors affecting practical adhesion strengths and the precautions which should be borne in mind for each technique are discussed. The techniques discussed are:

1. Qualitative tests.
2. Tension tests (use of solders, adhesives and electroformed grips; the Ollard method and its subsequent modifications).
3. Ring shear tests.
4. Peel tests.
5. Knife and scribing tests (knife test, scratch, or stylus test).
6. Ultracentrifugal test.
7. Nonmechanical tests.

In addition, the ultrasonic and blister methods for measuring the practical adhesion of organic coatings are included.

Dini and Johnson [*36*] have recently reviewed the following tests for measuring the practical adhesion of electrodeposits: ring shear, conical-head tensile, modified Ollard, I-Beam, and flyer plate. Modified Ollard and I-Beam are tensile type tests, and flyer plate tests are shock-wave tests used for evaluating parts under dynamic loading conditions. All have been shown to be quite effective in providing consistently reproducible data. Details are presented for each test, such as the shape and size of the specimen, preparation of specimens, and the manner in which the test is performed; data for various substrate-electrodeposit combinations are included. Most recently, Dini et al [*37*] have discussed in detail some of the techniques for measuring practical adhesion of electrodeposits.

The most comprehensive review on the practical adhesion measurement of thick or bulk coatings is that by Bullett and Prosser [*38*]. For an earlier review on this topic, the reader should refer to Lewis and Forrestal [*39*]. These two reviews discuss in great detail the state of the art with respect to practical adhesion of thick coatings. Primary emphasis is on organic coatings. There has not been any significant development since these reviews were published.

Future Developments

It appears that there is a definite need to expend efforts in the following areas:

1. Development of nondestructive techniques for measuring practical adhesion.
2. Thorough understanding of the factors which influence practical adhesion measurement as determined by a given technique.
3. The existing promising techniques for measuring practical adhesion should be standardized, so that results from different investigators can be directly compared.
4. Cross comparison of various practical adhesion measurement techniques. Some adhering systems should be tested for practical adhesion by various techniques.
5. Thorough understanding of the noninterfacial contribution to prac-

tical adhesion strength. If these noninterfacial contributions are understood and quantified, then there is the possibility of determining basic adhesion from measured practical adhesion, provided there is a clear-cut interfacial separation. It should be noted that in practical situations, one is concerned with practical adhesion strength, not with basic adhesion. However, in the case of clear-cut interfacial separation, the practical adhesion could be improved by improving basic adhesion by suitable interfacial treatments.

Summary and Conclusions

1. Generally speaking, an adhering system may consist of one or more interfaces and one or more interphases, in addition to the two bulk phases. When an adhering system is disrupted, or separated by applying external stresses, the separation may take place at an interface, in an interfacial region (also known as interphase), or in the bulk of the weaker of the adhering phases. If the separation occurs at an interface or in an interphase, then it is suggested that the stresses required be taken as a measure of practical adhesion. And since the practical adhesion value depends upon the locus of failure, the failure surfaces should be analyzed for the locus of failure. The information regarding the site of failure is important in understanding the mechanism of failure, and the modern surface analytical instruments should be of immense help in examining the failure surfaces.

2. Basic adhesion signifies the summation of all interfacial intermolecular interactions. In case of a clear-cut interfacial separation, practical adhesion is a function of basic adhesion and many other factors (intrinsic stresses, techniques used to measure practical adhesion, etc.). Unless the contributions due to these "other factors" can be accounted for, one cannot determine basic adhesion from practical adhesion.

3. In the case of failure in an interphase, the practical adhesion depends upon the intermolecular or interatomic bonding within the interfacial region and other factors (stresses, easy failure mode, mode of applying external stresses, etc.)

4. Practical adhesion may not vary directly with changes in basic adhesion. For example, attempts to improve basic adhesion may not result in increased practical adhesion, because of the contribution from "other factors." On the other hand, two adhering systems may have the same basic adhesion, but their practical adhesion could be quite different.

5. Different techniques yield different practical adhesion strengths. Practical adhesion, for a given adherend-adherate combination, as determined by different techniques may not be directly comparable. Furthermore, different techniques yield values of practical adhesion in different units (for example, peel strength in terms of force/length vis-a-vis tensile strength as force/area).

6. As there are many factors affecting practical adhesion, the experimenter, while quoting values of practical adhesion, must specify the experimental conditions, specimen size and geometry, and other relevant parameters. For example, unless the rate of peel and the angle of peel are kept constant, different experimenters will measure different peel strengths for the same adhering system.

7. Some practical adhension tests are qualitative in nature. In such tests, there is always some subjective interpretation. Unless the test yields some numbers for practical adhesion, its usefulness is restricted. However, qualitative tests can be used profitably in discerning cases of very poor practical adhesion.

8. The choice of the test for measuring practical adhesion should be based upon the type of stresses the test specimen is going to encounter in practice.

References

[*1*] Mittal, K. L., *Journal of Adhesion,* Vol. 7, 1974, pp. 377-378.
[*2*] Bikerman, J. J., *The Science of Adhesive Joints,* 2nd edition, Academic Press, New York, 1968.
[*3*] Nenadovic, T., Bibic, N., Kraljevic, N., and Adamov, M., *Thin Solid Films,* Vol. 34, 1976, pp. 211-214.
[*4*] Lee, L. H., Ed., *Recent Advances in Adhesion,* Gordon and Breach, New York, 1973.
[*5*] Lee, L. H., Ed., *Adhesion Science and Technology,* Vols. 9A and 9B, Plenum Press, New York, 1975.
[*6*] Proceedings of Recent Conferences on Adhesion and Adhesives held in London, England. These are chronicled in the series *Aspects of Adhesion,* D. J. Alner, Ed., CRC Press, Cleveland, Ohio.
[*7*] Good, R.J., *Journal of Adhesion,* Vol. 8, 1976, pp. 1-9.
[*8*] Allen, K. W. in *Aspects of Adhesion,* D. J. Alner, Ed., Vol. 5, CRC Press, Cleveland, Ohio, 1969, pp. 11-24.
[*9*] Raevskii, R. J., *Journal of Adhesion,* Vol. 5, 1973, pp. 203-210.
[*10*] Mittal, K. L., *Journal of Vacuum Science and Technology,* Vol. 13, 1976, pp. 19-25.
[*11*] Mittal, K. L. in *Adhesion Science and Technology,* L. H. Lee, Ed., Plenum Press, New York, 1975, Vol. 9A, pp. 129-168.
[*12*] Mittal, K. L., *Polymer Engineering and Science,* Vol. 17, 1977, pp. 467-473.
[*13*] Huntsberger, J. R. in *"Treatise on Adhesion and Adhesive,"* R. L. Patrick, Ed., Marcel Dekker, New York, 1967, Vol. 1, pp. 119-149.
[*14*] Sharpe, L. H., *Journal of Adhesion,* Vol. 4, 1972, pp. 51-64.
[*15*a] Chapman, B. N., *Journal of Vacuum Science and Technology,* Vol. 11, 1974, pp. 106-115.
[*15b*] Mattox, D. M., this publication, 54-62.
[*16*a] Bikerman, J. J., *Industrial and Engineering Chemistry,* Vol. 59, No. 9, 1967, pp. 40-44.
[*16*b] Bikerman, J. J., *Journal of Paint Technology,* Vol. 43, 1971, pp. 98ff.
[*17*] Crocker, G. J., *Rubber Chemistry and Technology,* Vol. 42, 1968, pp. 30-70.
[*18*] Bikerman, J. J., this publication, pp. 30-40.
[*19a*] Good, R. J., *Journal of Adhesion,* Vol. 4, 1972, pp. 133-154.
[*19b*] Good, R. J., this publication, pp. 18-29.
[*20*] Baun, W. L., *Journal of Adhesion,* Vol. 7, 1976, pp. 261-267.
[*21*] Wyatt, D. M., Gray, R. C., Carver, J. C., Hercules, D. M., and Masters, L. W., *Applied Spectroscopy,* Vol. 28, No. 5, 1974, pp. 439-444.

[22] Murr, L. E., this publication, pp. 82-98.
[23] Campbell, D. S. in *Handbook of Thin Film Technology,* L. I. Maissel and R. Glang, Eds., McGraw-Hill, New York, 1970, Chapter 12, pp. 12-3-12-47.
[24] Taylor, D., Jr., and Rutzler, J. E., Jr., *Industrial and Engineering Chemistry,* Vol. 50, 1958, pp. 928-934.
[25] Mittal, K. L., *Electrocomponent Science and Technology,* Vol. 3, 1976, pp. 21-42.
[26] Mittal, K. L. in *Properties of Electrodeposits: Their Measurement and Significance,* R. Sard, H. Leidheiser, Jr., and F. Ogburn, Eds., The Electrochemical Society Princeton, N.J., 1975, Chapter 17, pp. 273-306.
[27] Jacobsson, R., *Thin Solid Films,* Vol. 34, 1976, pp. 191-199.
[28] Goldstein, L. F. and Bertone, T. J., *Journal of Vacuum Science and Technology,* Vol. 12, 1975, pp. 1423-1426.
[29] Ahn, J., Mittal, K. L., and MacQueen, R., this publication, pp. 134-157.
[30a] Stephens, A. W. and Vossen, J. L., *Journal of Vacuum Science and Technology,* Vol. 13, 1976, pp. 38-39.
[30b] Vossen, J. L., this publication, pp. 122-133.
[31] Krongelb, S., this publication, pp. 107-121.
[32] Hitch, T. T. and Bube, K. R., "Basic Adhesion Mechanisms in Thick and Thin Films," Report from RCA Laboratories, Princeton, N. J., 31 Jan. 1975.
[33] Jacobson, L., *Proceedings,* IEEE and EIA Electronic Components Conference, 1971, pp. 474-479.
[34] Anjard, R. P., *Microelectronics and Reliability,* Vol. 10, No. 4, 1971, pp. 269-275.
[35] Hitch, T. T., this publication, pp. 211-232.
[36] Dini, J. W. and Johnson, H. R., *Metal Finishing,* March 1977, pp. 42-46; April 1977, pp. 48-51.
[37] Dini, J. W. and Johnson, H. R., this publication, pp. 305-326.
[38] Bullett, T. R. and Prosser, J. L., *Progress in Organic Coatings,* Vol. 1, 1972, pp. 45-73.
[39] Lewis, A. F. and Forrestal, L. J. in *Treatise on Coatings,* R. R. Myers and J. S. Long, Eds., Marcel Dekker, New York, 1969, Vol. 2, Part 1, pp. 57-98.

DISCUSSION

H. E. Ashton[1] (*written discussion*)—First, your statement that the scratch adhesion method was developed in the 1950s is not correct with regard to organic coatings. Bell Laboratories developed the Bell Scrape Adhesion Tester in the 1930s which led to Method A of D 2197. See *ASTM STP 500—Paint Testing Manual,* p. 320.

Second, the term "hesion" was used before Dr. Asbeck used it in the 1960s. I heard Dr. Reed Brantley of Occidental College use it in 1957. Try Brantley et al.[2]

K. L. Mittal (*author's closure*)—First, please note that I was referring to the use of the scratch test for *thin* films, and to my knowledge it was

[1] Division of Building Research, National Research Council of Canada, Ottawa, Ont., Canada.
[2] Brantley, L. R., Stabler, R., and Bills, K., *ACS Paint and Plastics Preprints,* American Chemical Society, Vol. 13, March 1953, p. 140.

first employed for a quantitative evaluation of "adhesion" of thin films in the 1950s. I fully agree with your comment that the Scrape Adhesion Tester was developed in the 1930s for organic coatings. These organic coatings were definitely thicker than the usual thin films for which I described the scratch test.

Thank you for bringing the Brantley et al reference to my attention. I used the term "hesion" during my talk, but it has not been used in the written account of the paper. The reason is that this term did not receive enthusiastic response from attendees I talked to.

H. E. Hintermann[3] (*written discussion*)—You made no mention of "acoustic emission" as a potential technique for adhesion measurement. Could you make a critical appreciation of this technique and summarize briefly the actual state of the art?

K. L. Mittal (*author's closure*)—I am not aware of any published work apropos of the use of the acoustic emission technique for determining practical adhesion for an adhering system (adherend-adherate combination). It has been used, however, for adhesive joints (see Curtis[4]).

APPENDIX

Some Conversion Factors and Units Commonly Used to Express Adhesion Strength

1 Pa (pascal) (that is, 1 N m^{-2})	= 10 dyn cm^{-2}
	= 1.4504×10^{-4} psi
	= 1.4504×10^{-2} gf cm^{-2}
	= 1.0197×10^{-2} gf cm^{-2}
	= 1.0197×10^{-5} kgf cm^{-2}
	= 1.0197×10^{-1} kgf m^{-2}
	= 1.0197×10^{-7} kgf mm^{-2}
	= 10 μbar
	= 7.5×10^{-3} torr
	= 0.9872×10^{-5} atmosphere
1 kg cm^{-1}	= 5.6 lb·$in.^{-1}$
1 J m^{-2}	= 10^3 erg cm^{-2}
1 N	= 10^5 dyn
1 torr	= 1 mm (Hg) = 133.3 Pa

[3]Laboratoire Suisse de Recherches Horlogeres, CH-2000 *Neuchatel,* Switzerland.
[4]Curtis, C. J., *Nondestructive Testing,* Vol. 8, 1975, pp. 249-257.

R. J. Good[1]

Locus of Failure and Its Implications for Adhesion Measurements

REFERENCE: Good, R. J., **"Locus of Failure and Its Implications for Adhesion Measurements,"** *Adhesion Measurement of Thin Films, Thick Films, and Bulk Coatings, ASTM STP 640*, K. L. Mittal, Ed., American Society for Testing and Materials, 1978, pp. 18–29.

ABSTRACT: The extreme importance of correctly establishing the locus of separation in an adhering system, such as a film on a solid, is pointed out. Two theories as to the level of possibility that true interfacial separation will occur are described and both are criticized. A logically rigorous formulation which draws upon both these theories is given.

KEY WORDS: adhesion, films, coatings, adhesives, bond strength, joint strength, fracture, cracks, flaws, fracture mechanics, interfaces, failure locus, interfacial separation, interfacial failure, cohesive failure, weak boundary layers

Two kinds of information are obtained in measurements of force or energy of adhesion. One is quantitative—the numerical value of the force or energy. The other is of a very different kind—it is the location of the failure, for example, whether or not separation occurred at the interface between two phases.

Scientists, engineers, and technicians commonly prize the first kind of information the most highly; quantitative results are always the best kind of results. Now it is true that generally a qualitative report such as "strong" or "weak" carries less information than does a numerical result. (But an isolated numerical result may carry little information, if an evaluative scale is not provided. A strength report of, say, 30 000 kPa may be regarded as "strong" for a system in which one or more phase is a polymer, but "weak" if both phases are metals.) The query as to where a failure locus is, is not a qualitative matter, however—certainly not in the sense that the terms "strong" and "weak" are qualitative. Indeed the question of whether failure starts at an interface can be put in the form of binary logic: "0"

[1]Professor, Department of Chemical Engineering, State University of New York at Buffalo, Buffalo, N. Y. 14214.

for at the interface, "1" for not at the interface. It is hard to call such a distinction anything but quantitative.

Practical Importance of Failure Locus

There is great practical importance to the correct identification of the failure locus. It is obvious that the measures which must be taken to remedy an interfacial failure are different from those that would remedy "cohesive" failure in either bulk phase. Moreover, lawsuits may turn on the question of where a failure occurred. For example, if there was cohesive failure within a primer phase (and if the primer was used according to manufacturer's directions), the maker of the primer might be held legally liable. If a coating separated from a solid, leaving upon the solid a detectable layer which had the chemical composition of the coating, then a case might be made for the proposition that the maker of the coating was liable. (This case will not always be valid; see the following paragraph, and an earlier paper by the author [*1*].[2]) If the separation was exactly at the interface, between primer and coating, this may have been due to a failure of *interaction* between the primer and the coating. Then there may be five possibilities as regards legal liability: neither party, both parties, one party or the other, or the party that made an inappropriate choice of primer-coating combination. It is all too easy to conclude that the liability should be shared equally by the two suppliers, but such a verdict would not necessarily represent scientific truth. (A similar argument could be made if we were discussing an adhesive, rather than a coating.)

An actual locus of failure may be in any of five different regions. See Fig. 1, which is drawn for a film on a solid, a system in which each phase is assumed to be uniform as to composition and molecular orientation. (The same argument can be made in the case of an adhering system in which both phases are thick.) Regions 1 and 5 of Fig. 1 are within the respective bulk phases, and are far from the interface. There is no sharp boundary, but a continuous gradient of properties and behavior into Regions 2 and 4, respectively. In Regions 2 and 4, local mechanical behavior is influenced appreciably by the presence of the interface and of the other phase. For example, if Phase B has a much higher Young's modulus than Phase A, then in Region 2 there is a serious constraint on the Poisson's contraction of the matter under tensile loading. This influence, of course, increases as the interface is approached.

The interface, Region 3, may be sharp on a molecular scale (or it may be molecularly diffuse to an appreciable degree) and the mechanical properties may exhibit a "jump," over a dimensional region equivalent to the distance between two atoms or molecules. Certain local properties that

[2]The italic numbers in brackets refer to the list of references appended to this paper.

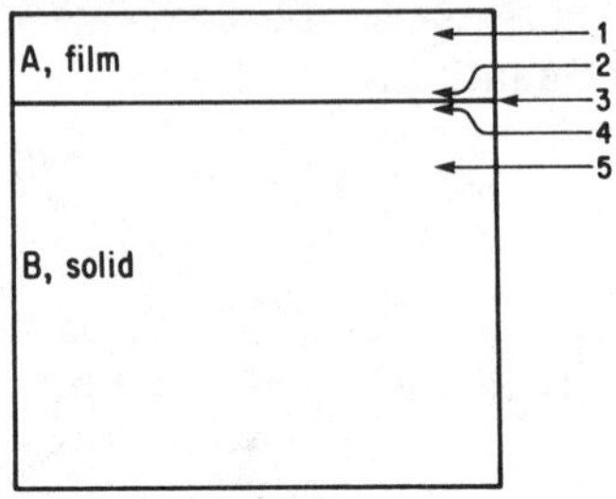

FIG. 1—*The five possible regions for the locus of failure.*

are most directly relevant to the strength of the two-phase system may change more gradually, even when the gradient of composition is the steepest possible. See Ref *1*, where it is pointed out that the dissipative processes associated with fracture [*2,3*], such as crazing, must occur in a zone whose thickness is 2δ; this thickness will be a function of the location of the separation surface.

We must, of course, take note of the fact that initiation of separation, and its propagation, will not necessarily have the same locus. Thus we should more properly refer, in the paragraphs above, to the processes of initiation of interfacial separation, and interfacial propagation of separation.

It is clear that any theory which forestalls the investigation of interfacial failure will make the remedies for interfacial failure inaccessible. It is also clear that an incomplete theory of interfacial separation could lead to the incorrect assignment of legal liability—and probably has done so already, many times.

We will now describe the two theories of interfacial separation; and in a subsequent section, we will criticize both. The criticism of the second theory, which has not been reported before, points the way to a new theoretical analysis which is needed.

Theories Regarding the Possibility of Interfacial Separation

The following quotation from a recent paper by Bikerman [*4*] shows a major component of a view which can be termed the Weak Boundary Layer (WBL) doctrine: "Rupture so rarely proceeds exactly between the adhesive and the adherend, that these events (that is, "failure in adhesion") need not be treated in any theory of adhesive joints." Bikerman, elsewhere, has explicitly included the adhesion between a film and a solid substrate in this generalization. In Ref *5*, the broad statement may be found, that if a system appears to have failed at a phase boundary, the failure must actually have occurred in an unsuspected layer of material at the interface, which had a low cohesive strength—a weak boundary layer. Such a layer might be a brittle oxide of a metal, or low-molecular-weight polymeric material which migrates to the surface of a high-molecular-weight polymer, or a minor component of one phase (for example, an antioxidant) which absorbs at the interface; or it may be contamination, such as foreign matter, which should not have been present at all.

Before discussing Bikerman's basis for his conclusions, we will mention the alternative theory which is presently available [*1*]. This is that failure is initiated according to the Griffith-Irwin criterion [*3,6*]

$$\sigma_c^2 = \text{constant} \times E\mathcal{G}/l \tag{1}$$

where σ_c is the critical stress needed to initiate failure, E the elastic modulas, $\mathcal{G}$ the (dissipated) work per unit area of crack extension, and l the length of the longest pre-existing crack (assumed to be perpendicular to the stress direction.[3]). To apply this to a two-phase system, Phases A and B, it was assumed that the stress was in the z-direction, perpendicular to the interface. For mathematical purposes, cracks of equal length, and parallel to the xy-plane, were assumed to exist, at various values of z. If, under stress, a crack starts to propagate, with the conversion of stored elastic energy into dissipated work (for example, heat), there must be a spatial transport of energy from regions around the crack, to the tip region (the effective thickness of which is 2δ) where the dissipative processes occur.

If the fracture occurs in the region near the interface, elastic energy is drawn from both phases. Therefore E in Eq 1 must be replaced by an effective modulus, which we will designate $\mathcal{E}(z)$—the average modulus, weighted according to the relative volumes of Phases A and B from which elastic energy is drawn when the crack starts to propagate. $\mathcal{E}(z)$ will be a continuous function, varying near the interface and joining the values of E_A and E_B. $\mathcal{G}$ is, likewise, an averaged quantity; the average for $\mathcal{G}(z)$ is taken over the region 2δ, around z. If the crack is at the interface, the region will extend the distances δ_A into Phase A and δ_B into Phase B. The gradient of $\mathcal{G}(z)$ in the region near the interface will, in general, be much steeper than that of $\mathcal{E}(z)$. We may write, for constant l

$$\delta_c^2(z) = \text{constant} \times \mathcal{E}(z)\mathcal{G}(z) \tag{2}$$

With this model, it was shown [*1*] that the location where fracture starts is at the value of z where the product $\mathcal{E}(z)\mathcal{G}(z)$ is a minimum. Various possibilities were worked out, such as the following: Assume that no weak boundary-layer material is present, and that strong intermolecular bonds across the interface exist. If $E_A > E_B$ and $\mathcal{G}_A < \mathcal{G}_B$, and if $\Delta \log E < \Delta \log \mathcal{G}$, then there may be a strong minimum in the value of $\mathcal{E}(z)\mathcal{G}(z)$, at a small

[3]Cracks at angles other than 90 deg to the stress direction have been given considerable theoretical attention in the past few years. See, for example, Ref. *7*.

distance from the interface, and on the A side, where $\mathcal{G}$ is lower. Failure will occur in such a way that a thin layer of the lower-$\mathcal{G}$ phase will be found on the surface of the other phase, even though as far as chemical composition (including molecular weight and structure) is concerned, both phases are homogeneous right up to the interface. This result could be misinterpreted as proving that there had been a chemical inhomogeneity, a weak boundary layer.

In a second example it was shown that if there are no strong bonds (such as would be developed by chemisorption or polymer grafting) across the interface, then there will be a strong minimum in $\mathcal{E}(z)\mathcal{G}(z)$ exactly at the interface, and interfacial failure will be the most probable mode of separation.

In a third example, if $E_A < E_B$ and $\mathcal{G}_A < \mathcal{G}_B$, and if interfacial bonding is strong, the interfacial region will be a highly improbable locus for failure, and failure will occur within Phase A, the weaker bulk phase. This case is in full agreement with Bikerman's doctrine; see the foregoing.

Critique of the Two Theories

Weak-Boundary-Layer (WBL) Theory

In Ref *1* a set of general criticisms was given of the four theoretical arguments offered by Bikerman. Subsequent to that paper, Bikerman described a fifth argument in favor of his theory, which needs to be dealt with: Consider a long and tortuous path across an ordinary sheet of paper, say 20 cm (8 in.) wide, generated by writing the script word *Massachusetts* from one side to the other. See Fig. 2. Start a crack at the script M, at the left side. What is the probability that the crack will follow the line, all the way across the paper? It is not negligible?

This model is a special case of Bikerman's general "probablistic" argument [5]: The probability that a crack will advance past one atom following an interface (as opposed to leaving the interface and traveling the same distance into Phase A or Phase B) is 1/3. The probability that it will continue at the interface, past n atoms, is $(1/3)^n$. If n is large, $(1/3)^n$ is a very small number. In the case just cited, for a tortuous path across a piece of matter 20 cm (8 in.) wide, the total path might be 100 cm (25 in.); and if an atom diameter is, say, 2Å, n will be 5×10^9.

But consider the fact that the bonds may not be of equal strength. Let us assume, for convenience, that the A-A energy is 70 kcal/mole, the B-B energy is 80 kcal/mole, and the A-B energy is 5 kcal/mole (5 kcal is a typical value for London dispersion bond energies, or dipole-dipole attractions). Let the interface contain 2×10^9 bonds, to fit the case just cited. What is the probability that interfacial separation will occur, at a succession of A-B bonds, without departure from the interface and the breaking of A-A for B-B bonds? A Boltzmann probability function will give us an

FIG. 2—*Bikerman's model for his locus-of-failure argument. Phases A and B are separated by a very tortuous interface.*

estimate of the frequency of breaking A-B bonds, relative to A-A or B-B bonds. If P(A-B) and P(A-A) are the relative probabilities of breaking these bonds,

$$\frac{P(\text{A-A})}{P(\text{A-B})} = e^{-[U_{\text{AA}} - U_{\text{AB}}]/RT} \tag{3}$$

With $U_{\text{AA}} = 70$ kcal/mole and $U_{\text{AB}} = 5$ kcal/mole, the value of the exponential, at room temperature, is approximately 5×10^{-55}. This is an exceedingly small number. (This is the probability of departing into Phase A. The probability of entering Phase B is even smaller.) We may estimate n (1 percent), the number of atoms the crack must pass before the probability is 1 percent that the crack will depart from the interface. This is the value such that $(1 - 5 \times 10^{-55})^n = 0.99$. With the aid of the binomial expansion, it is found that $n = 2 \times 10^{51}$. We have assumed one atom diameter to be 2×10^{-8} cm, so this value of n corresponds to a distance of a 1.5×10^{38}, a distance vastly larger than the diameter of the Universe. The answer to Bikerman's example, in the first paragraph of this section, is: the odds that the crack *will* follow the tortuous path are overwhelming for example, (of the order of 10^{50} to 1) if the path represents a phase boundary between two solids in which the cohesive energies are large, and if the adhesive energy corresponds to "physical" forces such as the London dispersion force or the Keesom dipole-dipole force.

Now, to refute the derivation of a theory which justifies a hypothesis does not necessarily disprove the hypothesis. In this case, the refutations, such as that just given, serve only to deny the *universality* of the WBL hypothesis.

The concepts of "necessary" and "sufficient" conditions, with respects to scientific proofs, must be considered here. Bikerman [*4,5*] contends that a WBL is both a necessary and a sufficient condition for failure close enough to a phase boundary that, on first inspection, it appears to be an interfacial failure. He explicitly says that the appearance of interfacial failure is proof that WBL material is present. This is fallacious.

On the other hand, there is no question that the *presence* of WBL material *is a sufficient condition* for failure which may appear, on first inspection, to be interfacial. There is ample experimental evidence for this proposition. But such evidence cannot, in logic, be generalized to prove the sufficient condition to be both necessary and sufficient. Counter-examples exist, for instance, Ref *8*, which prove that the presence of a WBL is not a *necessary*-and-sufficient condition, while not detracting from the conclusion that it is a *sufficient* condition.

Theory Based on Fracture Mechanics

The criticism of the theory proposed in Ref *1* has three components. The first is a matter of interpretation, the second is one of microscopic mechanism, and the third involves fracture mechanics per se.

1. The theory in Ref *1* was based on the Griffith criterion for fracture *initiation*. The value of $\mathcal{G}$ is known to depend on the rate of crack propagation [*9,10*]. Therefore the most probable locus of propagation of a running crack is not necessarily at the same value of z as the most probable locus of initiation. And propagation at an angle to the interface is not excluded. The observation that a failure locus of propagation is at some distance, large or small, from the interface cannot be taken as disproof of interfacial *initiation*.

2. The fracture mechanics theory of the relative probability of failure initiation at an interface did not, originally [*1*], take into account the fact that, in a glassy polymer under load, a microcraze region exists ahead of a crack tip. This defect was partially taken care of in Ref. *2*. We say partially, because the problem of crazing at or very near an interface has not yet been fully elucidated. It was pointed out [*2*] that, on the basis of our current understanding of crazing, it seems that a craze which bridges between two phases is a very improbable structure, or one which should contribute very little strength to the system. Experimental evidence for or against this proposition has not yet been obtained.

3. Fracture mechanics originally [*3*] took a thermodynamic approach, and related the energy dissipated [*3,6*] to the energy available; see Eq 1. Later, the theory was developed in terms of a model of crack shape [*11*]. Typically, a crack is considered to be bullet-shaped at its tip. For a known tip shape, it is in principle possible to compute the field of stress concentration around the tip, for an applied stress normal to the crack direction.

The result is a graph such as those in Fig. 3. The analysis also leads to the estimation of critical stress concentration factors, such as K_{Ic}, where I refers to the crack-opening mode.

Several workers have attempted analyses of this type for interfacial cracks. See, for example, Refs *12–14*. (These studies have not, however, included the question of an interfacial craze.) A very important component of this analysis, which has not been done as yet, is to carry out the analysis of local angular dependence for cracks near an interface, that is, at distances which are small compared with the crack length. The symmetry of the stress field about the crack direction is lost, for such cracks, and the degree to which the lobes in Fig. 3 are distorted will depend strongly on the distance from the interface.

It should be possible to conduct an analysis of this problem, by the finite-element method. Such an analysis would go far toward solving the problem of the locus of propagation of a crack, referred to above.

Conclusions

Fracture mechanics, consideration of intermolecular forces and energies, and experiment lead to a rigorous reformulation of Bikerman's conclusions in regard to actual interfacial failure, which we now propose. This formu-

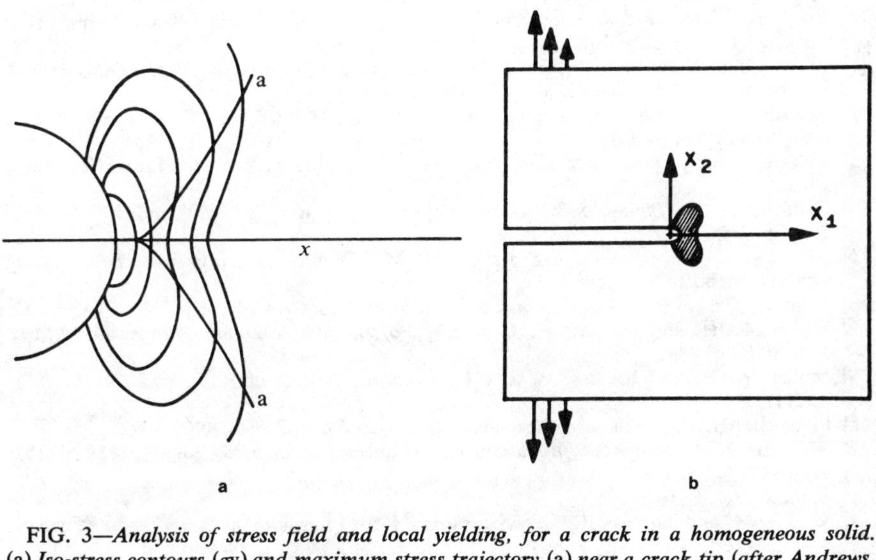

FIG. 3—*Analysis of stress field and local yielding, for a crack in a homogeneous solid.* (a) *Iso-stress contours* (σy) *and maximum stress trajectory* (a) *near a crack tip* (*after Andrews, E.H.,* Fracture in Polymers, *Oliver and Boyd, Ltd., Edinburgh and London, 1968*). (b) *Small-scale yielding near a narrow notch or crack* (*after Rice, J. R. in* Fracture, *H. Liebowitz, Ed., Vol. 2, Academic Press, New York, 1968*).

lation is applicable to thin films on solids, to thick films, to bulk phases in contact, and systems in which an adhesive layer exists between two solids.

1. Failure which, on first examination, appears to be at an interface, very often turns out not to be interfacial, when a closer examination is made, using sensitive methods such as ESCA, Auger spectroscopy, etc.

(*a*) The failure may be (and very often is) caused by WBL material.

(*b*) A case sometimes exists, which mimics WBL failure in systems where no such material is present. Failure propagates close to, but not at, the phase boundary, because of mechanical reasons. [*1*,*15*].

(*c*) True interfacial separation can occur when the cohesive strengths of both phases are large (for example, due to covalent bonds, as in polymers, or to ionic or metallic interactions) and when the adhesive forces across the interface are of the London dispersion type, or dipole-dipole type, or other types that are significantly weaker than the cohesive bonds within either phase.

2. The existence of WBL material at an interface is a sufficient condition for weakness in an adhering system. (It is not a necessary condition, however.) Any scientist, engineer or technician who is seeking the cause of failure or weakness in an adhering system would, in general, do well to investigate the possibility of a weak boundary layer first.

References

[*1*] Good, R. J., *Journal of Adhesion*, Vol. 2, 1972, p. 133.
[*2*] Good, R. J. in *Adhesion Science and Technology*, L. H. Lee, Ed., Plenum Press, New York, 1975, Part A, p. 107.
[*3*] Irwin, G. in *Fracturing of Metals*, American Society for Metals, 1947; *Transactions*, American Society for Metals, Vol. 40, 1948, p. 147.
[*4*] Bikerman, J. J. in *Recent Advances in Adhesion*, L. H. Lee, Ed., Gordon and Breach, New York, 1973, p. 351.
[*5*] Bikerman, J. J., *The Science of Adhesive Joints*, 2nd edition, Academic Press, New York, 1968.
[*6*] Griffith, A. A., *Philosophical Transactions of the Royal Society*, London, Vol. 221, 1920, p. 163.
[*7*] Swedlow, J. L. in *Cracks and Fracture, ASTM STP 601*, American Society for Testing and Materials, 1970, p. 506.
[*8*] Huntsberger, J. R., *Chemical and Engineering News*, Vol. 42, Nov. 2, 1964, p. 82.
[*9*] Muller, K. H. and Knauss, W. G., *Journal of Applied Mechanics*, Paper 71-APM-B, June 1971, p. 483.
[*10*] Muller, K. H. and Knauss, W. G., *Transactions*, Society of Rheology, Vol 15, 1971, p. 271.
[*11*] Liebowitz, H., Ed., *Fracture*, Academic Press, New York, 7 vols., 1968–1972.
[*12*] Williams, M. L., *Bulletin of the Siesmological Society of America*, Vol. 49, 1959, p. 199.
[*13*] Wang, T. T., Kwei, T. K., and Zupka, H. M., *International Journal of Fracture Mechanics*, Vol. 6, 1970, p. 127.
[*14*] Wu, E. M. and Thomas, R. L., *Proceedings*, Fifth International Congress on Rheology, S. Onogi, Ed., University Park Press, Baltimore, Md., Vol. 1, 1969, p. 575.
[*15*] Bascom, W. D., Timmons, C. O., and Jones, R. L., *Journal of Materials Science*, Vol. 10, p. 1037.

DISCUSSION

W. D. Bascom[1] (*written discussion*)—There are three points that I wish to make. I believe that apparent interfacial failure is well recognized by most workers in adhesion technology. In fact, advantage is taken of the fact that for certain test geometries the failure near the interface occurs normally. For example, the climbing-drum peel test is used deliberately in the aerospace (and other) industries to test the interfacial region. This points up the fact, as we have shown for brittle adhesives[2] and Gent, Kaelble, and others have shown for ductile adhesives, that the apparent locus of failure is strongly dictated by the combination of tensile and shear loading acting at the bond line.

Secondly, your reasoning that $\mathcal{G}$ should go through a minimum near an interface is based on a proportionality between $\mathcal{G}$ and K^2. However, such a linear relationship holds only for an isotropic, homogenous, linear elastic material. These conditions are not met for a crack near an interface and especially not for the rubbery (plastic) deformation you illustrate in your diagram.

Finally, there is considerable experimental evidence that a flaw, deliberately placed, at resin/metal interfaces grows *away* from the interface under tensile loading; for example, the work of Wu and Thomas.[3] Also, in the work of Mostovoy et al and ourselves the locus of failure is center-of-bond in adhesive double cantilever beam fracture tests *even if the initial crack is deliberately placed at the interface.*[4]

R. J. Good (*author's closure*)—(1) The pseudo-interfacial failure which Dr. Bascom refers to is exactly the kind of effect which I am considering in my paper. He, and Gent, and Kaelble, have shown experimentally that the *true* locus of failure is strongly influenced by the loading geometry. My theoretical studies have shown, in principle, why this is so. The detailed mechanical analysis (in terms of the stress tensor, the isostress contours, and local plastic yield in complex geometries) has not been completed, as yet.

[1]Naval Research Laboratory, Washington, D. C. 20375.

[2]Bascom, W. D., Timmons, C. O., and Jones, R. L., *Materials Science,* Vol. 10, 1975, p. 1037.

[3]Wu, E. M. and Thomas, R. L., *Proceedings,* 5th International Congress on Rheology, Vol. 1, 1969.

[4]Mostovoy, S., Ripling, E. J., and Versch, C. F., *Adhesion,* Vol. 3, 1971, p. 125.

However, the main point in the present paper is in regard to Bikerman's generalization that, in any "proper joint," true *interfacial* separation *practically never* occurs. Up to the time of my 1972 paper, Ref *1*, a growing number of workers in the field of adhesion were coming to accept this fallacious conclusion as being *universally* true.

(2) The reasoning, that $\mathcal{G}$ should (for certain kinds of systems) go through a minimum near an interface, is independent of all arguments based on the stress intensification factor, K^2. The $\mathcal{G}$ argument is thermodynamic, and independent of crack geometry; but the K^2 argument, used by writers on fracture mechanics, is based on a specific, ideal crack shape in a *homogeneous*, linear-elastic solid. Dr. Bascom's remark amounts to a reinforcement of my own comments, Item 3 in my section on the critique of the theory based on fracture mechanics. I am grateful for his support in this matter.

(3) Dr. Bascom's comment amounts to an elaboration on the first point which I have made, in the critique of fracture mechanics.

It does not matter where a crack *would* propagate, if the crack never starts to propagate. Therefore, we should be much more interested in the *location* of the pre-existing crack which does start to propagate, than in the locus where it propagates. And if we cannot trace a crack back to its origin, we cannot draw inferences with any extreme confidence, about where it originated. See Andrews, *Fracture in Polymers,* Oliver and Boyd Co., Edinburgh and London, 1968, Chapters 4 and 5.

K. L. Mittal[5] (*written discussion*)—You have mentioned that there is separation parallel to the interface, *not* at the interface. Does not the separation parallel to the interface signify failure in cohesion in one of the bulk phases? In other words, if separation is parallel to the interface, then are we talking about failure in adhesion? Would you clarify this point?

R. J. Good (*author's closure*)—A separation may occur at an interface, or in one or another bulk phase, far from the interface. There is also a third possibility. A region near the interface (on one side of it, of course) may be of the same chemical composition as the bulk phase. But mechanically, it is under important constraints. Matter within the lower-modulus phase, but close to the interface, cannot deform as freely as matter that is far from the interface. The presence of the harder phase causes behavior roughly equivalent to what would be expected if the softer phase had a region where its elastic modulus increased and became anisotropic. The effective stress at a point in such a region may exceed the effective yield

[5]IBM Corporation, Hopewell Junction, N. Y. 12533.

point, and plastic behavior may start, even though the stress at a point in the bulk phase was below the elastic limit. This plastic, irreversible behavior may take the form of failure, that is, that a pre-existing crack may start to propagate. It is clear that, mechanically, this failure will be different from failure in a region far from the interface.

A major point of my paper is, that we must cease referring to "failure in cohesion" as if all that mattered were the chemical composition of the region where failure is initiated. That is an oversimplification, which interferes with important tasks such as the diagnosing and remedying the weaknesses of adhering systems.

J. J. Bikerman[1]

Problems in Adhesion Measurement

REFERENCE; Bikerman, J. J., **"Problems in Adhesion Measurement,"** *Adhesion Measurement of Thin Films, Thick Films, and Bulk Coatings, ASTM STP 640,* K. L. Mittal, Ed., American Society for Testing and Materials, 1978, pp. 30–40.

ABSTRACT: Clinging of preformed films to other solids can be measured unequivocally, and the two most common causes of it are outlined. The real adhesion of films, applied as a liquid and then solidifed on the substrate, cannot be measured because mechanical separation of the film from the substrate exactly along their interface is almost impossible. Recent confirmations of this impossibility are reviewed. In the tests intended for measuring this adhesion, very often the tensile strength of the film material is determined, albeit in an indirect manner. In many other systems, the experimental breaking stress depends on a layer spread along the above interface. The use of the term adhesion for the strength of coatings is not advisable.

KEY WORDS: adhesion, breaking stress, capillary attraction, electrostatic attraction, films, improper joints, proper joints, rupture, weak boundary layers

Two entirely different fields are covered by the customary appellation of "film adhesion." In one, preformed solid films are seen to cling to another solid. In the other, liquids are spread over a solid surface, solidify on it, and thus form laminates whose destruction requires a measurable force.

Solid-to-Solid Adhesion

The first phenomenon is known as solid-to-solid adhesion [*1*].[2] Two effects usually are responsible for its existence: electrostatics and capillarity.

When a film F is pressed to a substrate S, the difference in the two work functions causes F and S to acquire opposite and equal charges. The mutual attraction of the latter can be determined by forcibly separating the two solids. However, even if F and S are perfectly isolated from the ground, they are surrounded by the same atmosphere and, consequently, lose their charges more rapidly the greater the electric conductivity of the gas phase. This conductivity increases with the relative humidity (RH). Hence, the

[1] Adjunct professor, Department of Chemical Engineering, Case Western Reserve University, Cleveland, Ohio 44106.

[2] The italic numbers in brackets refer to the list of references appended to this paper.

electrostatic adhesion is encountered more frequently in dry, than in a humid air.

Capillary attraction is observed when liquid water (or another liquid) is present between F and S, but not around the laminate. Suppose, for instance, that droplets, such as pictured in Fig. 1, are confined in the narrow slit between the two solids. If the distance between F and S is h centimeters,

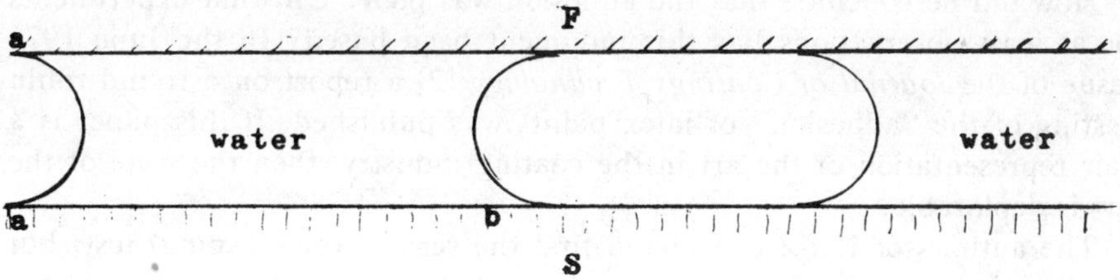

FIG. 1—*Capillary attraction between Film* F *and substrate* S, *caused by water droplets.*

and both F and S are wettable by the liquid, then the radius of curvature of the profile aa is 0.5 h, and the capillary pressure in the drop is $-2\gamma/h$, if γ is the surface tension of the liquid, and the length ab, that is, $2l$, is much greater than h. As the area of contact between the solid and the liquid is πl^2 for each drop, the force (per drop) pressing the solids together is $2\pi\gamma l^2/h$ g·cm/s (or kg·m/s). If n such drops exist in the slit, the total attractive force is

$$N = 2\pi n\gamma l^2/h \tag{1}$$

and may exceed the force exerted by the atmospheric pressure.

The multitude of droplets assumed in the derivation of Eq 1 cannot form or cannot exist for long, if the RH of the atmosphere is too low. Hence, the adhesion of a film to a substrate may have a minimum at a moderate RH. When the RH increases from this middle value, capillary attraction is intensified; when RH descends below this magnitude, then electrostatic attraction increases.

This minimum of attraction can be observed also when the whole space between F and S is filled with liquid. In these systems, the attraction per unit area is $2\gamma/h$, and the total attractive force is $2\gamma A/h$, A being the area of the liquid-solid contact.

Adhesion of Coatings

It appears, however, that this publication deals mainly with films which were applied in the liquid state and then solidified on the support. What is meant by the "adhesion" of such films? Is it advisable to speak of "film adhesion"? Are expressions such as "The adhesion of Film *A* is good, but the adhesion of Film *B* is poor" clear and instructive?

If, as I believe, the rules valid for adhesive joints are true also for coatings, then the answers to the second and third questions are in the negative. Whoever says, for instance, that "Film *B* has a poor adhesion" may effectively block his own progress.

How did he conclude that the adhesion was poor? On what experiments or at least observations has this judgment been based? In the June 1976 issue of the *Journal of Coatings Technology* [*2*] a report on a round-robin testing of the "adhesion" of latex paints was published. If this paper is a fair representation of the art in the coating industry, then the state of the art is deplorable.

The authors of Ref *2* considered first the scratch (or scraping) test, but finally agreed to use the cross-cut (or cross-hatch) method. A scheme of the first is indicated in Fig. 2. A film attached to a support (or substrate) is being scraped by a blade (knife) moving in the direction of the arrow. The force needed for maintaining this motion is measured and is said to be a measure of the adhesion of the film. It is obvious, however, that the knife simply crushes the film substance, and the crushing force has no relation to the forces acting between the film and the substrate.

If, to someone, this is not obvious, and if the doubter has access to the equipment required, I hope he will perform the scraping test on a free film, that is, on a film prepared as similarly to that of Fig. 2 as possible, but not attached to any solid and simply clamped at the ends. I believe that the force needed for scratching the free film will be very similar to the force required for scratching an attached film and that the thicker the free film, the smaller will be the difference between the two.

In the usual scraping test, the knife is immersed deeper than is indicated in Fig. 2 and removes also a thin surface layer of the support. This modifi-

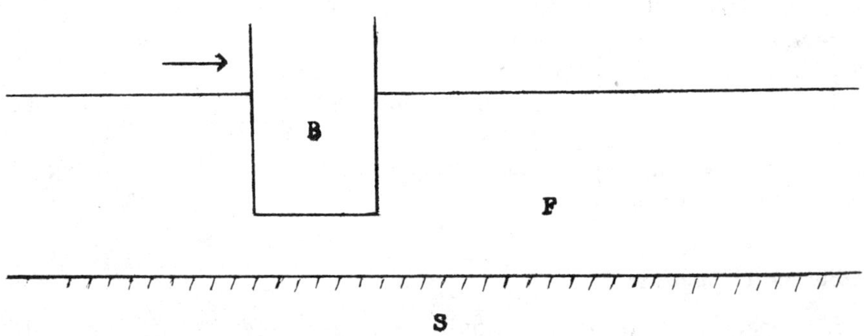

FIG. 2—*Scraping test.* S *is a solid coated with film* F. *Blade* B *is pushed to the right by a measurable force* (arrow).

cation does not alter the essential mechanism, but, instead of one, two materials are crushed, and the measured force is the sum of those needed to deform F and S separately; no term attributable to real adhesion is present.

The cross-hatch test consists of two operations. First, the film is cut in two perpendicular directions so that it is transformed into a multitude of plates which may be, say, 0.05 mm thick and 1 mm along the edge. It seems to me that this operation alone should disqualify the method. People are interested in the properties of a sound and useful film, not in those of a damaged coating which cannot protect and has no esthetic value.

In the second operation, an adhesive tape is pressed to and then peeled off the scratched film surface.

Peeling is commonly used for testing adhesive joints, and the first-order approximation for the minimum peeling tension Γ (that is, peeling force per unit width of the tape or ribbon) is

$$\Gamma = k\sigma_1 \left(\frac{E_2}{E_1}\right)^{0.5} h_1^{0.25} h_2^{0.75} \tag{2}$$

k is a number depending on the Poisson ratios of the two solids (the adhesive and the ribbon) and usually is not far different from 0.5, σ_1 is the tensile strength of the adhesive (which is supposed to fail), E_1 and E_2 are the moduli of elasticity of the adhesive and the ribbon (both elastic solids), and h_1 and h_2 are the thicknesses of the adhesive and the ribbon, respectively. If Film B "has a poor adhesion," its σ_1 may be too small, or the backing may be too flexible, and so on, and quantitative statements concerning the properties represented in Eq 2 would be much more helpful than simply abandoning the film in despair.

It should be emphasized that neither the approximation (2) nor the more sophisticated approximations contain any term or any quantity which can serve as a measure of adhesion. All quantities seen in (2) refer to the two phases separately; none is related to the interfaces between the adhesive and the ribbon or between the former and the rigid support.

The two preceding paragraphs are an "aside" because the customary cross-cut test does not even measure the peeling tension required to remove the adhesive tape; only the additional damage caused by the tearing away of the tape is qualitatively assessed. For instance [2], a coating is said to be of quality No. 3 if "small flakes of the coating are detached along edges and at intersections of cuts; the area affected is 5 to 15 percent of the lattice." As indicated in Fig. 3, the aformentioned "small flakes" (*N*) are, obviously, broken off the main body of the film (*M*). Thus, separation occurred wholly in the film material; the tensile (or shear) strength of this material was overcome by the peeling force, but the interface between the film and the support was not involved at all. Hence, the test affords no information on the adhesion between the two.

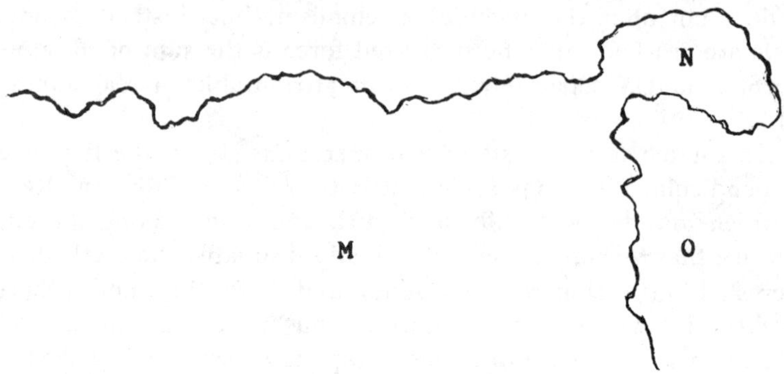

FIG. 3—*Cross section of a cut film.* M = *a unit of film remaining after cut* O *has been made.* N = *a protuberance produced by cutting.*

Failure in Adhesion

The information about adhesion just referred to in the previous section could be obtained, of course, if the adhesive tape would remove square plates (produced by cutting) completely, that is, exposing the initial surface of the support. As pointed up long ago [*1*], p. 137, such a failure in adhesion is so unlikely to occur that it need not be considered when discussing the strength of adhesive joints. This conclusion, derived from the probability theory of rupture, was confirmed by the crude experiments possible then. Now, however, much more powerful methods for analyzing surfaces are available. Whenever they are applied to the problem of "adhesion failure," they demonstrate the presence of the adhesive (or coating) on the adherend after mechanical rupture, or the presence of the adherend material on the separated coating, or both. Thus so far the improbability of true interfacial separation has been consistently verified in all instances in which true molecular contact between the two solids is achieved in the first place.

The most popular of these methods are electron spectroscopy for chemical analysis (ESCA) [*3,4*], Auger electron spectroscopy (AES) [*3–5*], and secondary ion mass spectroscopy (SIMS) [*5,6*]. The electrons or ions studied in the analytical half of the instruments all originate in the external layer of the irradiated solid, usually less than 20 Å thick, and thus give information on the chemical composition of this layer.

Some instances of the recent findings are mentioned in the following. Films of a copolymer of ethylene and acrylic acid, solidified on a tetrafluoroethylene copolymer, showed the presence of fluorine on their surfaces after separation [*7*]. Collodion films attached to and then peeled off an etched tetrafluoroethylene copolymer, were covered with a fluorine-containing layer [*8*]. When molten tetrafluoroethylene copolymers were spread on a metal (for example, copper) surface, permitted to solidify, and then peeled off, thin residues of tetrafluoroethylene were left on the support [*9*]. Poly (tetrafluorethylene) films pressed at a temperature above 300°C (572°F) between two aluminum foils were found to be coated with alumina after separation, and the oxide contamination seemed to be, on the average, less than 10 Å thick [*3*].

When the joint between two titanium plates bonded with an epoxy glue containing magnesium silicate was broken in shear, both magnesium and silicon, in addition to other elements, could be detected on the adherend surfaces. When the filler in the glue was titanium dioxide (TiO_2) instead of magnesium silicate, and the adherend was aluminum (covered, of course, with alumina), then titanium appeared in the alumina layer. When no molecular contact was achieved in the preparation of the joint, no transfer of material was seen; this apparently was the case in the system of iron adherends and Nylon-12 [*10*].

As both experiment and theory now show that a rupture exactly along the interface occurs very rarely, if at all, only two *loci* of rupture seem to remain: in the coating or in the substrate. In reality, however, numberless coated solids break in a stratum not even mentioned in the description of the system, namely, in a weak boundary layer (WBL) which separates the (strong) substrate from its (strong) coating. Adhesive joints containing WBL are denoted as improper joints; and a proper joint contains no WBL.

Weak Boundary Layers

According to the provenience of the material of WBL, seven kinds of these layers may be differentiated: (1) air or vapor, (2) coating material, (3) substrate, (4) air and coating, (5) air and substrate, (6) coating and substrate, and (7) all three.

WBL of the First Kind

To obtain a well-adhering coating, the liquid paint (or lacquer, etc.) has to displace the air from the substrate surface. This displacement takes time, especially because all solid surfaces have microscopic valleys or indentations which can be only slowly filled by the liquid. If the paint solidifies before the displacement is complete, a net of air (or vapor) bubbles remains between the coating and the support; when a disruptive force is applied, cracks propagate from one to the neighboring bubbles, and only a weak force is needed for separation.

Is it wise to speak of a "poor adhesion" in this instance? Obviously not. The breaking force is small because the rate of setting of the paint was too great, or its viscosity was excessive. These statements immediately indicate possible remedies: either the setting rate or the viscosity should be lowered. Thus, in many instances, a small and rationally selected alteration of Composition *B* or, even simpler, a temperature change, can transform "poor adhesion" into "perfect adhesion." Whoever considers the phrase "Compositon *B* has a weak adhesion" to be a final summary of the test results effectively blocks his own progress.

Many improper joints of the first kind are caused by a completely different effect. Air can be displaced from a solid surface only by liquids which wet the latter, that is, which form very small contact angles θ with it. If the value of θ is large (for example $\pi/2$ or 90 deg), no waiting, however long, will result in satisfactory contact between the coating liquid and the substrate. Here again, a statement such as "Composition *B* forms a large contact angle with the solid in question and therefore is not suitable in this case" is much more informative than the sigh "*B* is no good because of its poor adhesion." Because the wetting behavior of so many liquids on so many solids is known, probably it would not be too difficult to alter the contact angle obtained with *B* without drastically changing its chemical nature.

WBL of the Second Kind

Commercial polyethylene is a suitable example of WBL of the second kind. It has been long considered to have "poor adhesion" because of its nonpolar nature; thus, again, a basic reason was alleged. In reality, this polymer contains impurities of a low molecular weight (such as oligomers, lubricants, and antioxidants) which are squeezed out during its solidification and, especially, during its crystallization. These impurities, consequently, gather along the surface of the specimen where they give rise to a WBL. Purified polyethylene, which is even less polar than the commercial one, formed proper joints with all (clean!) solids tested. Thus, Composition *B* may have "weak adhesion" because it contains an unsuspected "oily" ingredient that spreads between the coating and the support.

To save space, the remaining six classes of WBL are not discussed here. Examples of them can be found, for instance, in Ref 1, p. 164.

Proper Joints

In proper joints, either the adhesive or the adherend breaks, although the rupture surface may also pass through both. If the substrate is a structural metal and the coating is polymer-based, the coating usually will rupture first. Consequently, the breaking stress of an attached coating is related to that of a free film of an identical composition; see also the scraping test discussed in the foregoing. In experiments on adhesive joints [*1,11*] it was found that the breaking stress of butt joints tends toward the breaking stress of the adhesive itself (tested as a rod), when the thickness of the adhesive layer in the joint increases; above the thickness of 1 mm the difference between the two stresses may be negligible.

Analogous results have been obtained on coatings [*12*], although a more complicated setup proved necessary. According to Sanzharovskii, the support must be deformed plastically at the stresses needed for breaking the coating. A thin copper foil was coated on both sides with, apparently, cellulose nitrate films. After aging for 120 h, the maximum strain of the coating decreased to that of the copper used, namely, 9 to 10 percent. When the 3-layer ribbon (coating-copper-coating) was extended longitudinally, all three layers ruptured simultaneously. Substracting from the total pull the force needed to snap the foil, the breaking stress of the coating proved to be equal to 910 bars (or 91×10^7 dynes/cm^2 or 91×10^6 N/m^2). Free, otherwise identical films needed breaking stresses of 930 bars, that is, not significantly different from those of attached coatings. After a longer aging (400 h), the breaking stresses of attached and free films were 740 and 780 bars, respectively; again, a reasonable agreement. It is clear that the "adhesion" of cellulose nitrate to copper has no effect on the experimental results, but the Sanzharovskii [*12*] managed to refer to adhesion also when reporting the foregoing data.

From time to time, the so-called adhesion of coatings is estimated from the breaking stress of butt joints. A rigid bar is glued to the coating so that the assembly consists of four materials, namely, support-coating-glue-bar. A pull perpendicular to the support-coating interface is applied, and the minimum force required for rupture is determined. If the rupture takes place in the glue layer, the result is discarded. If the coating breaks, the result is said to be a measure of the adhesion between the coating and the substrate.

In reality, all rules earlier derived for adhesive butt joints must be valid for these *ad hoc* joints as well. If a WBL is present between the support and the film, the minimum pull is a measure of the tensile strength of the WBL; and, if the joint is proper, then, as a rule, the coating splits and the breaking stress is a function of the tensile strength of the coating material. Why is this stress not simply identical to the tensile strength? Because, first, the distribution of stresses in a thin film generally is different from that in a specimen used for strength measurements on unattached solids, and, secondly, because the probability of a particularly dangerous flaw is greater in a large rather than in a small specimen.

The dependence of breaking stress on specimen dimensions points out yet another possibility of mistake when the term adhesion is employed. Tests on Composition *A* may have been performed on very thin coatings. The breaking stress was high, and the test results were summarized as "Composition *A* has a good adhesion." In practice, the coatings are thick and rupture at smaller stresses. Hence, "well-adhering" coatings may prove to be unsatisfactory in large-scale applications.

When it is understood that we are measuring the breaking stresses of a particular system rather than the mythical adhesion (supposed to have a general validity), and, moreover, when it is remembered that the science of the strength of materials is needed for comprehending the results of mechanical measurements, then the strength and other properties of coatings will be judged with more confidence than is possible at present.

References

[*1*] Bikerman, J. J., *The Science of Adhesive Joints*, 2nd edition, Academic Press, New York, 1968, p. 92.
[*2*] Montreal Society for Coatings Technology, *Journal of Coatings Technology*, Vol. 48, No. 617, 1976, p. 52.
[*3*] Clark, D. T. and Feast, W. J., *Journal of Macromolecular Science—Reviews in Macromolecular Chemistry*, Vol. 12, 1975, p. 191.
[*4*] Czanderna, A. W. and Wolsky, S. P., *Modern Methods of Surface Analysis*, American Elzevier, New York, 1975.
[*5*] *CRC Critical Reviews in Solid State Sciences*, Vol. 4, 1974, p. 333.
[*6*] Baun, W. L., *Preprints*, Organic Coatings and Plastics Division, American Chemical Society, April 1976, p. 344.
[*7*] Azrak, R. G., *Journal of Colloid and Interface Science*, Vol. 47, 1974, p. 779.
[*8*] Herglotz, H. K. and Suchan, H. L., *Advances in Colloid and Interface Science*, Vol. 5, 5, 1975, p. 79.
[*9*] Dwight, D. W. and Riggs, W. M., *Journal of Colloid and Interface Science*, Vol. 47, 1974, p. 650; Dwight, D. W., later (still unpublished) work.
[*10*] See Ref *6* and later (still unpublished) work.
[*11*] Bikerman, J. J. and Huang, C. R., *Transactions*, Society of Rheology, Vol. 3, 1959, p. 5.
[*12*] Sanzharovskii, A. T., *Methods of Determining Mechanical and Adhesion Properties of Polymer Coatings* (in Russian), Nauka, Moscow, 1974, p. 41.

DISCUSSION

R. J. Good[1] *(written discussion)*—I have two questions.

1. You say that even when interfacial forces are weak, if atomic contact between two phases exists, a true interfacial separation is "just *too* improbable" to be given any consideration.

I would like to refer you to my paper.[2] Boltzman probability weighing must be introduced into the model which you discuss in Ref *1* of the paper, pp. 139–40. When this is done, it is found that if the bonds across a sharp interface are only 20 or 30 kcal/mol weaker than the cohesive bonds in either phase, the probability is infinitesimal that an interfacial crack will

[1] Department of Chemical Engineering, State University of New York at Buffalo, Buffalo, N.Y. 14214.

[2] Good, R. J., *Journal of Adhesion*, Vol. 4, 1972, p. 133.

depart from the interface. (In my paper in this publication[3] I give an example of this probability estimate.) Now, have you found any logical flaws in my theory, or in the other arguments in my paper? If so, I would like to learn of them, so that we can discuss the question and reach a conclusion.

2. If, in setting out to investigate an example of failure which, on casual inspection, looks like interfacial separation, a scientist brings to the problem the closed-minded preconception that actual interfacial separation is "just too improbable," does not that condition prevent him from investigating true interfacial effects?

J. J. Bikerman (*author's closure*)—To answer your first question, it would be necessary for me to re-read the paper you refer to. It should be remembered, however, that a crack tends to move along the path of the least resistance, that is, from one defect to another. Consequently, values calculated for a perfect lattice are not to be employed in estimating the probability of a given path.

The answer to the second question is: No! When a failure looks like interfacial separation, the experimenter has the duty to find out where and why the separation really took place.

H. E. Ashton[4] (*written discussion*)—You stated that one should not use the term "adhesion" because it is such a complex mixture of properties. How can you then recommend "scratch resistance," because this is also a mixture of properties—adhesion, hardness, tensile properties—which generally leads to poor reproducibility if not repeatability?

J. J. Bikerman (*author's closure*)—The term "scratch resistance" identifies the method used and, hence, is infinitely preferable to the term, "adhesion." Scratch resistance is not affected by molecular adhesion and is a function of the mechanical properties of the coating, the geometry of the scratching tool, and so on.

T. T. Hitch[5] (*written discussion*)—While it is agreed that the fine points of interpreting the knife-blade and cross-hatch tests are open to question, both of these qualitative tests have value in distinguishing poorly adherent films from films of good adherence. Such use requires experience with the test and the material under test, however.

J. J. Bikerman (*author's closure*)—The knife-blade and the cross-hatch tests give no information on the true film adherence; I hope I showed the

[3] This publication, pp. 18–29.
[4] National Research Council of Canada, Division of Building Research, Ottawa, Ont., Canada.
[5] RCA Research Laboratories, Princeton, N.J. 08540.

validity of this opinion in my paper. The results of these tests depend on the brittleness and other mechanical properties of the coating. When the scientists and technicians in the industrial control laboratories realize this, the value of their tests, for both industry and science, will be greatly enhanced.

W. L. Baun[1]

Experimental Methods to Determine Locus of Failure and Bond Failure Mechanism in Adhesive Joints and Coating-Substrate Combinations

REFERENCE: Baun, W. L., "**Experimental Methods to Determine Locus of Failure and Bond Failure Mechanism in Adhesive Joints and Coating-Substrate Combinations,**" *Adhesion Measurement of Thin Films, Thick Films, and Bulk Coatings, ASTM STP 640,* K. L. Mittal, Ed., American Society for Testing and Materials, 1978, pp. 41-53.

ABSTRACT: The strength of an adhesive bond is assessed by means of several physical tests in which a load is placed on the joint until failure occurs. The durability of adhesive joints is evaluated by still other techniques where the bond is tested, usually under load, in salt spray, high humidity and temperature, or other environments deleterious to the bond. Regardless of the strength of the joint, a complete description of joint performance must include how and where the bond failed.

Ion scattering spectrometry (ISS) and secondary ion mass spectrometry (SIMS) combined with scanning electron microscopy (SEM) are capable of determining the true locus of failure in an apparent interfacial bond separation. Methods using an electron beam, such as Auger electron spectroscopy (AES), are useful for analyzing evaporated metal films and the metal or alloy surface in adhesive bonds but are of limited use for examining the adhesive side of a failure, because of heating, which rapidly destroys the specimen surface. X-ray photoelectron spectroscopy (XPS), sometimes called electron spectroscopy for chemical analysis (ESCA), provides not only elemental analysis but also gives information on how the elements are combined in the adhesive, adherend, and interfacial regions. Results using these and other techniques are discussed for failure surfaces obtained by lap shear, peel test, and wedge test methods. Advantages and limitations of various methods are discussed.

KEY WORDS: adhesion, adhesive bonding, thin films, failure locus, bond failure, adhesive joints, surface characterization, ion scattering spectrometry, secondary ion mass spectrometry, scanning electron microscopy, Auger electron spectroscopy, X-ray photoelectron spectroscopy

The strength of an adhesive joint is assessed by means of physical tests such as single lap shear, double lap shear, and peel in which an increasing load is placed on the joint until failure occurs. Following failure, visual

[1]Research chemist, Air Force Materials Laboratory, Wright-Patterson Air Force Base, Ohio 45433.

(or sometimes microscopic) examination of the surfaces is made to determine the mode of failure. If adhesive remains on each adherend and the joint appears to have failed in the adhesive itself, the failure is customarily termed "cohesive" failure. If the failure appears to have occurred at the interface between the adhesive and adherend, the failure is termed "adhesive." More comprehensive background on failure modes and mechanisms may be found in other contributions to this publication [*1–3*].[2]

Bikerman [*4*] says "rupture so rarely proceeds exactly between the adhesive and an adherend, that these events (that is 'failure in adhesion') need not be treated in any theory of adhesive joints." He points up that apparent failures in adhesion are quite common but they take place so near the interface that the adhesive remaining on the adherend after the rupture is not visible. This is the weak-boundary-layer (WBL) theory, first advanced in 1947 [*5*], which states that failure is always cohesive rather than adhesive in nature. These sometimes controversial ideas are critiqued by Good [*6*]. Although Good finds a practical usefulness to the WBL theory, he feels *experimental* proof has never been shown for the *universality* of the theory. Regardless of whether or not the WBL theory accounts for the mechanism of failure, it remains that an essential feature of the mechanical strength of an adhesive/adherend system is the site at which the failure is initiated and propagated, or the "locus of failure." Characterizing a joint by strength alone is not enough. We must also determine the mechanism of failure. Fracture mechanics indicates that the factors influencing bond strength are a product of terms describing interfacial separation, viscoelastic behavior of the materials, and fracture mechanisms such as crazing, divided by a function of initial distribution of original failure nuclei. Detailed investigation of the failed surfaces can give an estimate of the influence of each mechanism.

Frequently it is not simple using visual or even microscopic examination to determine after testing whether an apparent adhesive failure occurred at the interface due to improper wetting or at some new interface, leaving behind a thin layer of adhesive on the adherend or oxide on the adhesive. There is a resolution limitation of about 100 Å for most scanning electron microscopes (SEMs) which makes very thin organic films difficult to detect, especially when the adhesive is a pure polymer containing no fillers of higher atomic number than the polymer to increase contrast. Optical and staining methods have been reported [*7*] to determine the presence of adhesive films. However, the optical technique uses the interference phenomenon, which is applicable only to fairly thick films, certainly not to films only a few molecules thick or boundary layers containing both adhesive and adherend components. Staining techniques are sensitive only to specific compounds present in the usually complex adhesive systems. Several in-

[2]The italic numbers in brackets refer to the list of references appended to this paper.

vestigative techniques on both sides of a joint failure are necessary to determine the locus of failure conclusively.

It is also important to determine the locus of failure of thin films subjected to many mechanical tests depending on peel, scratch, abrasion, and deceleration as reviewed by Mittal [*8*]. Generally, thin polymer films on metals present the same problems as encountered in adhesive bonding. On the other hand, evaporated metal films, by nature of their simpler composition and more ideal interfaces, present fewer difficulties. There are usually differences in color between the evaporated film and the substrate which make it easier to determine the mode of failure. However, when both metals are the same or nearly the same color or when a metal is deposited on a thin oxide film or on a polymer, there can be problems of interpretation of visual results.

Just as in adhesive bonds, it is necessary to describe where a thin film failed following service or test as well as how well the film adhered to the substrate.

Characterization Methods

Each characterization method is based on an intrinsic property of the atoms and molecules on the surface. Such atomic and molecular surface properties are listed in Table 1 along with a partial list of techniques which use these properties. Park [*9*] has reviewed these methods and others and discusses the limitations as well as the advantages of each. He emphasizes in his critical review that in a field as new as surface characterization it

TABLE 1—*Properties of surface atoms and molecules and examples of characterization techniques utilizing these properties.*

Properties	Technique
Atomic mass	ion scattering spectrometry secondary ion mass spectrometry thermal, electron, ion desorption
Vibrational states	infrared absorption, emission Raman spectrometry inelastic electron tunneling
Valence states and	X-ray emission Auger electron spectrometry
Core states	photoelectron spectrometry soft X-ray appearance spectrometry
Crystallography	electron diffraction field ion microscopy

is not surprising that much remains to be learned about the effective use of these tools. In the present paper we consider those techniques which show promise for determining the locus of failure when an adhesive bond breaks or a coating separates from a substrate. We have largely ignored techniques exploiting vibrational states and emphasized those using atomic properties because the latter appear more sensitive and easier to interpret. This is not to say that vibrational spectroscopies are not useful. They are. For now, however, the discussion is limited to the four techniques listed in Table 2, ion scattering spectrometry (ISS), secondary ion mass spectrometry (SIMS), Auger electron spectrometry (AES), and X-ray photoelectron spectrometry (XPS), sometimes called electron spectroscopy for chemical analysis (ESCA). Facts on operating parameters are presented in this table along with opinions concerning advantages and disadvantages.

Ion Scattering Spectrometry

When an ion approaches a surface, it is most likely that the ion will become neutralized by electrons at the surface. Some ions do survive, however, and are back scattered from the surface after undergoing a simple elastic binary collision [*10*]. The masses of the atoms at the surface can be determined by measuring the energy lost by the incoming ion beam. The major components of the equipment as shown in Fig. 1 consist of an inert gas ion gun and a simple 127-deg electrostatic analyzer. This scheme was the basis for the original commercial ISS unit (3M Company, St. Paul, Minnesota). Only very recently was the system greatly improved by the addition of a cylindrical mirror analyzer which increased sensitivity and allowed data to be displayed in real-time fashion [*11*]. The ISS technique is the only one which analyzes the first atomic layer at the surface.

Secondary Ion Mass Spectrometry

The momentum transfer from the probing ion beam to the surface atom results in some atoms at the surface being dislodged. Although many of these sputtered species are neutral, some become ionized and can be analyzed by conventional mass spectrometry [*12*]. The components in ultrahigh vacuum are seen in the same figure, Fig. 1, as illustrated the ISS technique since they are commercially available in such a package and use the same ion beam to probe what leaves the surface (SIMS) and what remains on the surface (ISS).

Secondary Ion Mass Spectrometry/Scanning Electron Microscopy

In addition to being combined with ISS, SIMS is used frequently in conjunction with one of the other analytical techniques and only recently a unique

TABLE 2—*Comparison of primary elemental surface characterization techniques to determine locus of failure.*

	Technique			
Parameter	Ion Scattering Spectrometry (ISS)	Secondary Ion Mass Spectrometry (SIMS)	Auger Election Spectrometry (AES)	X-Ray Photoelectron Spectrometry (XPS)
Principle	elastic binary collision with surface ion	sputtering of surface atoms by ion beam	ejection of Auger electron upon recombination	ejection of photo-electons by photon
Probe	~1 to 3 keV ions	~1 to 3 keV ions	~1 to 3 keV electrons	0 to 2 keV photons
Signal	ion current versus energy	ion current versus mass	derivation e' emission versus energy	electron emission versus energy
Applicable elements	$Z \geq 3$	all (if positive and negative SIMS)	$Z \geq 3$	$Z \geq 3$
Surface sensitivity	high	variable	variable	high
Elemental profiling	yes	yes	yes, with ion beam	yes, with ion beam
Image-spatial analysis	yes	yes	yes	no
Spectral shift	possible, but generally no	no	yes	yes
Information on chemical combination	yes, in fine features but generally no	in some cases (finger-print spectra)	yes	yes
Quantitative analysis	yes	probably no, maybe with similar standards	yes, in principle but difficult	yes
Influence of operating conditions and matrix	no	yes	yes	yes
Isotopic analysis	yes, in principle but generally no because of resolution limits	yes	no	no
Beam induced surface changes	yes, sputtering damage	yes, sputtering damage	yes, due to sputtering and electron beam heating	no, (except when profiling)

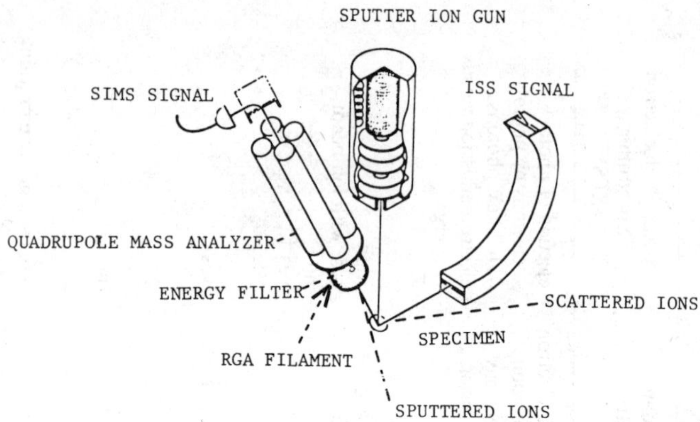

FIG. 1—*Essential components in ultrahigh vacuum for ISS-SIMS surface characterization.*

combination was developed between SEM and SIMS [*13*] which should prove very useful in adhesive bonding research. A diagram of the SIMS-SEM apparatus is shown in Fig. 2. A leak valve admits suitable gas into the ion source region where ions are produced and accelerated at voltages up

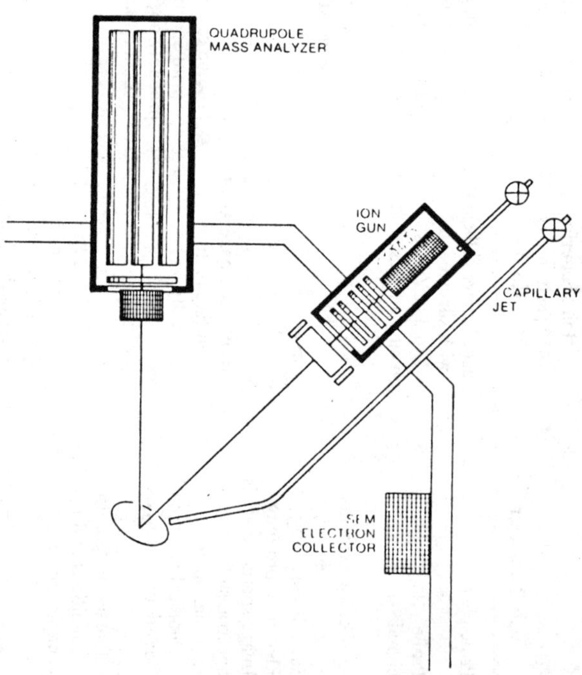

FIG. 2—*Diagram of a combination SIMS-SEM instrument.*

to 5 ke V. Lens assemblies permit focusing of the ion beam from less than 100 μm to a few millimeters diameter on the specimen surface. Deflection plates enable scanning of the ion beam for imaging or depth profile [*10*] purposes. Low-energy sputtered ions from the specimen surface are admitted to a quadrupole mass analyzer through an energy filter that eliminates the high background and peak broadening caused by high-energy sputtered or scattered particles. A gas capillary jet of oxygen directed at the specimen surface provides an "active gas" on the surface to maintain high secondary ion yields.

Auger Electron Spectrometry

If a material is irradiated with an electron or photon beam of sufficient energy, an electron is ejected from an inner level, placing the atom in an excited state. When the atom reverts to the ground state, it may do so by emitting characteristic X-rays or by radiationless transitions, the so-called Auger transitions. The emission of X-rays has been studied for three quarters of a century while the radiationless transitions largely have been ignored even though the Auger transitions outnumber the X-ray events for many elements. A system as usually assembled for AES uses a beam of several kilovolt electrons to excite the specimen in an ultrahigh vaccum. The Auger electrons are dispersed by a cylindrical mirror analyzer and detected by an electron multiplier. An elemental profile through the specimen may be obtained by placing the electron beam inside a larger beam of ions and slowly eroding away the surface as previously described. In this way one can deduce how thick a film was present and how it changed in composition with depth. One major disadvantage of AES for adhesive bonding research is that organic adhesives are very unstable under the electron beam due to localized heating. Naturally AES does not suffer such limitations for metal thin films. Information on chemical bonding is included in spectral details and is useful as a "fingerprint" but is extremely difficult to interpret from a theoretical point of view.

X-Ray Photoelectron Spectroscopy (Electro Spectroscopy for Chemical Analysis)

In X-ray excited photoelectron spectroscopy a specimen is irradiated with a beam of X-rays (obtained by electron bombardment of, typically, aluminum or magnesium) exciting electrons from the target that are energy analyzed. X-ray photoelectron spectra contain chemical information similar to that mentioned for Auger spectra, but it is easier to interpret because it involves only one simple transition rather than several. As the chemical environment of an atom changes, electron transfer from that atom to its surroundings varies and the binding energies of core levels change accordingly

[14]. Information such as this is invaluable in determining if chemical reactions take place in an adhesive joint following use in the field or after accelerated testing. In thin-film technology, changes in electron spectra allow the study of diffusion and alloying between thin films or thin-film/substrate combinations. Inert gas sputtering may also be used with XPS to provide elemental profile information.

Determining Locus of Failure

When we use these spectrochemical tools to determine species on the surface, we probably use them slightly differently each time, especially in adhesive bonding. Even when chemical and morphological information has been collected, interpretation may be difficult. Just how do we decide where failure has occurred? In the typical complex adhesive bonded system, we have several interfacial regions as shown in Fig. 3. Each of the materials

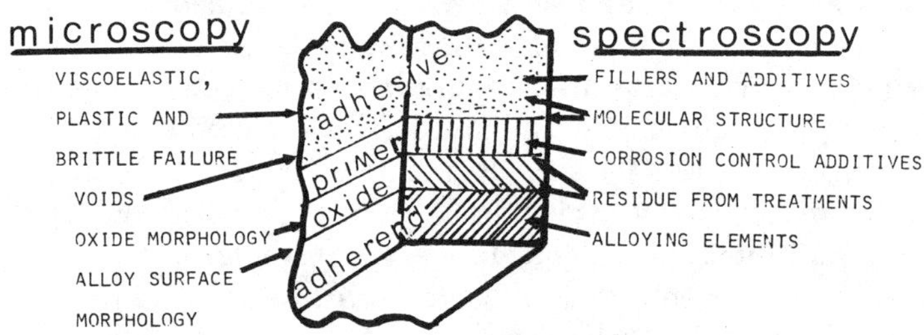

FIG. 3—*Typical complex adhesive bond and features which may be studied by microscopy and spectroscopy.*

coming together to form these interfaces has its own individual chemical signature. The substrate for instance usually contains alloying elements which vary in content between the surface and bulk. In addition to alloying elements, surface treatments leave behind elements characteristic of each treatment. For instance the popular FPL (Forest Products Laboratory) etch for aluminum alloys consists of sulfuric acid and sodium dichromate in distilled water and leaves a detectable amount of chromium on the alloy surface. Primers often contain anions and cations which can be followed by spectrochemical methods. These additives (such as strontium chromate) are usually placed in the primer to provide corrosion protection in the coating. The adhesive often contains fillers to provide conductivity or match coefficient of thermal expansion. A pure polymeric adhesive with no additives can cause problems since the elements present may look just the same to some characterization methods as organic contamination. It is here that vibrational spectroscopy or XPS provides important information on *molecular* configuration. The capability of differentiating isotopes with the SIMS method allows doping of the adhesive with an isotope not normally found in the adhesive such as ^{13}C or ^{15}N.

In addition to determining what elements exist on failure surfaces, it is most important to *look* at these surfaces using optical microscopy and SEM. Dwight and co-workers [*15*] have used the SEM extensively, especially in all cohesive failures, to attempt to understand the mechanism by which failure occurs. Plastic and brittle failure mechanisms are easily differentiated on the polymer surface. Initial and final flaws and voids may be determined and are of importance in evaluating joint performance. Some of the features observed by microscopy and the areas analyzed by spectroscopy are summarized in Fig. 3.

Discussion of Spectrochemical Characterization of Failure Surfaces

Failure surfaces from thin and thick-thin adherend wedge tests have been analyzed by the ISS-SIMS technique. The double cantilever beam wedge test provides information about adherend surface preparation in that it is sensitive to different surface treatments and can discriminate between bonding processes that give good and poor service performance [*16*]. The specimen consists of two thin adherends or a thick-thin configuration to concentrate stresses along the interface. A wedge is driven into the bondline as depicted in the drawing in Fig. 4. The position of the crack leading edge is determined microscopically, and then the specimen is subjected to various external stimuli such as changes in temperature and relative humidity. The propagation of the crack tip is followed with time. Sometimes when the wedge is driven into the bondline, separation of the specimen occurs over a portion of the bondline as in the photographs in Fig.4. Here the wedge was driven in as shown, first causing cohesive failure in the adhesive at the left, then apparent adhesive failure between *a* and *c* during testing at 71 °C (160 °F) and 95 percent relative humidity, followed at the right again by cohesive failure when the specimen is opened following the test. The areas *abc* ("clean adherend") and *def* ("clean adhesive") are matching regions from specimens which failed. The adherend shows no indication of adhesive either visually or in SEM, although there are slight reflectivity differences across the surface. Perhaps staining techniques such as described by Brett [*7*] would be effective in outlining areas which contained thin films of adhesive when they exist. With extremely thin films, however, it is doubtful if staining would provide any information or locate the locus of failure where a primer is used and therefore several interfaces exist. ISS and SIMS data for the DE-46 "clean" adherend at positions *a,b*, and *c* are shown in Fig. 5. These spectra indicate substantially adhesive failure at the oxide-primer interface at *a* and *b*. The increase in chromium content at *c* may indicate some cohesive failure in the oxide or adhesive

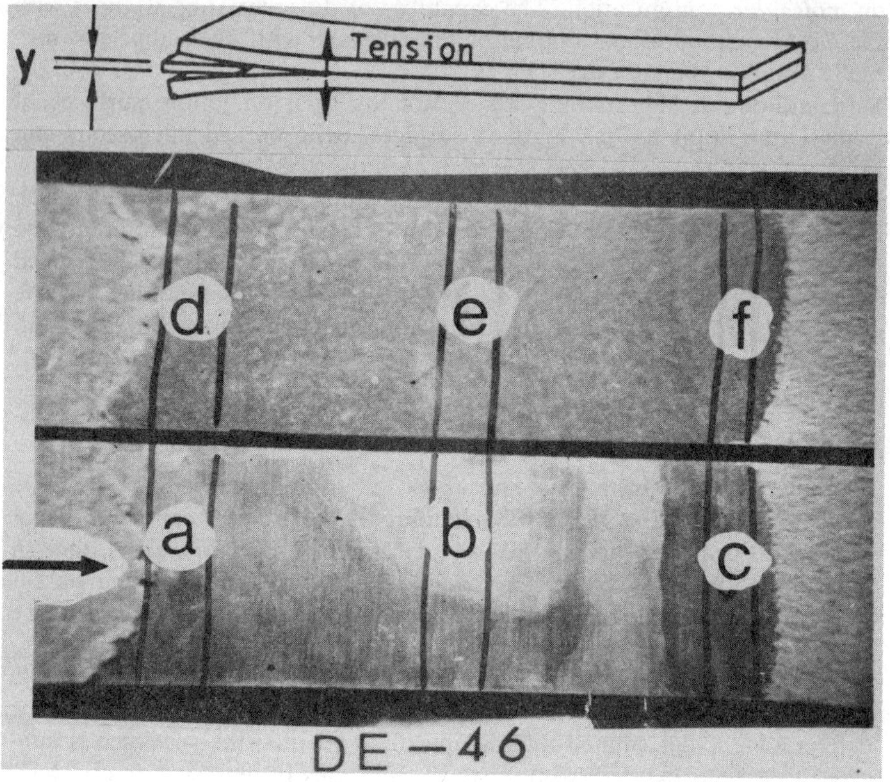

FIG. 4—*Diagram of wedge test for adhesive bonds* (top) *with both sides of test specimen following failure.*

failure at the oxide-alloy since the original adherend was prepared in a standard FPL etch, which contains chromate ions. The matching "clean" adhesive side gave the ISS-SIMS data shown in Fig. 6. Spectra at *d* and *e* appear to be contaminated primer or a mixed primer-adhesive zone, while *f* looks like it comes from a contaminated film of oxide which has adhered to the adhesive. The appearance of aluminum even at position *d* seems to indicate some cohesive failure in the oxide or in the oxide-metal interface all the way across the "clean" region of the specimen. ISS-SIMS data from other similar failures also show convincingly that the failure was similar to the DE-46, with failure near the end of the crack occurring in the oxide or at the oxide-metal interface. Analysis of several such fractures by ISS-SIMS after exposure to high humidity and elevated temperature indicates early crack propagation at the primer-oxide interface with large amounts of impurity ions (Na^+, K^+) present in this region. Studies were carried out with commercial adhesives and several alloy adherends to determine

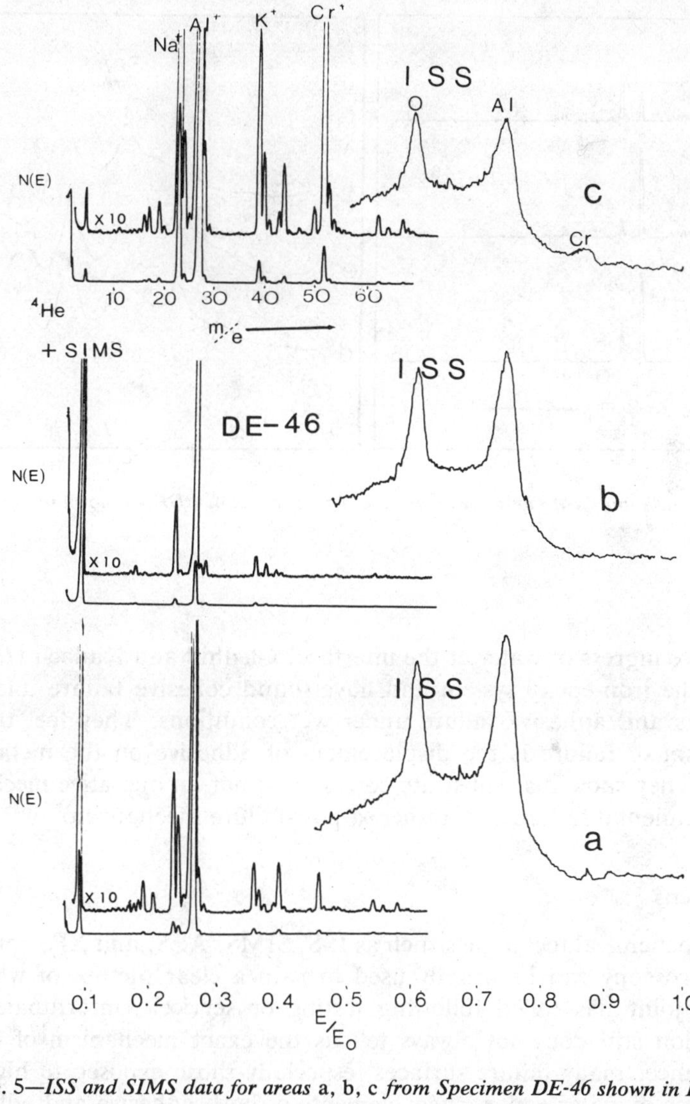

FIG. 5—*ISS and SIMS data for areas* a, b, c *from Specimen DE-46 shown in Fig. 4.*

the origin of these ions [*17*]. It was found that all materials contained small bulk amounts of numerous common elements. When the temperature of these materials was raised to approximately the adhesive cure temperature (121 to 177°C) (250 to 350°F), the very mobile ions such as Li^+, Na^+, and K^+ were concentrated on the surface. Perhaps if the conditions for bond failure under water attack were those for hydrolysis, the diffusion of alkali ions to the surface would increase the osmotic potential and enhance the

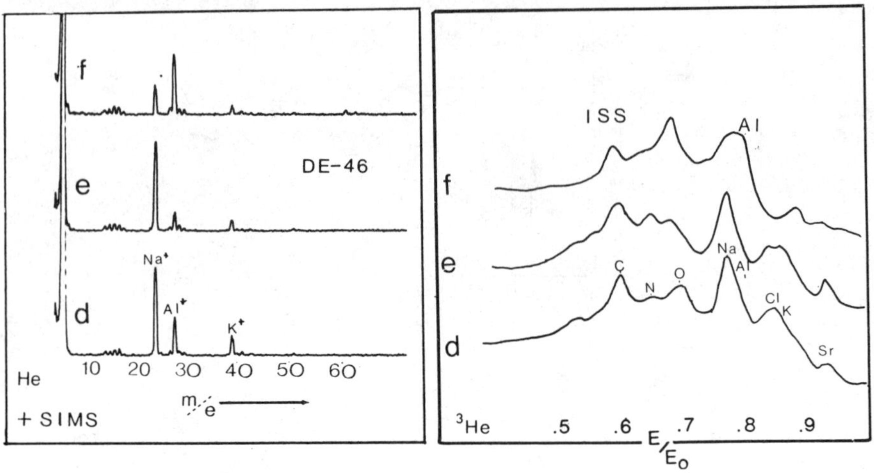

FIG. 6—*ISS and SIMS data for areas* d, e, f *from Specimen DE-46 shown in Fig. 4.*

destructive ingress of water at the interface. Gledhill and Kinloch [*18*] have studied the iron-epoxy system and have found cohesive failure under dry conditions and adhesive failure under wet conditions. They feel that the mechanism of failure is the displacement of adhesive on the metal-oxide surface. They show that substrate corrosion is not an operative mechanism in environmental failure but, rather, a post-failure mechanism.

Conclusions

Spectrochemical techniques such as ISS, SIMS, AES, and XPS combined with microscopy can be usually used to gain a clear picture of where an adhesive joint has failed following testing or service. Unfortunately this information still does not always tell us the exact mechanism of failure. For instance, many failure surfaces, especially those exposed to high concentrations of water vapor, show elements of both adhesive and adherend. Such a result would suggest the presence of a weak boundary layer, but Good [*1*] shows that the transfer of some material from one phase to the other does not prove the existence of a WBL at or adjoining the interface before failure. He does not rule out the WBL, but states only that it is not a universal concept which explains all apparent interfacial failures. Regardless of these interpretive difficulties, chemical information about the surface certainly will help clarify the mechanism of failure in adhesive bonds and evaporated films.

References

[1] Good, R. J., this publication, pp. 18–27.
[2] Bikerman, J. J., this publication, pp. 30–38.
[3] Mattox, D. M., this publication, pp. 54–62.
[4] Bikerman, J. J. in *Recent Advances in Adhesion*, Lieng-Huang Lee, Ed., Gordon and Breach, New York, 1973, pp. 351–356.
[5] Bikerman, J. J., *Journal of Colloid Science*, Vol. 2, 1947, p. 163.
[6] Good, R. J., *Journal of Adhesion*, Vol. 4, 1972, p. 133.
[7] Brett, C. L., *Journal of Applied Polymer Science*, Vol. 18, 1974, p. 315.
[8] Mittal, K., *Electrocomponent Science and Technology*, Vol. 3, 1976, pp. 21–42.
[9] Park, R. L. in *Surface Analysis Techniques for Metallurgical Applications, ASTM STP 596*, American Society for Testing and Materials, 1976, pp. 3–17.
[10] Smith, D. P., *Journal of Applied Physics*, Vol. 38, 1967, p. 340.
[11] Analytical Systems Technical Notes, 3M Company, St. Paul, Minn.
[12] Benninghoven, A., *Surface Science*, Vol. 28, 1971, p. 541.
[13] Leys, J. A. and McKinney, J. T., Extended Abstract of Paper Presented at Tenth Annual Meeting of the Microbeam Analysis Society, Las Vegas, Nev. 11–15 Aug. 1975.
[14] Palmberg, P. W., "New Technology for Surface Characterization," AFML-TR-73-69, Air Force Materials Laboratory, Dayton, Ohio, April 1973, p. 319.
[15] Dwight, D. W., Counts, M. E., and Wightman, J. P., "Surface Analysis and Adhesive Bonding, II Polyimides," ICSS, San Juan, P.R., June 1976; *Proceedings* to be published by Academic Press.
[16] Marceau, S. A. and Scardino, W., AFML-TR-75-3, Air Force Materials Laboratory, Dayton, Ohio, Feb. 1975.
[17] Baun, W. L., unpublished results.
[18] Gleadhill, R. A. and Kinloch, A. S., *Journal of Adhesion*, Vol. 6, 1972, p. 315.

D. M. Mattox[1]

Thin-Film Adhesion and Adhesive Failure—A Perspective*

REFERENCE: Mattox, D. M., "**Thin-Film Adhesion and Adhesive Failure—A Perspective,**" *Adhesion Measurement of Thin Films, Thick Films, and Bulk Coatings, ASTM STP 640,* K. L. Mittal, Ed., American Society for Testing and Materials, 1978, pp. 54-62.

ABSTRACT: A practical definition of "good adhesion" is that the interfacial region (or nearby material) does not fail under service conditions nor at unacceptably low stress levels under test conditions. Adhesion is a macroscopic property that depends on the chemical and mechanical bonding across the interfacial region, the intrinsic stress and stress gradients, and the adhesive failure mode. The failure mode depends on the interfacial structure and the stress to which the interface is subjected. Good adhesion is promoted by: (1) strong bonding across the interfacial region, (2) low stress gradients, either from intrinsic or applied stress, (3) absence of easy fracture modes, and (4) no long-term degradation modes. Adhesion testing must take into account the possible deformation and fracture modes, and a testing program must be devised to test all aspects of adhesion.

KEY WORDS: adhesion, interfaces, adhesion tests, interfacial failure, failure, deformation

The interfacial region between a thin film and a surface determines many physical and electrical properties of the couple. These include contact resistance, contact noise, acoustic coupling, electronic trapping and recombination, thermal conductance, and film adhesion. Measurements of these properties may be used to define the state of an interfacial region.

A practical definition of "good adhesion" is that the interfacial region (or nearby material) does not fail under assembly or service conditions nor at unacceptably low stress levels under test conditions. Adhesion or adhesive strength is a macroscopic property that depends on the bonding across the interfacial region, local stresses, and the adhesive failure mode. The failure mode depends on the type of stress to which the interfacial region

*Work supported by U. S. Energy Research and Development Administration.

[1]Division supervisor, Surface Metallurgy, Sandia Laboratories, Albuquerque, N. Mex. 87115.

is subjected. Typical stresses encountered in assembly and service may result from mechanical loading (tensile, shear, cyclical), thermal (high and low temperatures, cycling), chemical environment (corrosion—both chemical and electrochemical), and electrical environment. Thus, the failure mode may be determined by the environment, the chemical and electrochemical properties of the film and substrate materials, film morphology, the mechanical properties, defect morphology of the interfacial region, and the manner in which external stresses are applied. In addition, adhesive failure may be time dependent.

Good adhesion is promoted by

1. Strong atom-atom bonding within the interfacial region.
2. Low local stress levels.
3. Absence of easy deformation or fracture modes.
4. No long-term degradation modes.

These characteristics depend on the nature of the interfacial region, which in turn depends on the interactions between the depositing material and the surface.

Interfacial Regions

Interfacial regions may be classed as [*1-3*][2] mechanical, monolayer-to-monolayer (abrupt), compound, diffusion, or pseudo-diffusion and combinations of these types. The type of interfacial region formed during deposition depends on the substrate surface morphology, contamination, chemical interactions, the energy available during interface formation, and the nucleation behavior of the depositing atoms.

When atoms impinge on a surface, they lose energy to the surface and finally condense by forming stable nuclei [*4*]. During condensation, the adatoms have a degree of mobility on the surface which is determined by their kinetic energy and the strength and type of interaction between the adatom and the surface. A strong surface-atom interaction will give a high density of nuclei and a weak interaction will result in widely spaced nuclei. These nuclei will grow to form a continuous film. It has been proposed that the nuclei density and the nuclei growth mode determine the effective interfacial contact area and the development of voids in the interfacial region [*5,6*]. Nuclei density and orientation formed during deposition can be affected by ion bombardment [*7*], electric fields [*8*], gaseous environment [*9*], contaminant layers [*10*], surface impurities [*11*], surface defects [*4*], and deposition techniques [*12*]. In addition to the effective contact area, the mode of growth of the nuclei will determine the defect morphology in the interfacial region [*6*] and the amount of diffusion and reaction

[2] The italic numbers in brackets refer to the list of references appended to this paper.

tantalum that small amounts of surface contamination will affect the crystallographic orientation of the gold nuclei and the reaction depth [*10*].

The mechanical interface is characterized by mechanical interlocking of the film material with a rough surface. The strength of this interface will depend on the mechanical properties of the materials. Surfaces may be deliberately roughened or made porous in order to increase the mechanical interlocking [*13*]. It should be noted that often deposition of a film on a rough surface gives a porous film due to geometrical shadowing effects [*14*].

The monolayer-to-monolayer (abrupt) type of interface is characterized by an abrupt change from the film material to the substrate material in a distance on the order of the separation between atoms (2 to 5 Å). This type of interface may be formed when there is no diffusion and little chemical reaction between the film atoms and the substrate surface. This lack of interaction may be due to the lack of solubility between materials, little reaction energy available, or the presence of contaminant layers. In this type of interface, defects and stresses will be confined to a narrow region.

The compound interface is characterized by a constant composition layer, many lattice parameters thick, created by the chemical interaction of the film and substrate material. The compound formed may be either an intermetallic compound or some other chemical compound such as an oxide. Compounds are usually brittle materials. In this type of interface there may be abrupt physical and chemical discontinuities associated with the abrupt-phase boundaries. Often during compound formation there are segregations of impurities at the phase boundaries [*15*] and stresses are generated due to lattice mismatching. Porosity may develop in the interfacial region.

In the diffusion type of interface there is a gradual change in composition, intrinsic stress, and lattice parameters across the interfacial region (graded interface) [*16*]. If there is a difference of diffusion rates of the film atoms and the substrate atoms, Kirkendall porosity may be formed in the interfacial region [*17*]. For thin films Kirkendall porosity may not develop because of rapid surface diffusion [*18*].

Under more energetic situations such as ion bombardment, codeposition, ion implantation, or melting/quenching, a "pseudo-diffusion" type of interface may be formed by materials which are normally insoluble. Ion bombardment prior to film deposition may increase the interfacial solubility by creating very high concentrations of point defects [*19*] or stress gradients [*20*] or both which will enhance diffusion.

Obviously, a combination of several types of interfacial regions is possible. Compound and graded types of interface may be formed by controlling the environment or film composition during the initial portion of the film deposition [*21*], as well as heating during and after film deposition [*16*].

Interfacial Bonding

Bonding between unlike atoms and materials may be due to chemical, electrostatic, or polarization bonding. In chemical bonding the interaction is due to the transfer or sharing of electrons. Electrostatic bonding is due to charge separation and the resulting electrostatic attraction. Polarization bonding is due to asymmetry in the electric field around atoms or molecules and the resulting attraction. Chemical bonds vary in strength depending on the degree of electron transfer or sharing. If the degree of charge transfer is high, compounds or ionic solids are formed which are usually strong but brittle. If electron sharing is the bonding mechanism, alloys or metallic-type materials are formed which are more ductile. Electrostatic bonding has been proposed as the principal adhesion mechanism for metal-polymer systems [*22*] and is important in metal-atom nucleation on an insulator surface [*23*].

One would expect that a single atom on a surface would be bound with a different energy than if it were in a nucleus or a film since the like and unlike atom coordination changes with coverage. The bonding across the interfacial region should be the summation of the individual adatom/substrate bonds less any bond energy lost due to strain or defect formation. Thermal desorption studies of an insoluble/noncompound-forming metal system (gold-tungsten) have shown that a single absorbed metal atom (gold) on a metal surface (tungsten) will be bound with an energy near the cohesive energy of the film material (3.5 eV/atom) [*24*]. Generally the adhesion between nonsoluble/noncompound-forming film substrate couples is poor unless energetic deposition processes are used [*5,25*]. To improve adhesion, an intermediate layer of a metal which will react with both the film and substrate is used as a "glue" layer [*5*].

The problem of poor adhesion between insoluble materials must not lie in the intrinsic strength of the atom-surface bond but rather in the contact area, bond strain, or some other interfacial characteristic. When gold is deposited on an oxide surface, the adhesion is normally poor; and this is attributed to the low energy associated with the gold-oxygen bond. By sputter depositing the gold in an oxygen discharge, very adherent gold films can be deposited on oxide surfaces [*26*]. The only detectable difference is an increased nucleation density of the gold sputter deposited in the oxygen discharge. This may lead to an increased effective contact area.

Interfacial Stress

Intrinsic stress may be generated in the interfacial region by a variety of mechanisms [*27*]. The stress may be compressive or tensile in nature. For reasons which are not well defined, deposited thin films may have intrinsic stresses approaching the yield stress of the material. These stresses have been found to depend on the film thickness, deposition parameters, and deposition technique, and may be related to impurity atoms, nuclei coalescence patterns or quenching effects. High intrinsic stress may lead to anomalous film behavior such as room temperature grain growth in metal films [*28*] or very low-temperature strain points of deposited glass films [*29*]. Stresses may arise at phase boundaries due to differences in their crystallographic lattices and to property variations of the materials. Some of these stresses may be relieved by the formation of interfacial dislocations [*4*]. Analysis of this intrinsic stress distribution is difficult [*30*].

Adhesion testing utilizes externally applied stresses which may vary from tangential shear [*31*] to tensile [*32*], though the stresses appear at the interfacial region in a very complex manner due to variations in the physical properties of the material and the usually nonhomogeneous nature of the film and interfacial material. The presence of pores and voids in the interfacial region will give stress concentration and alter the values of the tensile and shear components of the interfacial stress.

The intrinsic and extrinsic stresses may be additive such that an interface with a high intrinsic stress may fail spontaneously or with little applied stress even though there is strong chemical bonding between film and substrate.

Interfacial Fracture

Fracture is composed of crack initiation and propagation. Initiation will begin at a flaw which allows stress concentration or bond weakening. Propagation of the fracture will occur by repeated bond breaking. Bonds may be broken by physical straining, by chemical effects, or by a combination of both. In ductile material, the stress at the tip of a propagating crack is relieved by plastic deformation of the surrounding material [*33*]. In a brittle material, there is little plastic deformation [*34*]. Fracture toughness [*35*] is a measure of the energy which must be added to the system to give rapid fracture propagation, and less energy is used to propagate a fracture in a brittle material than in a ductile material. A fracture will preferentially propagate in a plane of weakness such as might be found in an abrupt interface, an interface with a large number of voids, or a region of high intrinsic stress.

In stress corrosion cracking [*36*] or static fatigue [*37*] a foreign species at the crack tip will strain or weaken the bond, allowing the fracture to propagate. In metals, the corrosive agent may convert the material at the crack tip from a ductile to a brittle material, thereby reducing the amount of energy necessary to propagate the crack. Corrosion products may expand the crack by wedging.

Pores or cracks in the interfacial region act not only as crack initiators and stress concentrations but allow interfacial cracks to be propagated parallel to the direction of the applied force [*38*]. Analysis of the fracture surfaces (fractography) may be used to determine failure modes in some cases.

Time-Dependent Interfacial Changes

It has been reported that in some metal film on oxide substrate systems the adhesion improves with time at low temperatures [*40*]. This is attributed to the migration of active gases to the interface to form transition compounds or to stress relief. In the solid phase bonding of materials, a similar increase of bond strength with time has been noted and is attributed to stress relief in the interfacial region [*41*].

Treatment at high temperatures may improve adhesion by forming a desirable type of interface [*14*] or may decrease adhesion by diffusion void formation [*17*], diffusing a reactive species away from the interfacial region. Thermal and mechanical cycling likewise may lead to a time-dependent failure by fatigue [*42,43*].

Corrosion [*44*] may give a long-term loss of adhesion. It has been shown that only a small amount of halide and moisture can result in electrochemical corrosion of some common thin-film metallization systems such as titanium-gold [*44,45*]. Extensive work has been done to identify metallization systems which are not susceptible to electrochemical corrosion. Corrosion not only depends on the environment and the materials involved but also on the availability of the corrosive media at the interface. For this reason, film porosity [*46*] and pinholes are important in that they not only serve to absorb and trap corrosive liquids and vapors but allow them access to the interfacial region. Usually film porosity is measured by corrosion [*47*] or dissolution techniques [*48*]

Adhesion Testing

There has been considerable effort devoted to developing various adhesion testing methods [*49*], but to be meaningful a *testing program* must be developed for each situation. This means that the testing methods must duplicate the stresses to which the components are subjected in assembly and service. In addition, aging of components in service environments must be performed in order to determine the long-term adhesion properties. [*43*]

There have been many attempts to use adhesion tests to give quantitative information about the chemical bonding across an interface, though the analysis is highly questionable [*50*]. In bulk materials we find that the practical strength of a material never approaches the theoretical strength because failure is controlled by deformation and fracture processes. The chemical environment can affect the mechanical properties of hard [*51*] and brittle materials [52]. For bulk materials, a useful concept is that of fracture energy (toughness) [35]. If we consider that adhesive failure is a fracture process, then the energy necessary to fail an interface may be a more useful concept than that of chemical bond strength.

Loss of adhesion is the culmination of interfacial failure. Other techniques may be useful in detecting the degradation of interfacial adhesion. These include contact resistance [53], contact noise, acoustic coupling, shock wave coupling [54,55], and other processes which depend on the physical and electrical properties of the interface. Acoustic emission is used to detect fracture propagation in bulk materials [56] and may be useful in interfacial studies. Other properties such as film porosity and stress may have an indirect effect on adhesion and, in particular, on the long-term stability of film-substrate specimens. Since these properties are very dependent on deposition conditions, they should be studied and controlled.

Summary

Adhesion is a macroscopic property of a film-substrate system and depends on a number of factors. In particular, the measured adhesion depends on the failure (fracture) mode. If easy deformation or fracture modes are present, failure will occur at low stress levels even if the chemical bonding is high. A useful concept may be that of fracture energy, which is used widely in failure studies of bulk material and to some extent in organic adhesive testing [57]. To be meaningful, an adhesion-testing program must be designed to subject the film-substrate couple to the stresses it would encounter in production and in service.

References

[1] Mattox, D. M., "Interface Formation and the Adhesion of Deposited Thin Films," presented to the 1964 Gordon Research Conference on Adhesion and published as Sandia Laboratories Report SC-R-65-852, Jan. 1965 (available from National Technical Information Service, U.S. Department of Commerce, Springfield, Va. 22161).
[2] Berry, R. W., Hall, P. M., and Harris, M. T. in *Thin Film Technology,* D. Van Nostrand, New York, 1968, p. 584.
[3] Chapman, B. N., *Journal of Vacuum Science & Technology,* Vol. 11, 1974, p. 108.
[4] Neugebauer, C., *Handbook of Thin Film Technology,* L. Maissel and R. Glang, Eds., McGraw-Hill, New York, 1970, Chapter 8.
[5] Mattox, D. M., *Thin Solid Films,* Vol. 18, 1973, p. 173.
[6] Lloyd, J. R. and Nakahara, S., *Journal of Vacuum Science & Technology,* Vol. 14, 1977, p. 655.
[7] Tarng, M. L. and Wehner, G. K., *Journal of Applied Physics,* Vol. 43, 1972, p. 2268.
[8] Zanghi, J. C., Gauch, M., Metois, J. J., and Masson, A., *Thin Solid Films,* Vol. 33, 1976, p. 193.
[9] Paulson, G. G. and Friedberg, A. C., *Thin Solid Films,* Vol. 5, 1970, p. 47.
[10] Elliot, A. G., *Surface Science,* Vol. 51, 1975, p. 489.
[11] Sunahl, R. C., *Journal of Vacuum Science & Technology,* Vol. 9, 1972, p. 181.
[12] Westwood, W. D., *Progress in Surface Science,* Vol. 7, 1976, p. 71.

[13] Jennings, C. W., *Journal of Adhesion,* Vol. 4, 1972, p. 25.
[14] Bland, R. D., Kominiak, G. J., and Mattox, D. M., *Journal of Vacuum Science & Technology,* Vol. 11, 1974, p. 671.
[15] Seah, M. P., *Surface Science,* Vol. 53, 1975, p. 168.
[16] Katz, Gerald, *Thin Solid Films,* Vol. 33, 1976, p. 99.
[17] Philofsky, Elliot, *Proceedings,* 9th Annual Reliability Physics Symposium, Institute of Electrical and Electronics Engineers, IEEE Catalog No. 71-C-9-Phy., 1971, p. 114.
[18] Selikson, Bernard, *Applied Physics Letters,* Vol. 14, 1969, p. 283.
[19] Kornelsen, E. V., *Radiation Effects,* Vol. 13, 1972, p. 227.
[20] Dearnaley, G., *Applied Physics Letters,* Vol. 28, 1976, p. 244.
[21] Hollar, E. L., Rebarchik, F. N., and Mattox, D. M., *Journal of the Electrochemical Society,* Vol. 117, 1970, p. 117.
[22] Mittal, K. L., *Journal of Vacuum Science & Technology,* Vol. 13, 1976, p. 19.
[23] Kasprzak, L., Laibowitz, R., Herd, S., and Ohring, M., *Thin Solid Films,* Vol. 22, 1974, p. 189.
[24] Godwin, R. P. and Lüscher, E., *Surface Science,* Vol. 3, 1964, p. 42.
[25] Karnowsky, M. M. and Estill, W. B., *Review of Scientific Instruments,* Vol. 35, 1964, p. 1324.
[26] Mattox, D. M., *Journal of Applied Physics,* Vol. 37, 1966, p. 3613.
[27] Campbell, D. S., *Handbook of Thin Film Technology,* L. Maissel and R. Gland, Eds., McGraw-Hill, New York, 1970), Chapter 12.
[28] Patten, P. W., McClanghan, E. D., and Johnston, J. W., *Journal of Applied Physics,* Vol. 42, 1971, p. 4371.
[29] Mattox, D. M. and Kominiak, G. J., *Journal of the Electrochemical Society,* Vol. 120, 1973, p. 1535.
[30] Röll, K., *Journal of Applied Physics,* Vol. 47, 1976, p. 3224.
[31] Lin, D. S., *Journal of Physics D: Applied Physics,* Vol. 4, 1971, p. 1977.
[32] Jacobson, R. and Kruse, B., *Thin Solid Films,* Vol. 15, 1973, p. 71.
[33] Smith, E., *Engineering Fracture Mechanics,* Vol. 6, 1974, p. 213.
[34] Pugh, S. F., *British Journal of Applied Physics,* Vol. 18, 1967, p. 129.
[35] Eftis, J., Jones, D. L., and Liebowitz, H., *Engineering Fracture Mechanics,* Vol. 7, 1975, p. 491.
[36] *Proceedings,* Conference on Fundamental Aspects of Stress Corrosion Cracking, R. W. Staehle, A. J. Forty, and D. Van Roozen, Eds., National Association of Corrosion Engineers, 1969.
[37] Wiederhorn, S. M. and Bolz, L. H., *Journal of the American Ceramic Society,* Vol. 53, 1970, p. 543.
[38] Kendall, K., *Journal of Material Science,* Vol. 11, 1976, p. 638.
[39] *Fractography and Atlas of Fractographs, Metals Handbook,* Vol. 9, 8th Ed. American Society for Metals, Metals Park, Ohio, 1974.
[40] Benjamin, P. and Weaver, C., *Proceedings of the Royal Society, Series A: Mathematical and Physical Sciences,* Vol. 255, 1961, p. 516.
[41] Jellison, J. L., "Kinetics of Thermocompression Bonding of Gold," presented to the Institute of Electrical and Electronics Engineers Electronics Components Conference, San Francisco, Calif., April 1976, to be published in the *Proceedings.*
[42] Hasselman, D. P. H., *Journal of the American Ceramic Society,* Vol. 52, 1969, p. 600.
[43] Plumbridge, W. J., *Journal of Materials Science,* Vol. 7, 1972, p. 939.
[44] Kolesar, S. C., *Proceedings,* 12th Annual Reliability Physics Symposium, Institute of Electrical and Electronics Engineers, IEEE Catalog No. 74 CH 0839-Phy., 1974, p. 155.
[45] Speight, J. D. and Bill, M. J., *Thin Solid Films,* Vol. 51, 1973, p. 325.
[46] Rehig, D. L., *Plating and Surface Finishing,* Vol. 61, 1974, p. 43.
[47] Frant, M. S., *Journal of the Electrochemical Society,* Vol. 108, 1961, p. 774.
[48] Morrissey, R. J., *Journal of the Electrochemical Society,* Vol. 119, 1972, p. 446.
[49] Mittal, K. L., *Electrocomponent Science & Technology,* Vol. 3, 1976, p. 21.
[50] Butler, D. W., Stoddart, C. T. H., and Stuart, P. R., *Journal of Physics D: Applied Physics,* Vol. 3, 1970, p. 877.
[51] Almond, E. A., May, A. T., and Roebuck, B., *Journal of Material Science,* Vol. 11, 1976, p. 568.

[52] *The Science of Hardness Testing and Its Research Applications*, J. H. Westbrook and H. Conrad, Eds., American Society for Metals, Metals Park, Ohio, 1973, Section 6.
[53] Speight, J. P., Gerstenberg, D., and Basseches, H., *Proceedings*, 9th Annual Reliability Physics Symposium, Institute of Electrical and Electronics Engineers, IEEE Catalog No. 71-C-9-Phy., 1971, p. 195.
[54] Anderholm, N. C. and Goodman, Albert, "Method and Apparatus for Measuring Ad-!hesion of Material Bonds," U.S. Patent No. 3,605,486, 20 Sept. 1971.
[55] Stephens, A. W. and Vossen, J. L., *Journal of Vacuum Science & Technology*, Vol. 13, 1976, p. 38 (abstract).
[56] Evans, A. G. and Linzer, M., *Journal of the American Ceramic Society*, Vol. 56, 1973, p. 575.
[57] Mostovoy, Sheldon, Crosley, P. B., and Ripling, E. J., *Journal of Materials*, Vol. 2, 1967, p. 661.

DISCUSSION

K. L. Mittal[1] (*written discussion*)—You seem to suggest that if the failure is in the "interfacial region" then the use of the term "adhesion' is O.K. Would you like to comment?

D. M. Mattox (*author's closure*)—A strict definition of adhesion would apply only to systems where there is a sharp, well-defined boundary between two different materials. Even in this case, the boundary may induce stresses and changes in the nearby material which will allow fracture near, but not at, the interface. Many interfaces do not consist of sharp boundaries, and hence the strict definition of adhesion is meaningless. I therefore propose that my "practical" definition is more meaningful to all except the purist.

[1] IBM Corporation, Hopewell Junction, N. Y. 12533.

W. D. Bascom,[1] *P. F. Becher,*[1] *J. L. Bitner,*[1] *and*
J. S. Murday[1]

Use of Fracture Mechanics Concepts in Testing of Film Adhesion

REFERENCE: Bascom, W. D., Becher, P. F., Bitner, J. L., and Murday, J. S., "**Use of Fracture Mechanics Concepts in Testing of Film Adhesion,**" *Adhesion Measurement of Thin Films, Thick Films, and Bulk Coatings, ASTM STP 640,* K. L. Mittal, Ed., American Society for Testing and Materials, 1978, pp. 63–81.

ABSTRACT: The constant-compliance and applied-moment double-cantilever beam tests for adhesive fracture energy have been adapted to the measurement of the adhesion of thick-film metallizations on alumina substrates. The thick-film tests involve beams soldered to metallization strips and measure the strain energy release rate, $\mathcal{G}_c$. The fracture results are compared with peel strengths from solder-wire peel tests of the same metallizations. Both the fracture and peel tests indicate failure in the thick films but near the film/alumina boundary. The constant-compliance and the peel test data exhibit a bimodal distribution. The magnitude of the fracture energies indicate that failure usually occurs in the glass phase of the film but that film adherence is greatly enhanced by interlocking between the metal and glass phases. This interlocking and thus the film adhesion are strongly dependent on firing time and temperature.

KEY WORDS: thick films, adhesion, fracture mechanics, adhesion testing, microelectronics, ceramic-metal bonding, fractography, thick film processing

Despite the importance of adherence to the performance of coatings, test methods currently in use are essentially empirical with the physical significance of the measurement often in doubt. These test methods such as peel and scratch tests are useful for comparison purposes and are valuable for ranking the adhesion[2] of different films for quality-control pur-

[1]Head, Adhesion and Polymer Composites Section, Polymeric Materials Branch, head, Ceramic Processing and Characterization Section, Ceramics Branch, research chemist, Adhesion and Polymer Composites Section, Polymeric Materials Branch, and head, Surface Analysis Section, Surface Chemistry Branch, respectively, Naval Research Laboratory, Washington, D.C. 20375.

[2]The term "adhesion" is used here in the broad sense in that it refers to the failure strength of separating an adhesive (adherate) from an adherend. Failure need not occur along the boundary between the two phases.

poses. The test data have no inherent significance, however, with respect to material properties or to failure mechanisms.

In the work reported herein, a fracture mechanics approach was taken to coating adhesion in which the fracture testing methodology developed for organic adhesives [*1*][3] was utilized. Considerable success has been realized in relating adhesive fracture behavior to resin composition [*1,2*], morphology [*2*], and joint geometry [*3*]. Moreover, the test data can be used as failure criteria in the design of structural adhesive bonds. With some modification, this approach should be applicable to coatings.

The merit of a fracture approach to materials testing is that it recognizes that the strength of a solid is characterized not only by material properties but by flaw size as well. This flaw dependence takes explicit form for the case of opening mode ("cleavage") fracture by

$$\sigma_c = g\sqrt{\frac{E\mathcal{G}_{Ic}}{a}}$$

where

σ_c = failure load,
E = tensile modulus,
a = flaw size,
g = dimensional parameter determined by specimen geometry, and
$\mathcal{G}_{Ic}$ = total fracture energy [*4*].

For ideally brittle materials, $\mathcal{G}_{Ic} = 2\gamma$ where γ is the surface free energy of the solid. This relationship holds when the only irreversible deformation is the formation of new surface. Flaw-dependent strength is certainly to be expected for brittle materials but is also characteristic of most of the thermosetting polymers (epoxies and polyesters) and even some thermoplastic polymers [polymethylmethacrylate (PMMA) and polycarbonate]. The fracture energy characterizes the flaw sensitivity and as such is a material property relatable to chemical composition and microstructure although it may show time-temperature dependence and be sensitive to environmental conditions. The fracture energy varies with stress condition (tensile versus shear loading) and is designated $\mathcal{G}_{Ic}$, $\mathcal{G}_{IIc}$, or $\mathcal{G}_{IIIc}$ for opening, in plane shear or torsional shear, respectively, or simply $\mathcal{G}_c$ for mixed-mode loading.

The work reported herein was on the thick-film, conductive metallization widely used in microelectronics technology. These films are formed on ceramic substrates by screen printing inks containing metal powders, glass frits of silicon, boron, lead, and bismuth, and other metals, and an or-

[3] The italic numbers in brackets refer to the list of references appended to this paper.

ganic vehicle. The printings are fired and in the process the vehicle is driven off, the metallic constituents are at least partially sintered together, and the glass forms a bonding layer between the metal conductive and the substrate. In microelectronics devices, connections to the thick film metallizations are made by solder bonding, or by compression or thermal compression bonding. The methods presently in use for testing thick films are the same as used for the thin, evaporated metal films and include, "Scotch-tape" peel, scratch, and cleavage tests.

Experimental

Two separate studies are reported herein, both of which used metallizations prepared by the same facility but tested using different specimen geometries: a constant-compliance double-cantilever beam specimen [*5*] and a constant-moment double-cantilever beam specimen [*6*]. The two designs have the useful feature that the fracture energy is independent of crack length, a parameter which is usually difficult to determine. The constant-compliance specimen has found considerable use in adhesive fracture testing [*4,5*] and is the ASTM Test for Fracture Strength in Cleavage of Adhesives in Bonded Joints (D 3433-75). The more recently developed applied-moment specimen [*7*] has been used for testing bulk ceramics and ceramic/ceramic bonds.

Constant-Compliance Fracture Test

The tapered double-cantilever beam adhesive specimen is illustrated in Fig. 1 (*top*). The two beams, usually made of aluminum, are bonded by the organic adhesive. The short strips of tape (polytetrafluoroethylene resin) at the ends maintain the desired bond thickness and the springs hold the specimen together during heat cure of the adhesive. The force required to open the beams is measured and used to calculate the opening-mode fracture energy. The general relationship for Mode-I fracture of a double-cantilever beam is

$$\mathcal{G}_{Ic} = \frac{P_c^2}{2b}\left(\frac{dC}{da}\right)_p \tag{1}$$

where C is the compliance and b the beam thickness. For a rectangular beam, Eq 1 becomes from simple beam theory

$$\mathcal{G}_{Ic} = \frac{4P_c^2}{b^2E_b}\left[\frac{3a^2}{h^3} + \frac{1}{h}\right] \tag{2}$$

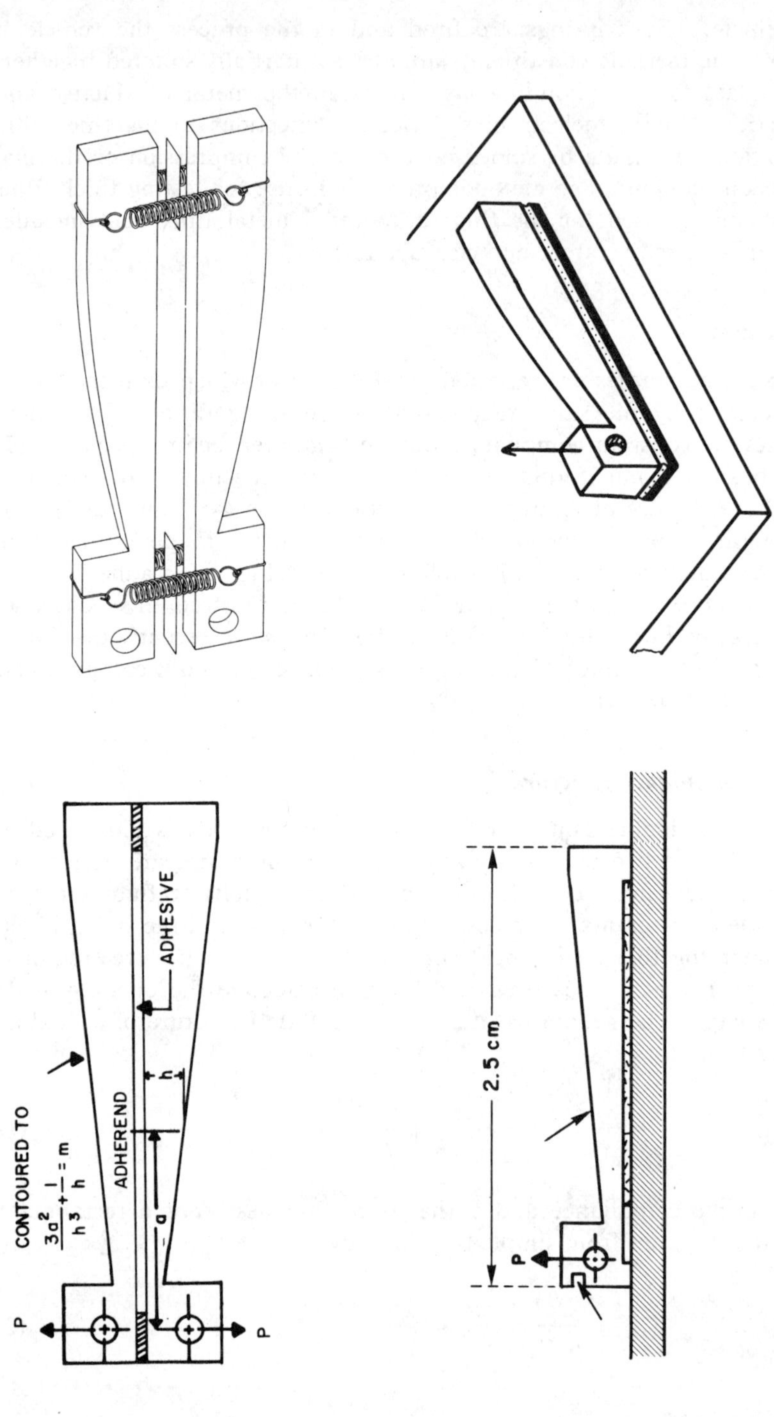

FIG. 1—(top) *Plan and schematic drawings of constant-compliance double-cantilever beam adhesive fracture specimen* [1,2]. (bottom) *Plan and schematic drawings of constant-compliance single beam specimen for thick-film adhesive fracture. Thin arrows indicate tension and compression points for precracking* [5].

where

E_b = bending modulus of the beam material,
a = crack length, and
h = beam height at the crack tip.

To simplify the test the beam is tapered [Fig. 1 (*top*)] such that the bracketed terms in Eq 2 are a constant, m, so that

$$\mathcal{G}_{Ic} = \frac{4P_c^2 m}{b^2 E_b} \tag{3}$$

Note that the fracture energy is independent of crack length, which is usually very difficult to measure.

To apply this technique to thick-film adhesion, a *single* beam test specimen was devised. The beam shape, tapered for $m = 100$, is shown in Fig. 1 (*bottom*) and was cut from brass plate ($E_b \approx$ tensile modulus, 108 GPa). These beams were 1 in. (2.5 cm) long, 0.25 in. (0.63 cm) high at both ends, and 0.08 in. (0.2 cm) thick, and were dip-soldered onto 80-mil-wide (2-mm), 0.75-in.-long (1.9-cm), thick-film strips printed onto 1 by 1 by 0.025-in. (2.5 by 2.5 by 0.06-cm) alumina plates. A schematic of a beam soldered to a thick film strip is given in Fig. 1 (*bottom*). The specimens were mounted in a universal test machine (Instron) by clamping the substrate in a holder on the lower cross arm and applying load to the beam via a hook through the loading hole. Testing was done at 21°C (70°F), 40 percent relative humidity and at a rate of 2×10^{-2} cm/s. The specimen holder was constructed so that the substrate was clamped firmly in a spring-loaded device such that the substrate would not experience any bending deformation [5].

It was found that consistent results could not be obtained using this technique unless a precrack was introduced before testing. A method was devised to initiate a crack and then arrest it by simultaneously applying a lifting force at the beam end and a compressive force at the top of the beam as illustrated in Fig. 1 (*bottom*).

Applied-Moment Fracture Test

The constant-moment specimen is illustrated in Fig. 2. The fracture energy ($\mathcal{G}_I$) of the double-cantilever beam specimen is obtained by equating $\mathcal{G}_I$ to the change in stored strain energy in the specimen when fracture occurs. This change in stored strain energy is based on the geometry of and the material in each cantilever arm. Thus for the thick-film substrate specimens, $\mathcal{G}_{Ic}$ is obtained from

$$\mathcal{G}_{Ic} = \frac{(P_cL)^2}{2t}\left[\frac{12}{E_1b_1h_1^3} + \frac{12}{E_2b_2h_2^3}\right]$$

where

P_cL = applied moment (with P_c the load on each arm and L the moment arm),
b and h = arm dimensions, and
t = thickness of the specimen in the crack plane (here the width of the thick-film conductor line).

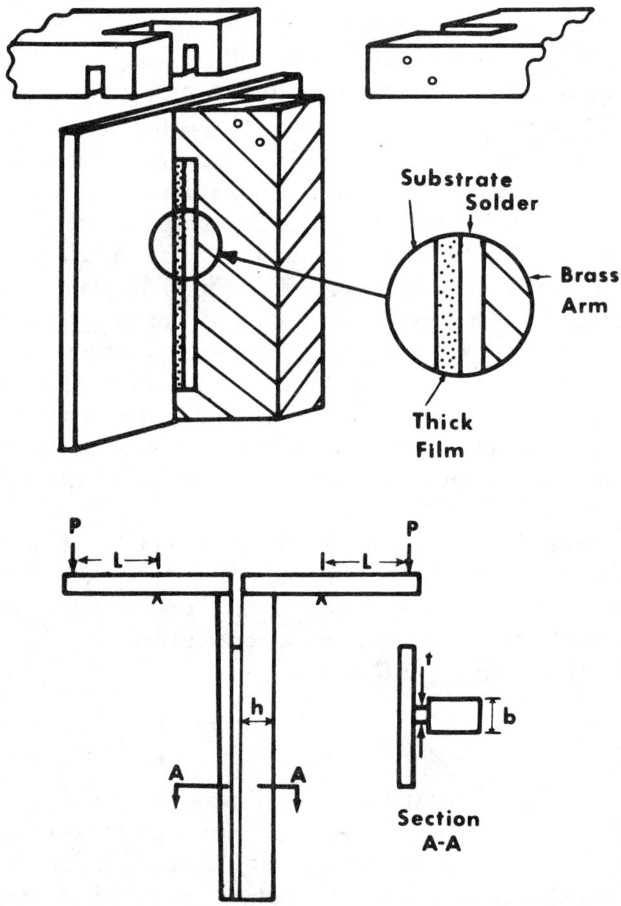

FIG. 2—*Applied-moment double-cantilever beam specimen for thick-film adherence.* (top) *Actual specimen used for thick film, showing load arm attachment and specimen cross section.* (bottom) *Schematic indicates specimen dimensions and how bending moment* (PL) *is applied to specimen* [6].

The subscripts 1 and 2 refer to the alumina and brass cantilever arms, respectively; see Fig. 2.

The test specimens consisted of thick-film strips printed on alumina substrates as used in the constant-compliance tests. The test specimens were then dip-soldered in order to attach the brass cantilever arm. An alumina (0.12 × 2.5 × 2.5 cm) backing substrate was then added and the loading arms attached. Precracks required to determine $\mathcal{G}_{Ic}$ were introduced on some specimens by immersion into liquid nitrogen (LN); however, such precracking was not required as the data for precracked and unprecracked specimens fell within 10 percent of each other, evidence that cracks already existed without the use of the LN_2 immersion.

Solder-Bond Peel Test

Measurements of thick-film peel adhesion strength were made using the method specified by the Electronics Products Division of E. I. duPont de Nemours and Co. This peel test consists of soldering No. 22 copper wire to small (0.2 × 0.2 cm) metallization pads as shown in Fig. 3.

The Metallizations

The two conductor inks examined in this study are listed in Table 1. The optimum firing temperature specified by the manufacturer is indicated. The solders used to bond the fracture test beams were 63Sn/37Pb for the platinum/gold films and 70Sn/18Pb/12In for both the platinum/gold and

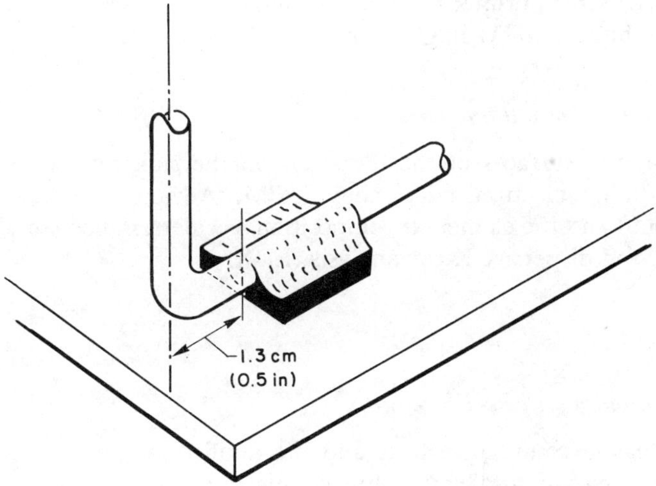

FIG. 3—*Schematic of wire bond peel test.*

TABLE 1—*Statistical analysis of fracture*[a] *energy data.*

Firing Temperature	$\mathcal{G}_{Ic}$(J/m^2) High Values	$\mathcal{G}_{Ic}$(J/m^2) Low Values	Number of Specimens
	Platinum/Gold (duPont 7553-63)Sn/37Pb		
755°C (1391°F)	74.0 ± 28.0 (25%)	17.6 ± 9.1 (75%)	24
[b]860°C (1580°F)	61.0 ± 14.0 (50%)	16.8 ± 4.4 (50%)	18
950°C (1742°F)	...	5.8 ± 3.0 (100%)	25
	Platinum/Gold (duPont 7553)—70Sn/18Pb/12In		
860°C (1580°F)	86.0 ± 30.0 (100%)	...	20
	Gold (duPont 8380)—70Sn/18Pb/12In		
800°C (1472°F)	64.0 ± 33.0 (63%)	12.6 ± 6.3 (37%)	19
[b]900°C (1652°F)[b]	64.0 ± 42.0 (85%)	7.6 ± 3.1 (15%)	30

[a]Constant-compliance fracture test method.
[b]Recommended temperature.

the gold films. The numerical values in the solder composition refer to the weight percent of each element. The tin/lead/indium solder was prepared by Semi-Alloys, Co., Mt. Vernon, N.Y. In the peel tests, only the platinum/gold metallization with 63Sn/37Pb solder was studied. The flux used in all soldering was LA-CO Nonacid (Lake Chemical Co., Chicago).

The ceramic substrates were 96 percent alumina (AISIMAG, American Lava Co., Tenn.) and were used as received without any cleaning. Printing of the metallizations was done using a Presco Model 100c (modified, Bound Brook, N.J.) screen printer (200 mesh), and the firing furnace was a 4-zone belt model built by BTU Engineering, Waltham, Mass.

Post-Failure Examination

The fracture surfaces of the films left on the substrate were examined using scanning electron microscopy (SEM, Advanced Metal Research, Model 1000) and the elemental composition was determined simultaneously using electron dispersive X-ray analysis (EDXA).

Results

Fracture Energy and Peel Strength

The constant-compliance tests and the applied-moment tests were conducted as separate studies and full details of the results are presented in Refs *5* and *6*, respectively. The intent of the research was the same in both instances—to develop fracture test methodology and apply it to study the effect of materials properties and processing. The standard deviations of the constant-compliance fracture results for the platinum/gold and gold films soldered with the tin/lead and the tin/lead/indium solders were judged to be unusually large and so the data were treated statistically using Gaussian probability plots. These plots revealed bimodal distributions for both the fracture data and the peel data. Typical examples of these plots are shown in Figs. 4 and 5, and the resulting averages are listed in Tables 2 and 3. Note that the high and low values of fracture energies and peel strengths for the three firing temperatures do not differ greatly in magnitude. Instead, the effect of firing temperature appears to be reflected in the ratio of the number of high and low $\mathcal{G}_{Ic}$ specimens.

The applied-moment test results are listed in Table 3 for the platinum/gold films soldered with tin/lead solder. They indicate decreasing adherence with increasing temperature and firing time. These data did not show the bimodal distribution found for the constant-compliance and peel data. In fact, the values for $\mathcal{G}_{Ic}$ in Table 3 appear to be comparable to the low $\mathcal{G}_{Ic}$ distribution values found using the constant-compliance test method.

Post-Failure Examination

The failed surfaces of the fracture specimens were examined using SEM and EDXA. These studies revealed that the overall thick-film structure consisted of an open network of glass-coated metal at the top of the film, beneath which was a glass matrix containing metal particles. This essentially glassy layer bonded the film to the alumina substrate. The effect of varying firing temperature at constant firing time on the film structure was determined from the fractured specimens of the constant-compliance and applied-moment tests. In the underfired films, failure occurred well away from the film/substrate interface, and the fractured surface showed evidence of incomplete sintering and flow of the metallization components. At the recommended firing temperatures, failure occurs primarily within the glass bonding layer situated between the metal conductor and substrate. The fracture path was through columns of glass and around metal particles (Fig. 6) and suggests a film microstructure as depicted by Fig. 7 (*top*) in which there is interlocking of the glass and metal. At the highest firing temperatures, the fracture mode shifts to failure at or near the glass-metal interface. The grain boundaries and other microstructural features of the metal phase could be seen replicated on the glass fracture surface. Figure 7 (*bottom*) depicts the development of a continuous layer of glass at the substrate because of excessive glass flow.

Extensive SEM examinations were made of the constant-compliance fractured specimens to try to distinguish differences between specimens

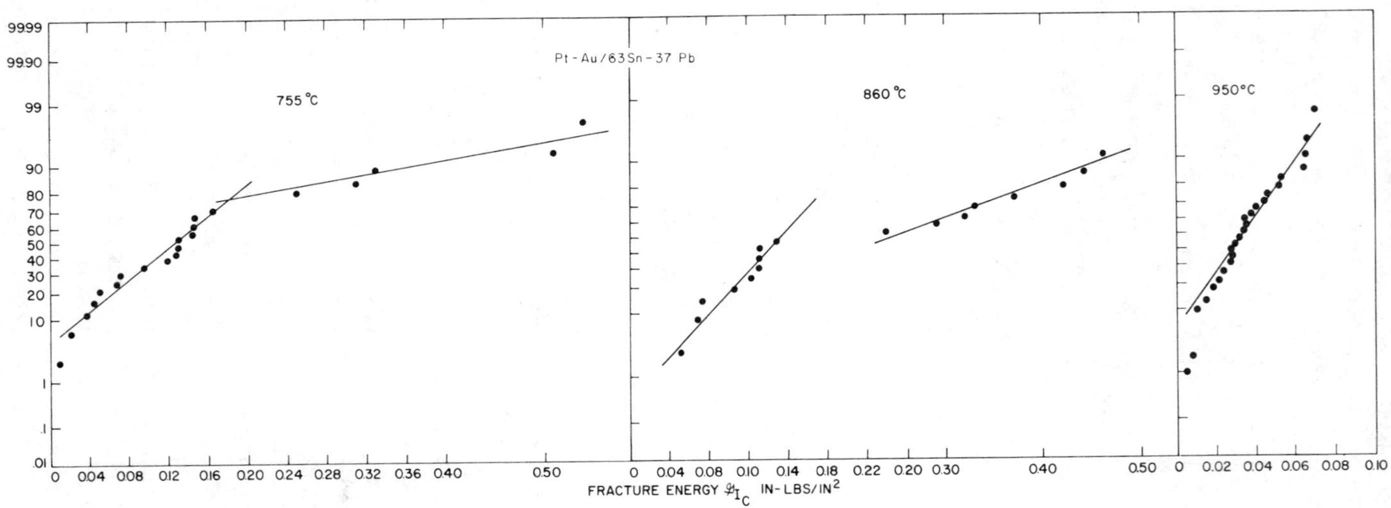

FIG. 4—*Probability plots of adhesive fracture data for the platinum/gold/lead/tin tests* [5]. *1 in · lb/in.^{2} = 174 J/m^2.*

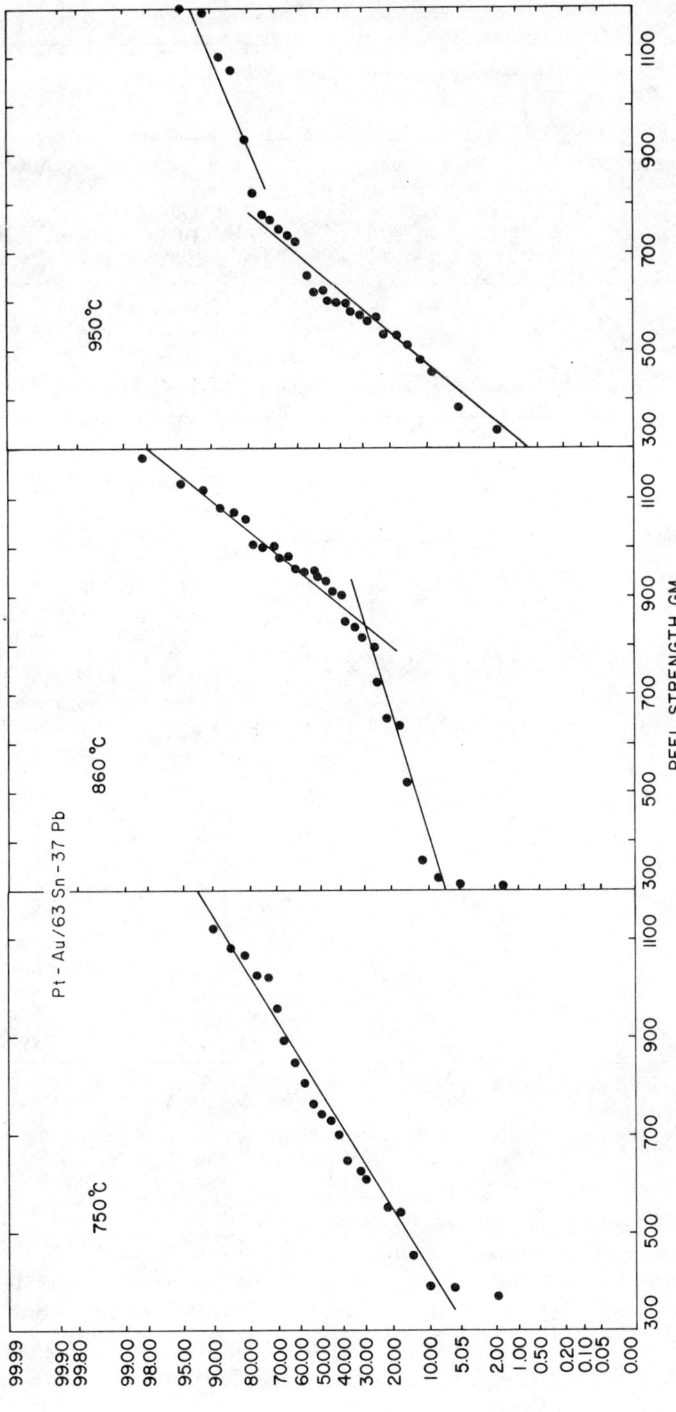

FIG. 5—*Probability plots of peel strength data for platinum/gold/lead/tin tests* [5].

TABLE 2—*Statistical analysis of peel strength data.*

Platinum/Gold (duPont 7553)—63Sn/37Pb			
Firing Temperature	Peel Strength (g) High Values	Peel Strength (g) Low Values	Number of Specimens
755°C (1391°F)	. . .	780 ± 250 (100%)	23
860°C (1580°F)	1000 ± 90 (63%)	570 ± 210 (37%)	30
950°C (1742°F)	1120 ± 120 (20%)	600 ± 120 (80%)	30

TABLE 3—*Applied-moment fracture test results.* [a]

Firing Conditions[b] Temperature	Firing Conditions[b] Time/min	$\mathcal{G}_{Ic}$, J/m^2	No. of Specimens	Locus of Failure
750°C (1382°F)	10	4.2 ± 2.2	11	glass
860°C (1580°F)	10	7.4 ± 3.2	21	glass
	30	7.4 ± 3.0	7	glass
950°C (1742°F)	10	5.8 ± 1.0	4	glass and glass/metal interface
	30	4.4 ± 2.0	4	glass and glass/metal interface
	60	5.4 ± 2.6	6	glass and glass/metal interface
	120	5.4 ± 3.2	14	glass and glass/metal interface
1050°C (1922°F)	10	5.8 ± 2.6	3	glass and glass/metal interface
	30	4.4 ± 1.2	4	glass and glass/metal interface
	120	< 1.0	3[c]	glass/metal interface
1100°C (2012°F)	10	5.2 ± 1.2	6	glass and glass/metal interface
	30	0.4 ± 0.2	6	glass/metal interface

[a]Platinum/gold film soldered with 63Sn/37Pb solder.
[b]Peak time and temperature.
[c]Upper limit value since most specimens broke in handling.

that gave high or low $\mathcal{G}_{Ic}$ values. No differences were found that could not be considered subjective. It was noticed that the film structure was very heterogeneous in that, on a given fracture surface, areas corresponding to more than one of the structures of Fig. 7 could be seen. In none of the studies was it possible to observe the point of crack initiation, and there was no evidence of ductile yielding of the gold, platinum, or solder metals.

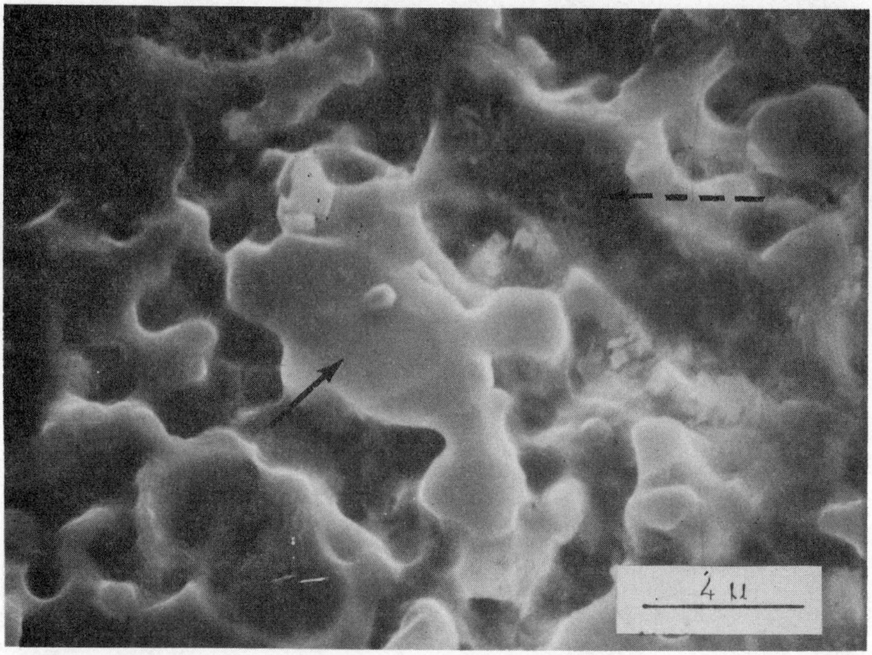

FIG. 6—*SEM photomicrograph of failure through region of interlocking metal and glass. Arrows indicate fracture through glass columns* (←) *and around sockets left by metal particles* (←---).

Discussion

It is reasonably evident that maximum adherence occurs when the film structure involves interlocking of the glass and metal phases. When this condition is attained, there has been sufficient sintering of the metal network to prevent its failure and, on the other hand, the firing time or temperature or both have not allowed accumulation of a continuous glass phase at the film/substrate boundary. The thick-film morphology and its effect on film adherence found here are in agreement with the conclusions of Hitch [*8,9*], who determined film structure by selectively removing the metal phase with mercury vapor and examining the remaining glass phase.

When it was judged that failure occurred through a glass layer, the magnitude of the fracture energies is in reasonable agreement with the known fracture energies of silicate glasses, ~4 to 10 J/m^2 [*10,11*]. For example, consider the constant-compliance results for the platinum/gold films fired at 950°C (1742°F) and for the low $\mathcal{G}_{Ic}$ values of the 900°C (1652°F) fired gold film and all of the applied-moment results of Table 3. The progressive decline in $\mathcal{G}_{Ic}$ with overfiring or prolonged firing occurs because of a loss of interpenetration at the glass/metal interface.

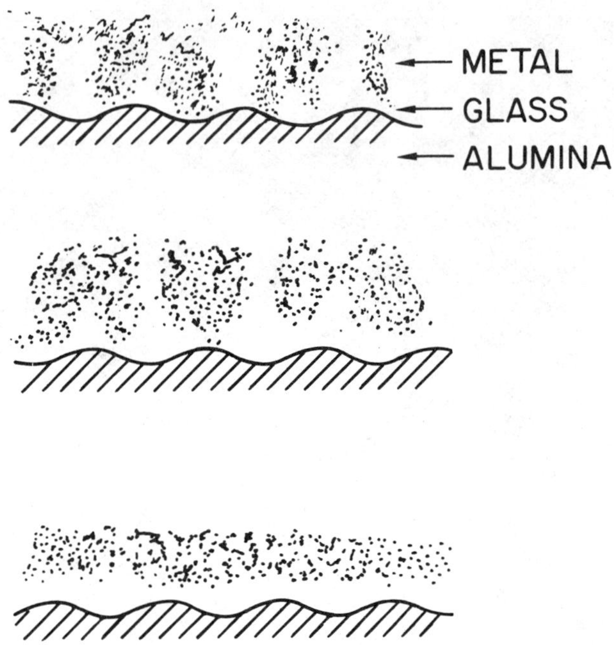

FIG. 7—*Schematic drawings of the metal/glass interface illustrating* (top) *strong interlocking,* (middle) *weak interlocking, and* (bottom) *metal-glass phase separation.*

The very high $\mathcal{G}_{Ic}$ values, 60 to 90 J/m^2, observed in the constant-compliance tests are surprising in that they obviously cannot be reconciled with glass fracture and yet are not high enough to be due to metal fracture ($\mathcal{G}_{Ic}$ of ductile metals $>10^3$ J/m^2). Moreover, there was no evidence of metal yielding from the SEM observations. A hypothesis has been advanced [5] that the bimodal distribution of $\mathcal{G}_{Ic}$ values is related to the crack tip deformation zone from which fracture propagates. The SEM study revealed that for a given thick-film specimen there were variations in morphology along the film, for example, good interlocking in some areas, poor interlocking in others. This variation probably developed during printing or firing or both. These variations are random, so a precrack may terminate in a region of good interlocking, and in order to advance when the specimen is tested it must first develop a large damage zone (that is, high $\mathcal{G}_{Ic}$) as illustrated in Fig. 8*a*. Alternatively, a precrack terminating in a region of poor interlocking need only develop a plane damage zone (low $\mathcal{G}_{Ic}$, Fig. 8*b*) before becoming unstable and the specimen fractures. The argument is then made that the "properly prepared" metallization has a high percentage of interlocked area, and a test sampling large enough to be statistically significant will show a larger percentage of high $\mathcal{G}_{Ic}$ data.

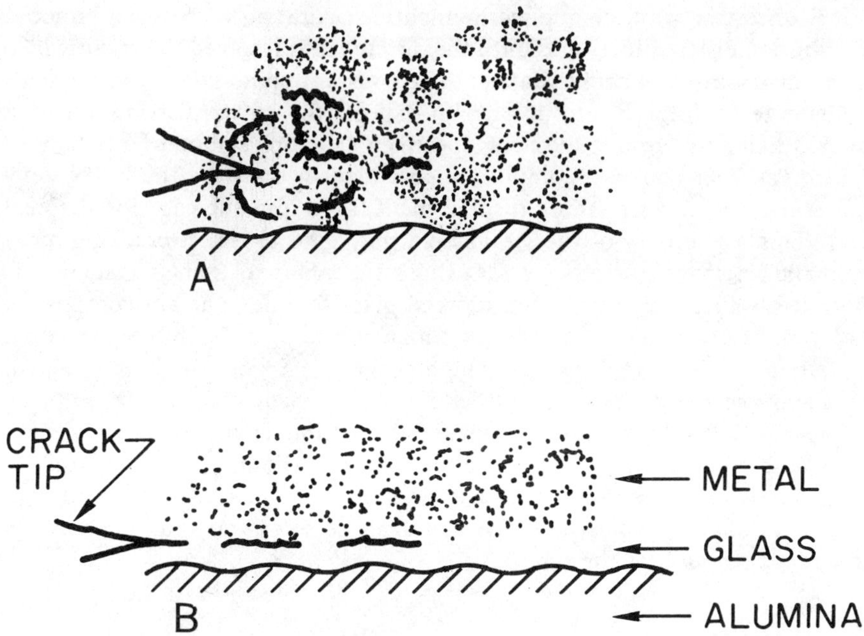

FIG. 8—*Crack-tip damage zones.*

Undoubtedly there are residual "thermal" stresses in the thick film and the solder bond due to cooling from the firing temperature and the soldering temperature. Clearly, these stresses are not overwhelming, otherwise the measured fracture energies would be much less than the 4 to 10 J/m^2 characteristic of glass fracture, and residual stress cannot explain the unusually high $\mathcal{G}_{Ic}$ values. Attempts were unsuccessful in observing any differences in $\mathcal{G}_{Ic}$ for normal cooling of the solder bonds versus slow oven cooling. As far as the metallizations were concerned, the cooling rates after reaching peak firing temperature were about 25°/min, which corresponds to normal technical practice and probably prevents excessive thermal stressing.

If the peel test is viewed as a fracture test with the horizontal segment of the wire a cantilever beam [of 1.3 cm (0.5 in.) length, Fig. 3], then Eq 2 indicates that the ratio of high to low $\mathcal{G}_{Ic}$ values should equal the ratio of the square of the corresponding high and low peel strengths. This correlation was met by all of the platinum/gold data [*5*]. Moreover, it was possible to show [*5*] that the effective crack length of the peel test is the distance from the bend in the wire to the edge of the metallization pad (Fig. 3). Therefore, following the same argument made for the constant-compliance test results, the peel strength would depend on whether the morphology of the film edge is highly interlocked (high peel strength) or glass-rich (low strength).

It is uncertain why the applied moment test data did not show a bimodal distribution of $\mathcal{G}_{Ic}$ values, and until the difference is resolved the significance of the distribution is problematical. The constant-compliance and applied-moment tests differ in one important aspect: the stress distribution at a crack tip in an applied-moment specimen is more symmetrical than in the *single beam* constant-compliance specimen in that, for the latter, only one arm experiences bending displacement (as is also the case for the peel test). This asymmetry of the strain field may tend to direct cracking away from the interface up into the film (high $\mathcal{G}$ region) so that propagation is more likely to occur in metal-reinforced glass for the constant-compliance and peel tests than for the applied-moment test. This difference could explain the high $\mathcal{G}_{Ic}$ values and the high peel strengths but does not establish the significance of these data to thick film adherence.

Looked at from the point of view of thick-film reliability, the test data from both fracture specimens agree on the lower-limit (worst case) failure criterion being essentially the fracture energy of glass. As far as design considerations are concerned, the metal phase offers no reinforcement to adherence strength of the films.

Conclusions

This study has demonstrated that a fracture mechanics approach can be successfully used to characterize the adhesion of thick-film metallizations. In comparison with the peel test, the fracture test is not sufficiently more discriminating to warrant its use for quality-control purposes, especially in view of its greater experimental complexity. Indeed, if the peel test were conducted with a sufficiently stiff wire, it could itself be used as a fracture test (see Discussion in Ref 5). In research work or in new ink development the fracture approach has the advantage of providing fracture energies for comparison with the fracture energies of the film constituents. Also, the fracture results have at least the potential of being used for design purposes.

In addition to demonstrating the feasibility of thick-film fracture testing, this study has added support to the morphological description of thick films developed by Hitch. Also, the question of bimodal strength distribution revealed by both the peel and fracture tests needs to be pursued.

Acknowledgments

It is a pleasure to note the helpful discussions of the work with R. W. Rice, S. W. Freiman, and D. R. Mulville of the Naval Research Laboratory (NRL) and T. T. Hitch and K. R. Bube of RCA Laboratories, as well as the work of L. A. Mann of NRL in the preparation of the thick-film specimens. The studies were supported by the Naval Air Systems Command, J. W. Willis contract monitor.

References

[1] Ripling, E. J., Corten, H. T., and Mostovoy, S., *Journal of Adhesion,* Vol, 3, 1971, pp. 107, 125, 145.
[2] Bascom, W. D., Cottington, R. L., Jones, R. L., and Peyser, P., *Journal of Applied Polymer Science,* Vol. 19, 1975, p. 2545.
[3] Bascom, W. D. and Timmons, C. O. in *Adhesion Science and Technology,* L. H. Lee, Ed., Plenum Press, New York, Vol. 9B, 1976, p. 501.
[4] Irwin, G. R. in *Handbuch des Physik,* S. Flugge, Ed., Vol. 6, 1958, p. 551.
[5] Bascom, W. D. and Bitner, J. L., *Journal of Materials Science*, Vol. 12, 1977, p. 1401.
[6] Becker, P. F. and Newell, W. L., *Journal of Materials Science*, Vol. 12, 1977, p. 90.
[7] Freiman, S. W., Mulville, D. R., and Mast, P. W., *Journal of Materials Science,* Vol. 8, 1973, p. 1527.
[8] Hitch, T. T., *Proceedings,* International Society for Hybrid Microelectronics Symposium, 1971, p. 7.7.1.
[9] Hitch, T. T., *Journal of Electronic Materials,* Vol. 3, 1974, p. 553.
[10] Coble, R. L. and Parikh, N. M. in *Fracture, An Advanced Treatise,* H. Liebowitz, Ed., Vol. 7, 1972, p. 243.
[11] Mecholsky, J. J., Rice, R. W., and Freiman, S. W., *Journal of the American Ceramic Society,* Vol. 57, 1974, p. 440.

DISCUSSION

T. T. Hitch[1] (*written discussion*)—Recent complications in our soldered wire, tension peel test studies of gold/platinum thick films were caused by the differential thermal expansion coefficients of the alumina substrate and the solder, wire, and thick-film pad. The same thing may be affecting your data. The use of a tapered cantilever beam of Kovar, Invar, etc. instead of brass would substantially reduce the mismatch of thermal expansion coefficients in the test part and thus reduce the possibility that thermally induced stresses might be affecting your fracture energy data.

W. D. Bascom (*authors' closure*)—We attempted to determine the influence of residual thermal stresses by using different cooling rates. Although we found no effects, there is no doubt that such stresses exist and your suggestion of using Kovar or similar metals is excellent and will be considered in future work.

S. Schroter[2] (*written discussion*)—It was pointed up that for completeness of presentation the data should include results after temperature cycling and after high-temperature storage.

[1] RCA Research Laboratories, Princeton, N.J. 08540.
[2] Raytheon Company, Quincy, Mass. 02169.

W. D. Bascom (authors' closure)—The work reported here is part of ongoing programs. We are looking into the effects of thermal cycling and many other variables that may effect thick-film performance.

K. L. Mittal[3] *(written discussion)*—Is your fracture mechanics approach applicable to thin evaporated films? Do you think that we can expect a relationship between the fracture mechanics results and the pull test for measuring adhesion? What are the main limitations of your fracture mechanics approach?

W. D. Bascom (authors' closure)—In principle, the fracture mechanics approach is applicable to characterizing any material which fails by flaw growth. It is simply a methodology for quantitatively measuring the flaw sensitivity. If one judges that flaw growth (brittle fracture) is the failure mode of the material (or component or bond, etc.), then it is a matter of devising a test which measures fracture resistance. In the specific case of thin film metallizations it is a matter of finding a means of attaching an arm to the film which does not alter the film properties, or, as was the case in the thick film study, the attachment (soldering) simulates what is done in practice. In this case one is characterizing a system (solder-bonded thick film).

The pull test is a good example of how not to test adhesion. Unless the film fails by a general ductile yielding across the bond area, then the pull test is a fracture test but without any measure of the flaw size from which failure initiates. Having this parameter uncontrolled can lead to an inexplicably wide data scatter and large differences in test results between laboratories. The fracture test and the pull test are similar insofar as technique is concerned but differ in that the former gives a material parameter independent of test geometry.

The disadvantage of fracture testing in general is its complexity. Also, the validity of the results being a true material parameter requires careful considerations of stress analysis, test procedure, etc. As we indicate in the paper, there are some unresolved questions about the single beam test. However, the organic adhesives industry has gradually accepted adhesive fracture testing and has found ways of simplifying the "ideal" test conditions to routine quality-control procedures.

J. J. Bikerman[4] *(written discussion)*—It would be interesting, I believe, to repeat your measurements using a metal foil instead of metal powder

[3] IBM Corporation, Hopewell Junction, N.Y. 12533.
[4] Department of Chemical Engineering, Case Western Reserve University, Cleveland, Ohio 44106.

which sinters (more or less) during the firing. Did you perform such experiments; and if you did, how do the results compare with those presented here?

W. D. Bascom (*authors' closure*)—You are quite correct in pointing up the complexity of the thick films and that we should address the validity of these test methods with better characterized coatings. We are indeed doing so but with epoxy polymer coatings, rather than metal foils, since we have fracture data on epoxy polymers from other, more conventional, tests.

L. E. Murr[1]

Techniques for Measuring Adhesive Energies in Metal/Ceramic Systems

REFERENCE: Murr, L. E., **"Techniques for Measuring Adhesive Energies in Metal/Ceramic Systems,"** *Adhesion Measurement of Thin Films, Thick Films, and Bulk Coatings, ASTM STP 640,* K. L. Mittal, Ed., American Society for Testing and Materials, 1978, pp. 82-98.

ABSTRACT: The adhesive energies of nickel on thorium dioxide (ThO_2), nickel/chromium (80/20) on ThO_2, and iron/nickel/chromium (304 stainless steel) on aluminum oxide (Al_2O_3) were determined at temperatures ranging from roughly 200°C (360°F) below the melting point of the metallic component to approximately 50°C (90°F) above. The technique involves the measurement of the average surface free energy for the solid metals and liquid metal drops utilizing the zero-creep and sessile-drop shape concepts, respectively. Measurements of equilibrium energetics and associated geometries as well as the contact angles at metallic particles in contact with the ceramic substrates are described using the techniques of scanning electron microscopy (SEM) and transmission electron microscopy (TEM). Measurements of surface free energy of nickel/chromium and stainless steel drops equilibrated at temperatures below the melting point and utilizing the sessile-drop method are compared with those determined by the zero-creep method. The results of this investigation represent the first systematic study of the variation of adhesive energy of a metal/ceramic system with temperature, and the comparison of surface free energies determined from equilibrium particle shapes with zero-creep measurements of wires of the same material in the solid state.

KEY WORDS: adhesion, adhesive strength, electron microscopy, sessile drop, zero creep, surface energy, contact angle, temperature coefficient, thin films

There have been few measurements of the work of adhesion (or adhesive energy) associated with thin films or coatings on ceramic substrates, even though this represents an important technological area in applications which range from integrated circuit component fabrication to certain aspects of cast metal and alloy production. More importantly, except for the preliminary work of Murr and Ahmad [*1*][2] there have not been any systematic measurements of adhesive energy as a function of temperature in a specific metal/ceramic system.

[1]Professor and head, Department of Metallurgical and Materials Engineering, New Mexico Institute of Mining and Technology, Socorro, N. Mex. 87801.

[2]The italic numbers in brackets refer to the list of references appended to this paper.

In the general case, the adhesive energy associated with the contact of a metallic coating (M) in thermal equilibrium with a ceramic (substrate) material (C) in the absence of any mixing or reaction at the contact interface is defined by

$$E_{Ad} = F_{S(M)} + F_{S(C)} - \gamma_{i(MC)} \tag{1}$$

where $F_{S(M)}$ and $F_{S(C)}$ are the solid surface free energies of the metal (M) and ceramic (C), respectively, and $\gamma_{i(MC)}$ is the solid-solid (metal/ceramic) interfacial free energy. If we assume that the metallic component of the couple is in the liquid state, Eq 1 can be written alternatively as

$$E_{Ad} = \gamma_{LV\,(M)} + F_{S(C)} - \gamma_{LS(MC)} \tag{2}$$

where $\gamma_{LV\,(M)}$ is the surface tension (liquid-vapor surface free energy) of the metal and $\gamma_{LS(MC)}$ is the liquid metal/ceramic interfacial free energy.

For a liquid metal drop or solid metal "particle" [*2*] resting on a solid ceramic substrate as depicted schematically in Fig. 1, the equilibrium contact angle, which can be measured for sufficiently large particles, can be utilized in relating the surface free energies to the interfacial free energies in the form

$$\gamma_i = F_{S(C)} - \gamma_{LV\,(M)} \cos\Omega_c \tag{3a}$$

and

$$\gamma_i = F_{S(C)} - F_{S(M)} \cos\Omega_c \tag{3b}$$

where Ω_c is the contact angle (Fig. 1). Substituting for γ_i in Eqs 1 and 2

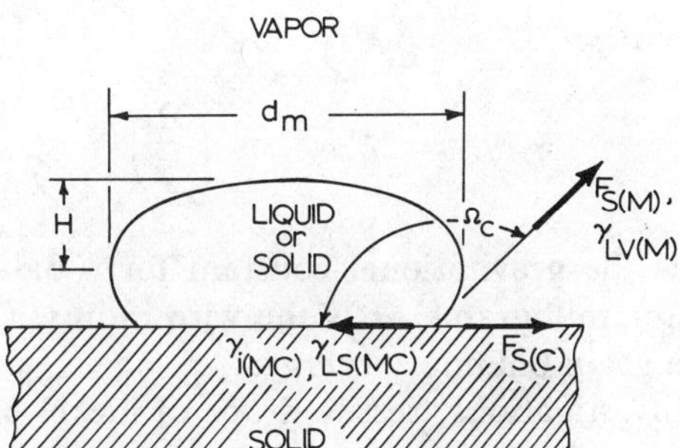

FIG. 1—*Metal sessile drop or particle in equilibrium with a solid ceramic substrate showing associated geometrical and energy parameters.*

then results in the following simple expression for the adhesive energy associated with a metal particle resting in equilibrium upon a ceramic substrate

$$E_{Ad} = \gamma_{LV(M)}(1 + \cos\Omega_c) \tag{4a}$$

and

$$E_{Ad} = F_{S(M)}(1 + \cos\Omega_c) \tag{4b}$$

where, as defined in the foregoing, $F_{S(M)}$ and $\gamma_{LV(M)}$ are the solid and liquid metal surface free energies, respectively, at some fixed temperature. Actually, Eqs 3*b* and 4*b* are the analogue of Eqs 3*a* and 4*a*, and do not strictly apply when the two phases are solid except near the melting point of the metal drop. These features have been discussed previously [*2*].

Experimental Procedure

It is apparent from Eqs 4*a* and 4*b* that by measuring the contact angle for a sessile "drop" of a metal (for which the associated surface free energy is known) resting upon a ceramic substrate, the adhesive energy can be determined at any specific (equilibrium) temperature. In the solid state, the surface free energy is difficult to determine at temperatures appreciably below the melting point, but near the melting point [$>0.8T_m$(T_m in K)] approximate (average) values of surface free energy can be obtained by the method of zero-creep [*2*] where, as illustrated in Fig. 2, the equilibrium conditions are described by

$$w_0 g = \pi r \left[F_{S(M)} - \gamma_{gb(M)} \left(\frac{r}{l}\right)\right] \tag{5}$$

and

$$\gamma_{gb(M)} = 2F_{S(M)} \cos\left(\frac{\Omega_S}{2}\right) \tag{6}$$

where g in Eq 5 is the gravitational constant for a balance load, w_0, expressed in grams (or milligrams), r is the wire radius, l the average grain length, and γ_{gb} the grain boundary energy.

In the present investigations, sessile drops of nickel on thorium dioxide (ThO_2) substrates, nickel/chromium (80 percent nickel) on ThO_2 substrates, and nickel/chromium/iron (9 percent nickel in 304 stainless steel) on aluminum oxide (Al_2O_3) substrates were equilibrated at temperatures above the melting point as described previously by Ahmad and Murr [*3*].

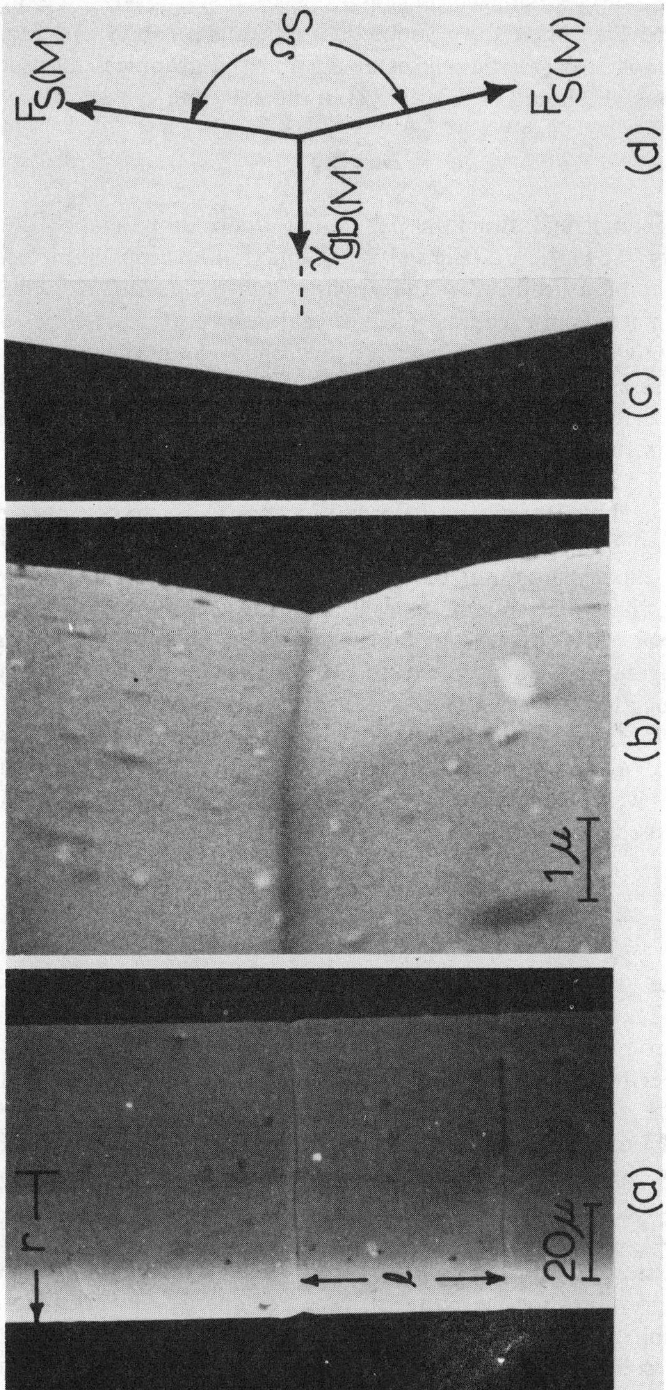

FIG. 2—*Determination of equilibrium, geometrical data for calculating surface free energies for crept wires by electron microscopy techniques.* (a) *304 stainless steel wire section crept in helium at 1060°C (1940°F) as observed in the SEM* (b) *Magnified SEM micrograph of groove at intersection of grain boundary with wire surface.* (c) *Groove angle similar to* (b) *magnified 10 times greater than in* (b) *as a shadowgraph in the TEM.* (d) *Schematic representation of equilibrium attained at grain boundary grooves.*

The surface free energies (surface tensions) for the metal drops (>0.5 cm in diameter) were calculated from geometical parameters shown in Fig. 1 utilizing the approach of Koshavnik et al [4] and the graphical modifications or extensions of Ahmad and Murr [3] in the equation

$$\gamma_{LV(M)} = D d_m^2 \rho g \tag{7}$$

where D is a geometrical function dependent upon drop shape [$D = f(dm/2H)$ in Fig. 1] [3,4], d_m (Fig. 1) is the maximum drop diameter, ρ the density at temperature, and g the gravitational constant. Drops were equilibrated for periods in excess of 0.5 h at each temperature at which the shapes were photographed [3]. A purified hydrogen gas environment was employed in all sessile drop experiments.

In several experiments on nickel/chromium and 304 stainless steel sessile drops, the drops were slowly cooled to temperatures below the melting point and then equilibrated for 200 h at specific temperatures, and the drop shapes were photographically recorded in an attempt to compare the surface free energies in the solid state utilizing Eq 7 with those determined from zero-creep experiments utilizing Eqs 5 and 6. The contact angles were also recorded in these experiments as well as in those conducted on liquid drops at temperatures above the melting point (Fig. 1) in order to determine the adhesive energy at temperature from a knowledge of the average value of the associated surface free energy utilizing Eqs 4*a* and 4*b*.

The solid surface free energies utilized in calculations employing Eqs 4 for nickel and 304 stainless steel were determined previously [2], while additional values were determined for nichrome (nickel/chromium) utilizing the same zero-creep procedures described by Murr et al [5,6].

Results

The examples of equilibrium angle measurements for stainless steel shown in Fig. 2 are typical of those used to determine the surface free energy for nichrome (80Ni/20Cr) at a temperature of 1400°C (2552°F) in the present experimental program utilizing the zero-creep concept. Figure 3 illustrates the results of zero-creep measurements on 80Ni/20Cr at 1400°C (2552°F) in a purified helium atmosphere (99.9999 percent). The balance load, w_0 in Eq 5 is determined as the load which effectively balances the surface tension of the 0.006-cm-diameter wires employed.

Figure 4 shows some examples of the sessile drops analyzed in both the solid and liquid state. By determining the drop shape parameters illustrated in Fig. 1, average values for the liquid surface free energy (surface tension) could be determined for temperatures above the melting point, while for particles in the solid state equilibrated at temperatures below the melting

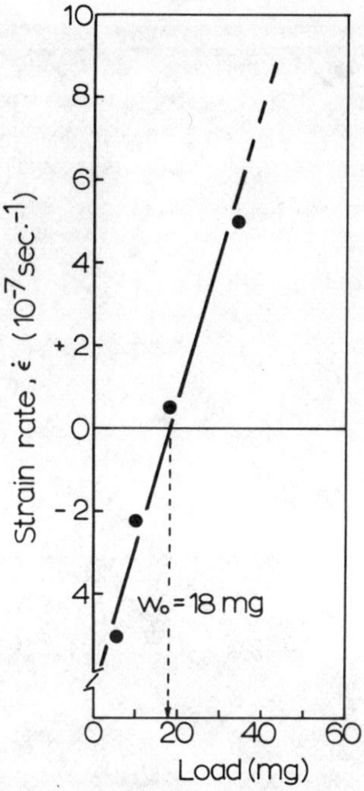

FIG. 3—*Plot of creep rate (for 90 h) against load in helium at 1400°C (2552°F) for 80Ni/20Cr alloy wire. The average wire radius was 0.0035 cm while the measured ratio* r/l *was 0.5. The load indicated at* $\dot{\epsilon} = 0$ *is the balance load* w_0*. The ratio* γ_{gb}/F_S *was determined to be 0.36.*

point, approximate values for the solid surface free energy could be calculated for comparison with values determined from the zero-creep measurements. As shown in Fig. 4*b*, experiments were undertaken in which more than one sessile drop or equilibrated particle could be analyzed simultaneously, and these have been extensively described elsewhere [*6, 7*]. Figure 5 illustrates another method which has been utilized in determining the contact angles for very small particles of nickel equilibrated on thoria substrates [*6*]. Because the equilibrated particles shown in Fig. 5 are extremely small, the gravitational influence on shape is negligible and the associated interfacial free energies cannot be determined from Eq 7 [*2*]. As shown in Fig. 5*d*, however, the use of the scanning electron microscope (SEM) [*7*] in the measurement of contact angles affords an opportunity to develop reasonably valid averages (at least 100 contact angle measurements have been made for each such average) for Ω_c which when substituted along with average values for $F_{S(M)}$ at temperatures into Eq 4*b* has allowed average values for the energy of adhesion to be determined at some specific temperature. It should be apparent of course that both the value of $F_{S(M)}$

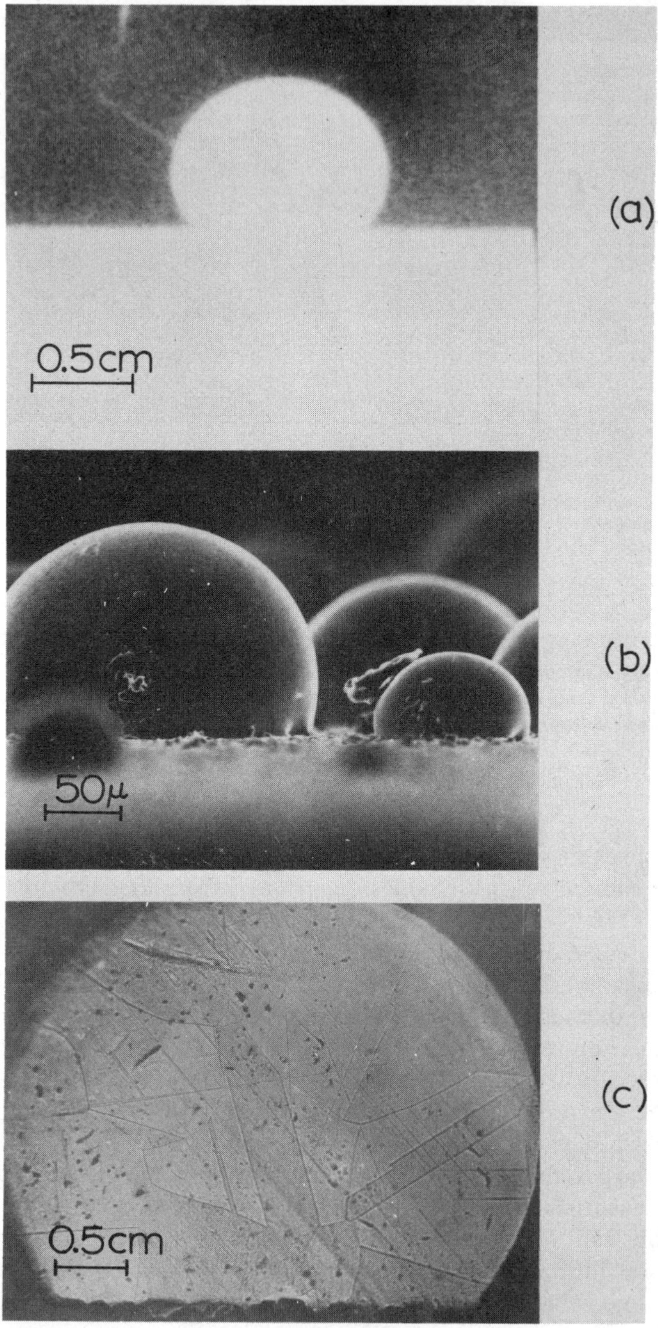

FIG. 4—*Sessile drops of nickel and 304 stainless steel on ceramic substrates.* (a) *Nickel liquid sessile drop on ThO_2 at 1470°C (2678°F),* (b) *304 stainless steel solid "sessile drop" at 1200°C (2192°F) equilibrated on Al_2O_3 for 200 h.* (c) *Polished and etched "drop" section of 304 stainless steel following equilibration on Al_2O_3 at 1420°C (2588°F) for 200 h. The particle is composed of numerous grains and annealing twins.*

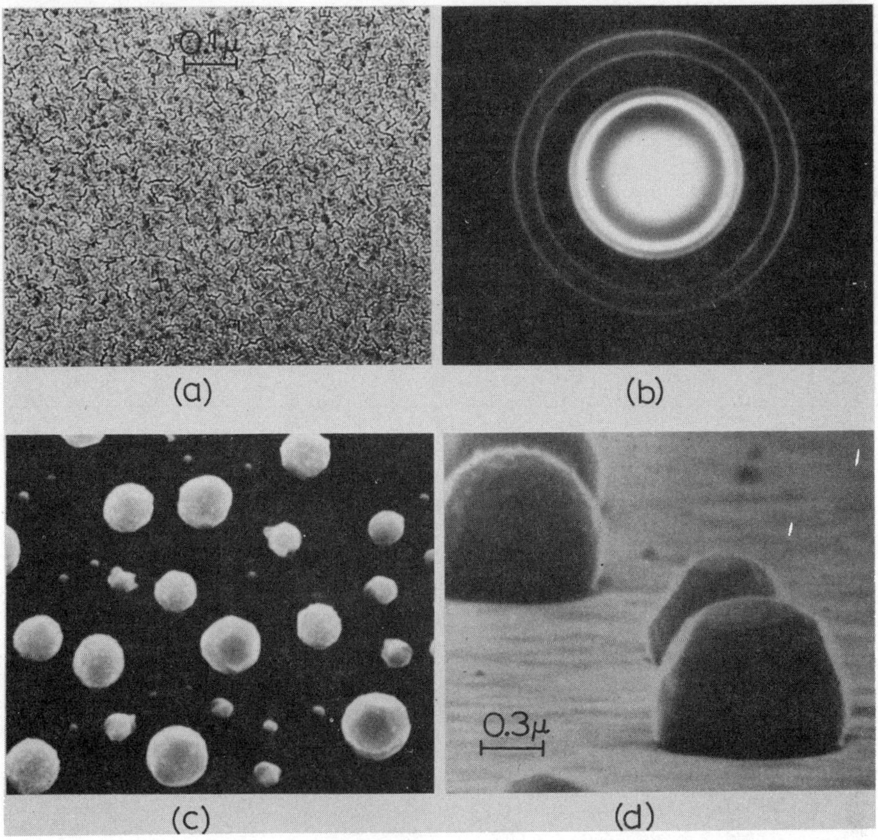

FIG. 5—*Development of equilibrated nickel particles on ThO_2 substrates for contact angle measurements.* (a) *TEM micrograph showing discontinuous thin film of nickel vapor-deposited onto sodium chloride having the same (001) surface orientation as ThO_2 single crystals which were also simultaneously vapor coated. The film thickness was approximately 400 Å.* (b) *Associated selected-area electron diffraction pattern of* (a) *showing small-grain polycrystalline nature of the vapor-deposited nickel layer on the sodium chloride, which presumably was the same on the ThO_2 crystal surfaces.* (c) *SEM micrograph showing equilibrated particles of nickel on the ThO_2 (001) surface following 200 h heat treatment at 1200°C (2192°F) in hydrogen. The viewing direction is the same as in* (a). (d) *Enlarged view at a more advantageous angle of particles in* (c), *showing the particles contacting the ThO_2 surface and illustrating the ability to accurately determine the contact angles.*

and Ω_c must be measured at a specific temperature in order that E_{Ad} be calculated with reference to the same temperature.

The results of measurements of surface free energies and contact angles at temperature determined in the present experiments on the NiCr/ThO_2 system, the Ni/ThO_2 system, and the stainless steel/Al_2O_3 system, as well as previous experiments [*1-3,5-7*] are listed in Tables 1-3 along with the calculations of the associated adhesive energies (at temperature) as de-

TABLE 1—*Surface and adhesive energy data for the Ni/ThO$_2$ systems.* [a]

Temperature, °C (°F)	$F_{S(Ni)}$ [b]	$\gamma_{LV(Ni)}$ [b]	Ω_c(deg)	E_{Ad} [b]
1200(2192)	2.200	...	120	1.100
1300(2372)	2.147	...	...	0.623[c]
1320(2408)	2.137	...	...	0.783[c]
1337(2436)	2.127	...	...	0.858[c]
1373(2506)	2.108	...	...	0.525[c]
1455(2651)	2.060	...	135	0.600
1455(2651)	...	1.776	135	0.515
1460(2660)	...	1.759	133	0.563
1470(2678)	...	1.729	132	0.571
1480(2696)	...	1.699	130	0.611
1490(2714)	...	1.682	129	0.621
1500(2732)	...	1.652	127	0.660

[a]Data from Ref *1*.
[b]Interfacial free energy in J/m^2.
[c]Values calculated from Eq 1 utilizing data in Ref *1*.
Solid-state densities are assumed to remain essentially constant at 8.86 g/cm^3.

TABLE 2—*Surface and adhesive energe data for the (80Ni/20Cr)/ThO$_2$ system.*

Temperature, °C (°F)	$F_{S(NiCr)}$ [a]	$\gamma_{LV(NiCr)}$ [a]	Ω_c(deg)	E_{Ad} [a]
1060(1940)	2.160[b]	...	...	...
1200(2192)	2.040[b]	...	131[d]	0.700
1400(2552)	1.960	...	138	...
1400(2552)	2.220[c]	...	138	...
1475(2687)	1.930	...	143	0.500
1475(2687)	...	1.700	143	0.360
1500(2732)	...	1.670	140	0.390
1525(2777)	...	1.670	142	0.340

[a]Interfacial free energy in J/m^2.
[b]Data from Refs *2,5,6*.
[c]Value determined from Eq 7 utilizing the following measured parameters: $d_m = 0.855$ cm; $2H = 0.754$ cm; $D = 0.37$ (see Fig. 1); $\rho = 8.40$ g/cm^3.
[d]Corrected value from Ref *5* determined in the present investigation.

termined from Eqs 4*a* and 4*b*. As a means of comparing these data, the results for each system are plotted in Figs. 6–8. Figures 7 and 8 also allow measurements of solid surface free energy of nickel/chromium and stainless steel equilibrated particles on ThO$_2$ and Al$_2$O$_3$ substrates, respectively, utilizing Eq 7, to be compared directly with the average-value curves determined from zero-creep measurements. While the values plotted are noticeably displaced from the zero-creep measurements, they are nonetheless higher than corresponding liquid-state sessile drop measurements, as would be expected and as demonstrated by the associated plots of zero-creep and sessile drop data in the solid and liquid states, respectively.

TABLE 3—*Surface and adhesive energy data for the 304 stainless steel-Al_2O_3 system.*[a]

Temperature, °C (°F)	$F_{S(304)}$[b]	$\gamma_{LV(304)}$[b]	Ω_c(deg)	E_{Ad}[b]
1200(2192)	1.944	...	104	1.440
1200(2192)	1.795[c]	...	104	...
1420(2588)	1.557	...	114	0.920
1420(2588)	2.020[c]	...	114	...
1475(2687)	1.460	...	119	0.780
1475(2687)	...	1.172	119	0.610
1480(2696)	...	1.115	118	0.590
1485(2705)	...	1.089	117	0.576
1490(2714)	...	1.025	116	0.560
1500(2732)	...	0.966	114	0.570

[a]Data from Ref *1* except for those marked *c*.
[b]Interfacial free energy in J/m^2.
[c]Values determined from Eq 7 utilizing the following measured parameters: at 1200°C (2192°F), $d_m = 0.510$ cm; $2H = 0.480$ cm; $D = 0.90$; at 1420°C (2588°F), $d_m = 0.515$ cm; $2H = 0.490$ cm; $D = 1.0$ (see Fig. 1); $\rho = 7.80$ g/cm^3.

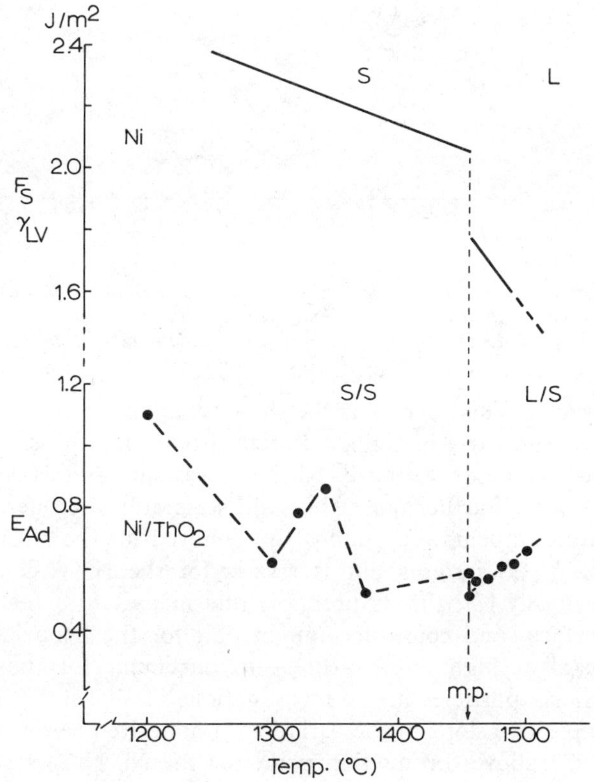

FIG. 6—*Variation of adhesive energy with temperature and comparison with the variation of surface free energy in the Ni/ThO_2 system. Solid and liquid phases are denoted* S *and* L, *respectively.*

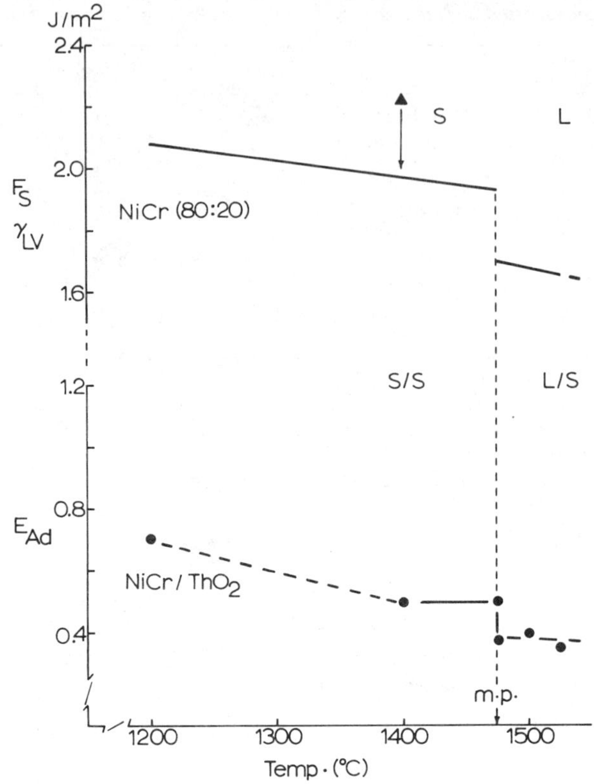

FIG. 7—*Variation of adhesive energy with temperature and comparison with the variation of surface free energy in the (80Ni/20Cr)/*ThO_2 *system. Solid and liquid phases are denoted S and L, respectively.* ▲ *denotes surface free energy values determined from Eq 7.*

It is observed in Figs. 6–8 that the greatest adhesion at any particular temperature (particularly in the solid state) occurs for the stainless steel/Al_2O_3 system, decreases with the Ni/ThO_2 system, and is least for the NiCr/ThO_2 system. Furthermore, the solid-state adhesive energy variation with temperature appears reasonably well behaved for the NiCr/ThO_2 and stainless steel/Al_2O_3 systems but is erratic for the Ni/ThO_2 system. As discussed previously [*1*], this is probably due in part to a reaction at the Ni/ThO_2 interface, and could account in part for the behavior of thoria-dispersed nickel at high temperatures—in particular for the decohesion observed at thoria particles in the investigations of Webster [*8*] and Olsen et al [*9*]. The positive slope of the variation of adhesive energy with temperature (dE_{Ad}/dT) above the melting point for the Ni/ThO_2 system plotted in Fig. 6 is also indicative of the formation of a reaction product or similar interfacial alteration at the liquid nickel-solid ThO_2 interface [*2*]. By comparison, the adhesive energy for the stainless steel/Al_2O_3 system in Fig. 8

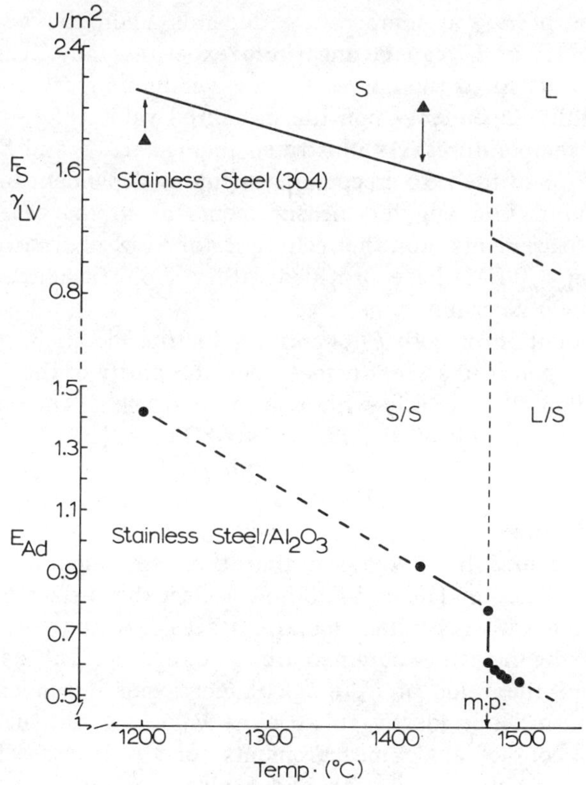

FIG. 8—*Variation of adhesive energy with temperature and comparison with the variation of surface free energy in the 304 stainless steel/*Al_2O_3 *system. Solid and liquid phases are denoted* S *and* L, *respectively.* ▲ *denotes surface free energy values determined from Eq 7.*

is observed to be well behaved, with a linear temperature coefficient in both the solid and liquid states for stainless steel. The value of $dE_{Ad}/dT \simeq -2.3$ mJ/m^2°C is in fact identical both above and below the melting point of the stainless steel. While the slopes of the solid and liquid-state curves for the NiCr/ThO_2 system in Fig. 7 can be interpolated in such a way as to make them essentially linear and equal, the magnitude of the slope, dE_{Ad}/dT, is equal to only approximately -0.2 mJ/m^2°C, or roughly an order of magnitude smaller than the corresponding value for the stainless steel/Al_2O_3 system.

Discussion

The technique for measuring adhesive energy in a metal/ceramic system as presented and elaborated upon in this paper is essentially summarized by, and predicated upon, Eqs 4. As a consequence, the reliability of calculated values of E_{Ad} at temperature depends upon the accuracy with which $F_{S(M)}$, γ_{LV}, or Ω_c can be measured experimentally. Therefore, the significance attached to the curves for E_{Ad} versus temperature plotted in Figs. 6–8 is fully dependent upon the measured values of $F_{S(M)}$, γ_{LV}, and Ω_c at various temperatures. As illustrated in Figs. 2, 4, and 5, the determination of $F_{S(M)}$ in the zero-creep experiments is dependent to a large extent upon equilibrium angular measurements as in the case of contact angle (Ω_c) measurements, and the techniques of SEM or transmission electron microscopy (TEM) have provided rather sophisticated and accurate methods for such determinations.

The determination of both $F_{S(M)}$ and γ_{LV} for the metals involved in this investigation depends to a great extent upon the purity of the experimental (furnace) system, particularly with regard to oxygen. The fact that the present value of $F_{S(M)}$ for 80Ni/20Cr at 1400°C (2552°F) is consistent with previous measurements adds some credibility to the values obtained, and recent zero-creep work on nickel at varying oxygen partial pressures has produced data corroborating the values for $F_{S(M)}$ listed in Table 1 [*10*].

It must of course be recognized that the experimentally determined values for $F_{S(M)}$ listed in Tables 1–3 do not reflect the surface energy anisotropy associated with crystalline metal surfaces [*2*], and this is an additional reason why the values obtained are to be regarded only as an average. However, since the value of $F_{S(M)}$ calculated is not dependent upon the creep mechanism, it is ideally an absolute value for the surface energy. Furthermore, contact angle measurements for the larger solid particles such as those depicted in Fig. 4*b* and *c* also represented averages from assemblages of polycrystals and suffer the same crystallographic limitations. Moreover, the variations in contact angle values measured show a considerable spread which seems to be more influenced by experimental error than surface or interfacial energy anisotropy effects.

In the development of Eqs 1 and 2 it was stated at the outset that they represent a somewhat idealized case where the contacting components of the system (metal/ceramic) bond adhesively, energetically, but do not mix. Obviously at elevated temperatures where diffusion can be important and where reactions at the interface can occur, this is invalid. This is exemplified perhaps in Fig. 6, and it is well known that high-temperature diffusion not only plays a part in particle stability at high temperature in the Ni/ThO_2 system but also particle agglomeration and reaction, even though it has been shown that equilibrium is rapidly achieved at a planar interface between nickel and ThO_2 at temperatures between 1250 and 1400°C (2282 and 2552°F) [*11*]. Certainly if the results shown in Figs. 6–8 are to be considered representative of the idealized case, extensive analysis of interfacial chemistry would have to be made. Ignoring these features, however, the data might still be of importance in the engineering evaluation of the high-temperature performance of such metal/ceramic systems. In this respect, the trend of the data in the solid state seems physically sensible, and the catastrophic loss of adhesion upon melting of the metal component of the contacting materials also seems to be physically as well as thermodynamically sensible.

The results of Figs. 6–8 would seem, however, to be somewhat limited to the case of a thick metal deposit or thick film on the respective ceramic substrates, the range of thick films being defined here as greater than about 0.2 μm. The reason for this is a rather fundamental if not thermodynamic one which can be recognized in principle from Fig. 5. In Fig. 5*a*, a thin, discontinuous film of nickel is shown vapor-deposited onto a ThO_2 crystal substrate. After heating this system in H_2 at 1200°C (2192°F) for 200 h, the thin film discontinuously but somewhat uniformly adhering to the ThO_2 substrate was changed into isolated particles having much larger dimensions than the initial film components. The part that surface diffusion plays in such a process is well known, but this has been shown to occur even in continuous thin films not only as a result of surface diffusion but also as a result of the system striving to achieve an energy minimum. In addition, small crystallites condensed upon a ceramic substrate possess—in association with the Wulff condition driving particles to equilibrium shapes—a thermal dependence on shape. The important (thermodynamic) feature of this dependence is that the effective melting point must always decrease with a reduction in the particle size at constant pressure [*2*]. As a consequence, polycrystalline aggregates are, under certain conditions, influenced thermodynamically. This is illustrated phenomenologically in the sequence of electron micrographs shown in Fig. 9. Figure 9 shows that, as suggested previously in this paper, the implications or applications of the adhesive energy data obtained (Figs. 6–8) may not be applicable to very thin films ($<$0.1 μm) on ceramic substrates. Figure 9*d* also illustrates that while the shapes of solid particles may approximate liquid drops, their crystal structure can be accommodated by a regular atomic array. These features have been described in detail in previous work by Murr [*12*].

Conclusions

The use of the sessile drop concept to measure adhesive energies in metal/ceramic systems provides a direct method for determining absolute values for the energy of adhesion. It should be noted that values of adhesive energy represent average values of temperature, that is, values independent of orientation and normally the arithmetic average of numerous measurements at the same temperature. This approach has allowed the variation of adhesion energy with temperature in a narrow range above and below the melting point of the metallic component to be determined in the Ni/ThO_2, NiCr/ThO_2, and 304 stainless steel/Al_2O_3 systems. This study represents the first systematic measurements of the variation of adhesive

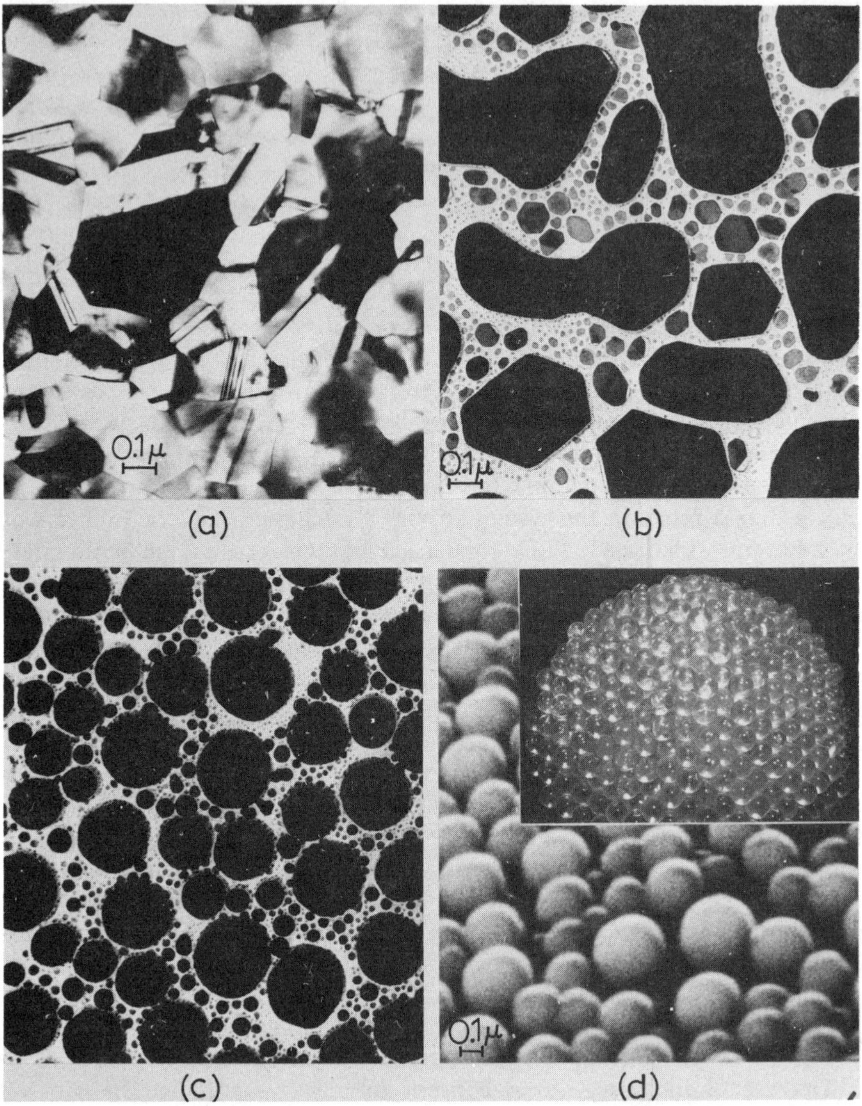

FIG. 9—*Structure and equilibrium morphologies of indium thin films vapor-deposited onto sodium chloride (001) substrates, illustrating the effect of thermodynamic and related phenomena on adhesion and continuity.* (a) *Continuous indium film (1150 Å thick). Substrate temperature of 75°C (167°F) [as compared with a melting point for indium at standard pressure of 156°C (313°F)].* (b) *Discontinuous indium film characterized by crystallographic islands (mean film thickness of 260 Å). Substrate temperature of 85°C (185°F).* (c) *Liquid-like growth of indium film (600 Å mean thickness) at a substrate temperature of 130°C (266°F).* (a–c *are TEM micrographs.*) (d) *SEM micrograph showing the liquid-like (sessile-drop) appearance of indium particles in* (c). *The insert shows a marble atom model for the particles, which have been demonstrated to be single crystals. The background pressure in all cases* (a–c) *was* 10^{-6} *torr and the evaporation rate was constant at* 10^2 *Å/s.*

energy with temperature, and demonstrates the unique capabilities of TEM and SEM in the determination of equilibrium energetics and geometries. A comparison of the surface energies of nickel/chromium and stainless steel measured from equilibrium drop shape in the solid state with values determined from zero-creep experiments indicates some deviation. The values obtained, however, are higher than those measured for liquid metal drops, and the differences observed may be due to crystallographic constraints imposed upon the solid drop shapes.

Acknowledgments

The author is grateful to Dr. U. M. Ahmad for his help in executing many of the experiments described and in determining a number of the experimental parameters. Early portions of some of the work described were supported by the Office of Naval Research and the Energy Research and Development Administration.

References

[*1*] Murr, L. E. and Ahmad, U. M., *Scripta Metallurgica,* Vol. 10, 1976, pp. 299–302.
[*2*] Murr, L. E., *Interfacial Phenomena in Metals and Alloys,* Addison-Wesley, Reading, Mass., 1975.
[*3*] Ahmad, U. M. and Murr, L. E., *Journal of Materials Science,* Vol. 11, 1976, pp. 224–230.
[*4*] Koshavnik, A. Y., Kusankuv, M. M., and Lubman, N. M., *Journal of Physical Chemistry,* Vol. 27, 1953, pp. 1887–1895.
[*5*] Murr, L. E., Horylev, R. J., and Wong, G., *Surface Science,* Vol. 26, 1971, pp. 184–196.
[*6*] Murr, L. E., *Journal of Materials Science,* Vol. 9, 1974, pp. 1309–1319.
[*7*] Murr, L. E., *Materials Science and Engineering,* Vol. 12, 1973, pp. 277–283.
[*8*] Webster, D., *Transactions,* American Society for Metals, Vol. 62, 1969, pp. 1309–1315.
[*9*] Olsen, R. J., Judd, G., and Ansell, G. S., *Metallurgical Transactions,* Vol. 2A, 1971, pp. 1353–1360.
[*10*] Stickle, D. R., Hirth, J. P., Meyrick, G., and Speiser, R., *Metallurgical Transactions,* Vol. 7A, 1976, pp. 71–74.
[*11*] Alcock, C. B. and Brown, P. B., *Metal Science Journal,* Vol. 3, 1969, pp. 116–120.
[*12*] Murr, L. E., *Thin Solid Films,* Vol. 20, 1974, pp. 81–89.

DISCUSSION

J. J. Bikerman[1] *(written discussion)*—I hope that the author will find time to consult two sources of information. First, some people, including myself, criticized the methods presently used to determine the surface

[1]Department of Chemical Engineering, Case Western Reserve University, Cleveland, Ohio 44106.

energy of solids; my conclusion was that this energy cannot be measured by any method existing now, not excluding the zero-creep experiment. This view was published in *Physica Status Solids* (1965) and in my book *Physical Surfaces* (1970).

Secondly, several people who studied the wetting of ceramic materials by molten metals came to the conclusion that the contact angle is small whenever the metal chemically reacts with an oxide in the ceramics; when there is no reaction, the contact angle is large.

L. E. Murr (author's closure)—The author is indeed familiar with both points raised by Dr. Bikerman. The second point is discussed as a caution within the text of this presentation. The first point is not a valid criticism. While it is certainly admitted that the zero-creep method of measuring surface energies is approximate, it is nonetheless valid in theory and principle, and this has been demonstrated in numerous cases over the past 15 years. The measurements are in fact consistent with theory, and the formulation has been verified from first principles. Dr. Bikerman correctly stresses that it is his "view" that the method is invalid, but it has been demonstrated to be a reasonable approach by many investigators. One might, for example, consult my Ref *2* for additional details.

W. D. Bascom[2] *(written discussion)*—I must agree with Dr. Bikerman's criticism of the techniques and analyses used in this work, but "it's the only ball game in town." Work is critically needed in this area of metal-ceramic interaction, especially for the new composite materials. Perhaps, with high-quality research such as has been presented here, some of the long-standing objections to determining solid/liquid and solid/solid interfacial energies can be removed.

L. E. Murr (author's closure)—At the risk of being somewhat patronizing, I too must agree that the techniques for directly measuring adhesive or surface energies can be criticized; but they are not claimed to be exact. We are simply looking for a way to make some sense of adhesion phenomena on a more fundamental basis. As I alluded in the initial response to Dr. Bikerman's comments, this work was executed on the basis of certain assumptions, mainly that no reaction takes place between the substrate and overlayer. Obviously on an atomic level or even larger this may not be true. The degree to which this deviation from the ideal is reflected in the results is, however, unknown. It may in fact not be large under certain conditions.

[2]Adhesion Section, Naval Research Laboratory, Washington, D.C. 20375.

Adhesion Measurement of Thin Films

L. W. Crane[1] *and C. L. Hamermesh*[1]

Adhesion of Thin Plasma Polymer Films to Plastics

REFERENCE: Crane, L. W. and Hamermesh, C. L., **"Adhesion of Thin Plasma Polymer Films to Plastics,"** *Adhesion Measurement of Thin Films, Thick Films, and Bulk Coatings, ASTM STP 640,* K. L. Mittal, Ed., American Society for Testing and Materials, 1978, pp. 101-106.

ABSTRACT: To date, no comprehensive study of the adhesion of plasma-polymerized films on plastics has been made. In the present work, 1400-Å films produced by the radio frequency discharge plasma polymerization of styrene and acrylonitrile were deposited on polyethylene, polystyrene, polypropylene, poly(ethyleneterephthalate) (Mylar), polytetrafluoroethylene (Teflon), polycarbonate (Lexan), polyimide (Kapton), poly(methyl methacrylate) (Lucite), polyamide (Nylon 6), poly(oxymethylene) (Delrin), and poly(vinyl fluoride) (Tedlar) substrates. Adherence of these films was then measured by applying a ½ by 2 in. (1.27 by 5 cm) piece of Scotch No. 810 tape to the film and then removing the tape in a 90-deg peel. This provides a lower limit of 190 g/in. for the adhesion of the polymer film to the substrate. It was found that both plasma polystyrene and plasma polyacrylonitrile were removed by this procedure from poly(oxymethylene), polypropylene, poly(methyl methacrylate), and polycarbonate. Both films remained adhered to the other seven plastics during this test. These results can be explained on the basis of molecular structure and can be correlated with literature data on bond strength results for activated gas-treated substrates. It was also concluded, considering the divergent properties of the two plasma polymers, that adhesion to any substrate is not dependent on the structure of the plasma polymer. Soaking films of plasma-deposited polyacrylonitrile on Mylar in solvents of varying polarity for long periods of time did not reduce the adhesion of the film to the substrate. It would seem that the film-substrate bond is either a physical one of a very high order or more likely a true chemical bond.

KEY WORDS: adhesion, plasma polymers, thin films, peel tests

Deposition of thin films via the polymerization of volatile reactive chemicals using the glow-discharge technique has been a subject of interest for several years. To a large extent, such studies have involved the structure of these species and its effect on polymer formation. Where polymeric substrates have been employed, most of the effort has been devoted to surface modification not by film deposition but rather by substrate reaction with

[1] Staff associate and member of technical staff, respectively, Science Center, Rockwell International, Thousand Oaks, Calif. 91360.

activated gas streams, the so-called CASING method [*1*]. In an earlier paper [*2*][2] data are reported on the deposition of thin films on plastic substrates and the marked changes in surface properties that occur under such conditions. The implications from this work in regard to improved bondability, altered wettability of the substrate, etc. have led to an investigation of the extent and nature of the adhesion of thin films deposited by the plasma polymerization technique on a variety of polymeric substrates.

Experimental

Substrates employed were low-density polyethylene, polypropylene, polystyrene, Mylar polyester, polycarbonate, Kapton polyimide, poly(methyl methacrylate), Nylon 6, Delrin poly(oxymethylene), Teflon poly(tetrafluoroethylene), and Tedlar poly(vinyl fluoride). All such commercial films or slabs were cleaned with a detergent solution, rinsed with water, and dried at ambient before deposition of the plasma films. The films were grown in the afterglow region of an electrodeless glow-discharge apparatus previously described [*3*]. The plasma was operated at 10-μm Hg monomer pressure and 10-W radio frequency power. Two monomers, acrylonitrile and styrene, were employed. Using a recently described ellipsometric technique [*3*], the deposition rates for plasma-polymerized polyacrylonitrile (PPAN) and polystyrene (PPS) were determined as ~17 and ~20 Å/min, respectively. Sufficient exposure time was allowed for the deposition of films of 1400 Å thickness.

Wettability studies were made by measuring the advancing contact angle of sessile drops at 23°C (73°F) with a Naval Research Laboratory (NRL) goniometer (Rame-Hart, Inc.). The test liquids used and the procedures followed in these measurements have been detailed previously [*4*].

Adherence of the plasma films to the substrates was measured by applying by hand a ½ by 2-in. (1.27 by 5-cm) strip of Scotch No. 810 tape to the film and immediately removing the tape by hand in a 90-deg peel. The effect of environment and aging on adhesion was measured similarly.

Results and Discussion

In Table 1 the wettability of five of the polymeric substrates on which PPAN had been deposited is reported as well as the substrate surface energetics prior to coating. As expected from prior work [*2*], γ^p, the polar component of surface tension, has been drastically increased by the thin PPAN. Also, regardless of the chemical structure of the substrate, all PPAN-coated surfaces give within experimental error the same γ^p and γ^d values. The fact that the sum of these two values equals the surface tension of

[2] The italic numbers in brackets refer to the list of references appended to this paper.

TABLE 1—*Wettability of polymeric substrate as a function of presence of PPAN coating.*

	Wettability			
	Uncoated		Coated	
Substrate	γ^p	γ^d	γ^p	γ^d
Teflon	3.4	20	59	15.9
Mylar	8.3	32.9	55.7	16.5
Polypropylene	0.9	30.7	52.4	17.4
Polycarbonate	8.45	30.9	55.5	16.4
Poly(vinyl fluoride)	7.93	32.0	48.1	18.2

water is a coincidence. Therefore, the wettability of the other PPAN and all the PPS-coated substrates was not measured.

In Table 2 the adhesion to the substrates of plasma films from both monomers is reported. The method described in the foregoing is a GO/NO-GO test. In all cases of NO-GO, failure was adhesive. When a film was not removed by the Scotch tape, it was calculated that the bond strength of the plasma film to the substrate must be at least 190 g/in.

From Table 2 it can be seen that the chemical structure of the plasma polymer does not influence adhesion to the substrate. Rather, it is the nature of the substrate that determines the adhesion. Thus, regardless of the film deposition, polypropylene, poly(methyl methacrylate), polycarbonate, and poly(oxymethylene) do not provide a surface to which a plasma can adhere strongly. The first three polymers contain the same unique chemical structure, a carbon atom bonded to at least three carbon atoms (that is, a tertiary carbon atom). It has been hypothesized that the first step in the

TABLE 2—*Adhesion of plasma-polymerized film as a function of polymeric substrate.*

	Adhesion Performance[a]	
	Plasma Polymer	
Substrate	PPAN	PPS
Polyethylene	adheres	adheres
Polystyrene	adheres	adheres
Polypropylene	no adhesion	no adhesion
Mylar	adheres	adheres
Teflon	adheres	adheres
Polycarbonate	no adhesion	no adhesion
Kapton Polyimide	adheres	adheres
Poly(methyl methacrylate)	no adhesion	no adhesion
Poly(oxymethylene)	no adhesion	no adhesion
Poly(vinyl fluoride)	adheres	adheres

[a] As per peel test described in text.

deposition of a plasma film is activation of the substrate by the plasma. It is possible that for these three polymers, the presence of the tertiary carbon in the structure may stabilize the species and prevent reaction with the monomer. This would result in no adhesion of the plasma film. An alternative possibility is preferential attack on the tertiary carbon by a free radical species accompanied by chain cleavage. This substrate degradation into small fragments could prevent adhesion.

Failure of the films to adhere to poly(oxymethylene) may be due to a different mechanism in which the ionizing radiation causes severe deterioration of the substrate, probably by removal of the stabilizing ester end groups. This mechanism is substantiated by weight loss studies [5] in which poly(oxymethylene) exhibited a very high loss rate.

The conclusions reached regarding the substrates which did not provide good adhesion for the plasma film are confirmed to some extent by a previous study [6] in which polymer films were treated with activiated gas (oxygen, helium, etc.), plasmas, and bonded with adhesive, and the bond strengths obtained compared with those for untreated films. It can be seen from Table 3 that the bond strengths of polypropylene, polycarbonate, and poly(oxymethylene) show little improvement as a result of this treatment using activated helium. This is true for oxygen except for polypropylene, where strength is significantly increased. In our experiments, when polypropylene was first treated with oxygen and then a film of PPAN was deposited thereon, no improvement in adhesion resulted.

In an attempt to determine the nature of the bond between substrate and plasma film, a series of tests was performed on PPAN films on Mylar in which the films were exposed to various gases or solvents for prolonged periods of time. Surface energetics as well as adhesion after exposure were determined.

The films were aged in air, nitrogen, oxygen, argon, vacuum, and hexane vapor. Regardless of the environment, the modest decrease in γ^p (Fig. 1) observed with time is essentially the same. Thus, the change in γ^p appears

TABLE 3—*Improvement in bond strength of polymeric substrates as a function of treatment with activated gas stream prior to bonding (from Ref* 5).

Substrate	Shear Strength, psi	
	No Treatment	Treatment
Polyethylene	372	1324
Polycarbonate	410	660
Polypropylene	370	450
Polystyrene	566	4015
Mylar	530	1660
Poly(vinyl fluoride)	278	1290
Poly(oxymethylene)	118	186

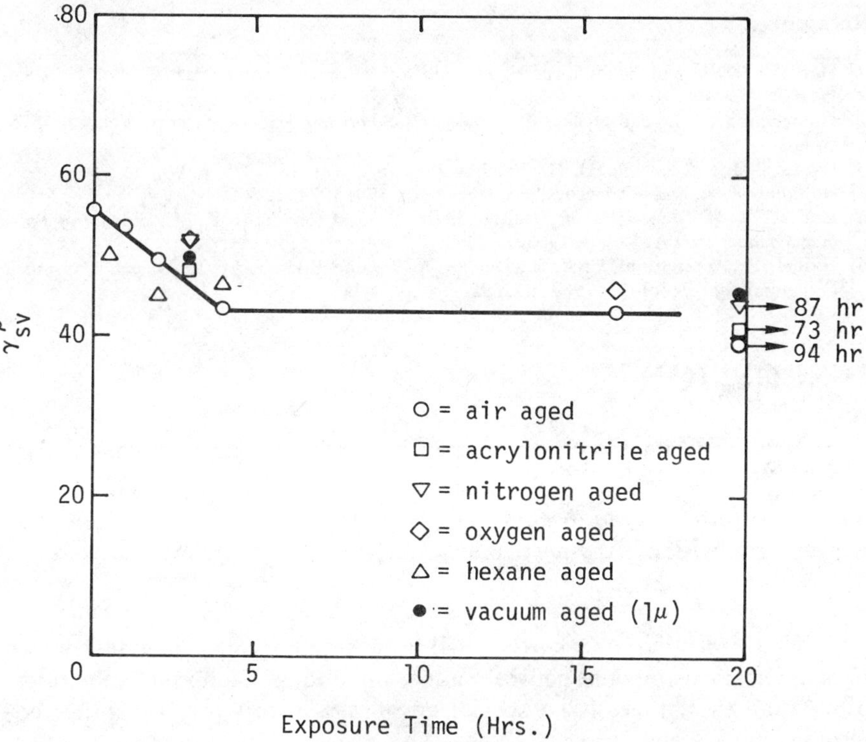

FIG. 1—*PPAN/Mylar aging in various gases.*

to be solely a function of time and not environment. Similarly, specimens of film were immersed in each of four solvents: water, dimethyl formamide, hexane, and polyethylene glycol E200. After various time intervals (2, 16, 96, and 120 h), the Scotch tape adhesion test was performed on dried specimens. In no case was there any evidence of adhesion deterioration of bond failure. Therefore, it was concluded that the bond between polymer substrate and the plasma polymerized film must be either a physical bond of a very high order or it is truly a chemical bond. Although we have not been able to prove this conclusively, the exposure tests lead us to believe that the latter hypothesis is more likely.

Conclusions

1. Adhesion of a plasma film to a polymeric substrate is a function of the chemical nature of the substrate and not that of the film.
2. The bond between film and substrate is probably chemical in nature.

References

[1] Hansen, R. H. and Schonhorn, H., *Journal of Polymer Science*, Part B, Vol. 4, 1966, p. 203.
[2] Hamermesh, C. L. and Dynes, P. J., *Journal of Polymer Science*, Letters, Vol. 13, 1975, p. 663.
[3] Dynes, P. J. and Kaelble, D. H., *Polymer Preprints*, Vol. 16, 1975, p. 98.
[4] Hollahan, J. R. and Wydeven, T., *Science*, Vol. 179, 1973, p. 500.
[5] Hansen, R. H., Pascale, J. V., DeBenedictis, T., and Rentzepis, P. M., *Journal of Polymer Science*, Part A, Vol. 3, 1965, p. 2205.
[6] Hall, J. R., Westerdahl, C. A. L., Devine, A. T., and Bodnar, M. J., *Journal of Applied Polymer Science*, Vol. 13, 1969, p. 2085.

DISCUSSION

K. L. Mittal[1] (*written discussion*)—Why did you select the peel test for measuring adhesion? Any particular advantage of this test over others?

L. W. Crane (*author's closure*)—It is a simple, convenient test which gives a clear distinction between adhering and nonadhering substrates. Other than that, there is no special advantage in this test over other adhesion tests.

[1] IBM Corporation, Hopewell Junction, N.Y. 12533.

Sol Krongelb[1]

Electromagnetic Tensile Adhesion Test Method

REFERENCE: Krongelb, Sol, **"Electromagnetic Tensile Adhesion Test Method,"** *Adhesion Measurement of Thin Films, Thick Films, and Bulk Coatings, ASTM STP 640,* K. L. Mittal, Ed., American Society for Testing and Materials, 1978, pp. 107–121.

ABSTRACT: A method is described for applying a known tensile stress to a thin-film structure without requiring any mechanical attachment to the film. The stress is developed by the interaction of an external magnetic field with an electric current through a suitably patterned specimen of the film to be evaluated. An apparatus for producing the required magnetic fields and electric currents is described, and procedures are given for preparing suitable test specimens using conventional electronic device fabrication technology. Results showing the use of this technique for measuring the adhesion of evaporated copper on thermally oxidized silicon are presented. The maximum stress which can be produced is limited by the available magnetic field, the fabrication of the test structure, and the heating effect of the current through this structure. These factors are discussed, and it is shown that useful stresses of at least 24 000 psi (165 600 kPa) are practical by the electromagnetic method.

KEY WORDS: adhesion, thin films, stresses, electric current, magnetic fields

The direct-pull test has long been an accepted procedure for measuring the adhesion between bonded specimens of bulk material where the specimen can be readily grasped by the test instrument. Attempts have also been made to apply the pull test to measure the bond between a thin film and its substrate [*1–3*].[2] In this application, however, difficulty is encountered in (*a*) providing some attachment to the thin film which allows the force to be applied but does not perturb the interface being tested, and (*b*) applying the stress uniformly and in a direction truly normal to the film.

This paper describes a method of using the interaction between an electric current in the film and an external magnetic field to produce a normal force on the film without requiring soldering or similar attachment to the film. The basic concept of the method is shown in Fig. 1. The thin film

[1] Research staff member, Thomas J. Watson Research Center, International Business Machines Corporation, Yorktown Heights, N.Y. 10598.

[2] The italic numbers in brackets refer to the list of references appended to this paper.

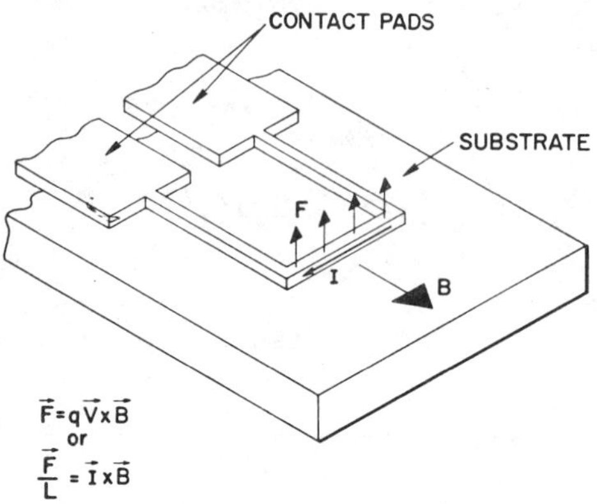

FIG. 1—*Basic configuration for electromagnetic test method.*

to be tested is patterned into a rectangular U-shaped form. Current leads are then clamped to the legs of the U, and the specimen is placed in a magnetic field oriented as shown. (By applying the magnetic field normal to the interface, a tangential stress can be also produced.) When a pulse of current is passed through the test specimen, a force perpendicular to both the direction of the current and the magnetic field will be developed in the lower segment of the U. Knowing the width of the interface between the strip and substrate and the magnitudes of the current and field, the force per unit area can be readily calculated.

For a given interface width and magnetic field, the maximum stress which can be developed is determined by the current passed through the line. This current is in turn limited by the I^2R heating of the line and the consequent temperature rise which one is willing to consider as creating a negligible perturbation on the measurement. The discussion herein will show that achieving a stress of 24 000 psi (165 600 kPa) with a temperature rise less than 20°C (11°F) requires a magnetic field of 200 to 400 kilo oersted (kOe) and a current pulse of about 20 to 40 A with a duration of a few tenths of a microsecond. The high magnetic field requirement is best met by a pulsed field magnet, and the apparatus constructed for these experiments employs such a magnet synchronized with a commercial high current pulse generator to produce a current pulse at the peak of the magnetic field. (Superconducting magnets can also achieve fields in excess of 100 kOe. Some of the initial experimental work reported here was done in a superconducting magnet; however, the pulsed field magnet is more convenient to use and is capable of achieving significantly higher fields.)

The following sections describe the experimental apparatus and the preparation of suitable specimens. Some results demonstrating the application of the techniques are then given. Factors relating to the maximum stress which can be achieved are treated in the discussion.

Apparatus

A schematic of the apparatus is shown in Fig. 2. The specimen is clamped to a holder (Fig. 3) which has provision for leads to feed the current pulse into the specimen. The holder is supported in the solenoid by a pivot point at the bottom and is clamped between a pair of adjustable screws at the top. These screws are used to align the specimen so that the magnetic field will be in the specimen plane, thus assuring that the stress will be normal to the surface. Since any misalignment also results in a voltage being induced in the specimen loop, a simple procedure for achieving alignment is to pulse the magnet and to adjust the screws until no induced voltage is detected at the current leads.

The magnet solenoid has an 0.5-in. (12.7-mm) bore and is of the helical construction described by Foner and Kolm [*4*]; the design is based on formulas given by Montgomery [*5*]. The magnet is powered by a pair of parallel-connected 240-μF capacitors which are charged from a CVC Type LC-031 d-c sputtering supply and discharged through the solenoid through a Type WL 7171 ignitron. The current limiting feature which is inherent in the sputtering supply is particularly useful for capacitor charging. The magnitude of the magnetic field is determined by the voltage to which the capacitor is initially charged. Our coil design produces approximately 60 kOe per 1000 V and reaches its peak field in about 50 μs.

The actual magnetic field at the specimen is determined for each test by means of a pickup coil which is wound around the specimen holder at the specimen location. The output of this coil is integrated and is displayed

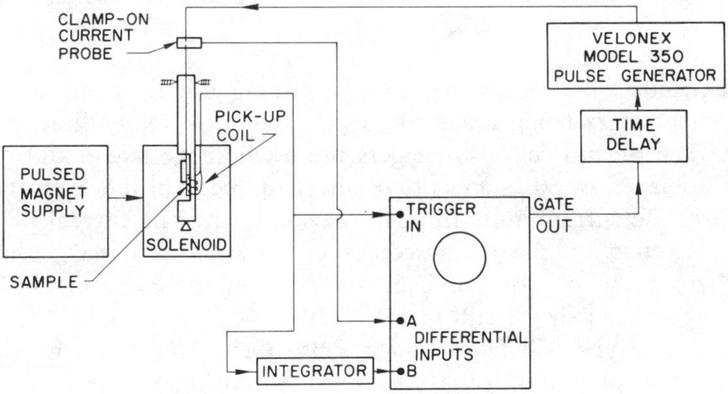

FIG. 2—*Schematic of apparatus.*

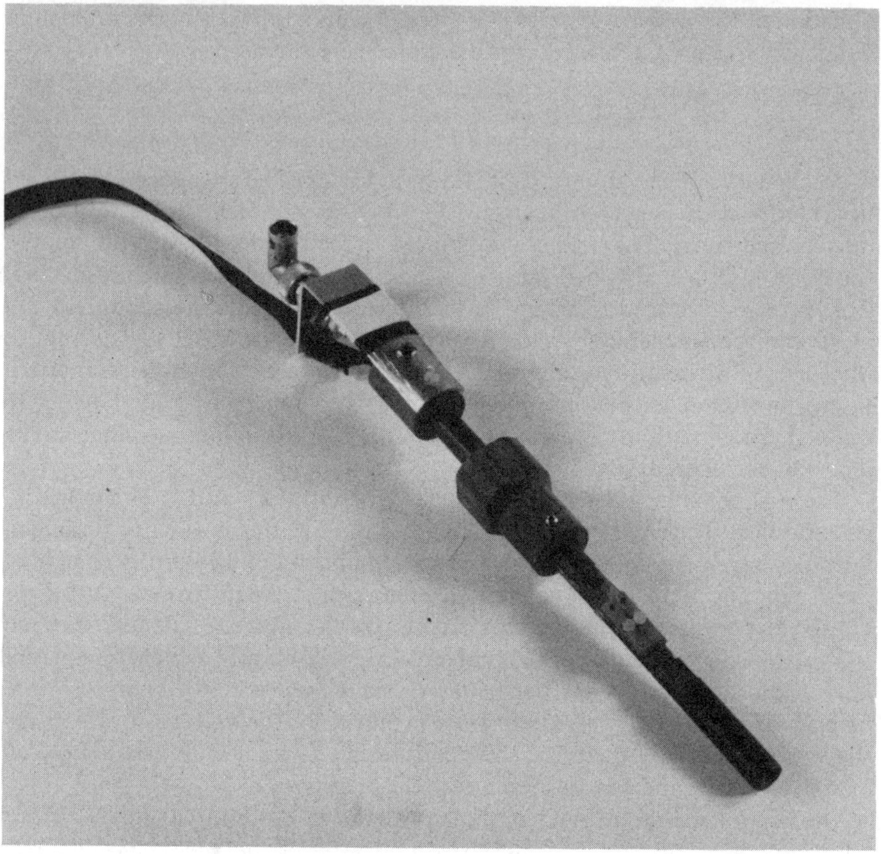

FIG. 3—*Photograph of specimen holder. Specimen is clamped in place on milled flat in lower part of holder. Brass strips under clamp blocks provide contact to strip line lead, which is seen as a ribbon in upper part of photo. Leads from field pickup coil terminate at BNC connector.*

on a Tektronix 545 oscilloscope as a plot of magnetic field versus time, the circuit parameters being adjusted so that the scale reads directly in kilo oersted. The pickup coil also triggers the oscilloscope sweep and provides a signal (delayed by 50 μs so as to occur at the peak of the magnetic field) which fires the current pulse through the specimen. The current pulse trigger is derived from the oscilloscope's gate output terminal rather than directly from the pickup coil because the steeper waveform of the gate pulse triggers the delay circuit more reliably.

A Velonex Type 350 High Power Pulse Generator fitted with a high current output plug-in unit that has 0.5 Ω output impedance is used as the source of the current pulse. The lead from the generator to the specimen is a piece of flexible stripline with a characteristic impedance of about 0.5 Ω

which was formed by etching copper-clad Mylar. Provision was also made for clamping a Tektronix Type P-6021 current probe around one of the leads so that the amplitude of the current pulse could be measured. By use of a differential input on the oscilloscope, this current pulse can be superimposed on the magnetic field traces as shown in Fig. 4. Knowing the scale factors for field and current, the magnitudes of each can be read from the oscilloscope trace and the applied stress calculated.

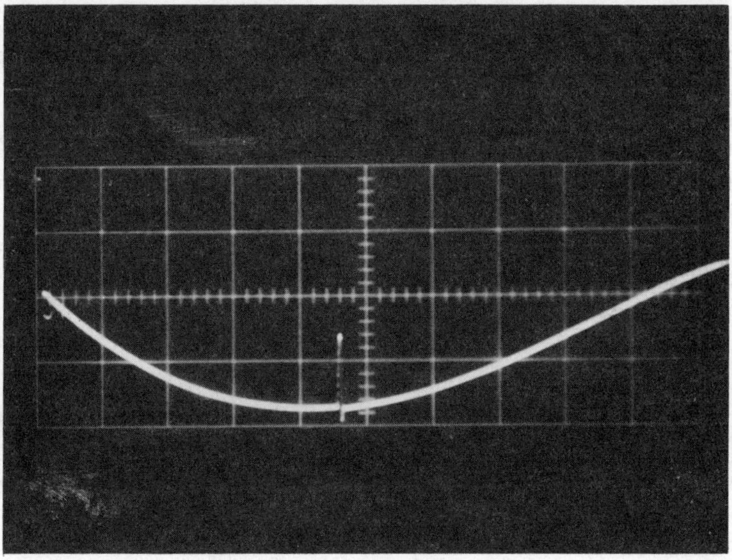

FIG. 4—*Typical oscilloscope trace showing simultaneous display of magnetic field and current pulse amplitudes.*

In testing a specimen, the structure is clamped in the holder and placed in the magnet solenoid. Approximate values of the peak magnetic field and current to produce the desired stress are selected by presetting, respectively, the charging voltage on the magnet power supply and the output power of the current pulse generator. The magnet supply is then discharged through the solenoid, creating the magnetic field and triggering the current pulse as described in the foregoing. This sequence is repeated using higher fields, higher currents, or both until the specimen fails. (Failure can often be detected without removing the specimen from the magnet because the line usually opens when it is pulled from the substrate). The failure stress can be calculated using the magnetic field and current as measured on the oscilloscope during the last stress pulse.

Specimen

The following discussion will show that, to minimize heating of the specimen by the current pulse, the line under test is required to have a substantial cross section. The interface between line and substrate should also be narrow so that maximum stress is developed for a given current through the line. These conditions can be achieved in structures fabricated using current device fabrication technology [*6, 7*].

The copper on silicon dioxide (SiO_2) specimens used in this work were prepared by first evaporating a few hundred angstroms of copper at the temperature specified for the particular specimen. A thin layer (about 1μm) of Shipley AZ 1350 J photoresist was applied and exposed to leave openings in the pattern of the test structure. Copper was then electroplated through the resist openings to a thickness of 10 or more μm. When plated to this thickness, the copper mushrooms over the resist to produce a line cross section as shown in Fig. 5 which provides the required current-carrying capability while retaining the narrow interface. As a final step, the photoresist was removed, and the unplated areas of evaporated copper were chemically etched. The lower portion of Fig. 6 shows a series of such structures before stressing.

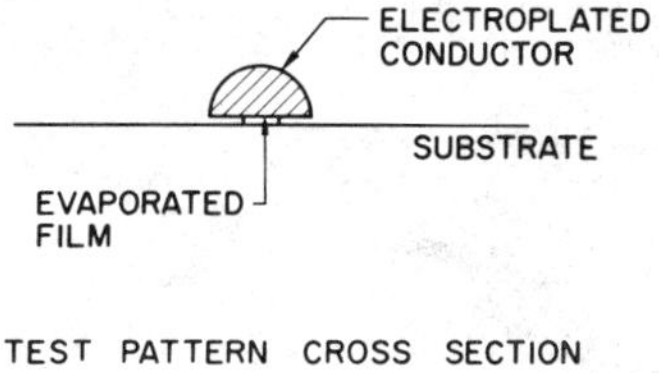

FIG. 5—*Mushroom cross section of plated conductor provides adequate current-carrying capacity.*

These structures had an interface width of about 12.5 μm as determined by examining the interface of specimens which had failed. The state of current technology is such that interface widths of less than 5 μm can be prepared in a similar manner. However, the use of chemical etching makes it somewhat difficult to control the actual interface width because invariably there is some undercut. An alternative process outlined in Fig. 7 avoids this problem. In this process a layer of photoresist having the pattern of the shaded area in Fig. 7*a* is formed on the initial film. The loop in the clear window will be the test line and therefore has the desired interface width. The window is next sputter etched using the resist film as a mask. A fresh layer of resist is then patterned as shown by the cross-hatched area of Fig. 7*b*, and the copper is electroplated through this mask. Plating within the window area takes place only on the preformed line so that the desired

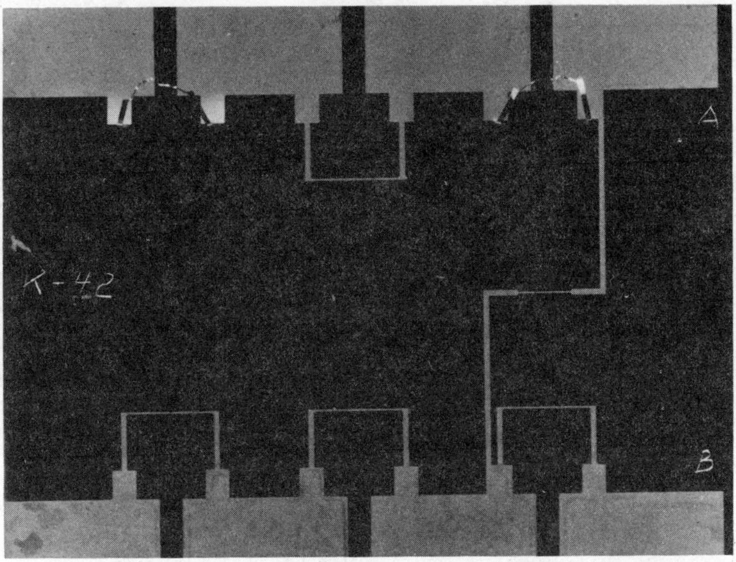

FIG. 6—*Photograph of specimen containing six test sites. Two patterns in upper portion have been lifted from substrate by electromagnetic stressing.*

mushroom structure with controlled interface width is formed. As a final step, the mask pattern of Fig. 7*c* is applied and the exposed evaporated layer is chemically etched. The interface of the test line is fully protected by the resist during this etching; slight undercut which may occur at the edges of the contact pads is of no consequence. Figure 8 shows a specimen with a 5-μm-wide interface made by this technique. (Only the center loop is the test line. The other lines are included to achieve better control of the plating process as discussed in Ref *6* and will be etched off in the final step.)

Experimental Results

As a demonstration of the technique, specimens with copper evaporated on SiO_2 at deposition temperatures ranging from room temperature to 300 °C (572 °F) were prepared and stressed until either failure occurred or the maximum achievable stress had been applied. This series was done using a superconducting magnet with a maximum field of 90 kOe. With this magnet, the maximum stress was limited to 9000 psi (62 100 kPa) to avoid undue temperature rise. The results are shown in Fig. 9. For the room temperature and 125 °C (257 °F) points, the average failure stress and standard deviation are plotted. These values are based on 12 test sites from four different specimens for each temperature. For the specimens prepared by 200 or 300 °C (392 or 572 °F) evaporation, detachment generally

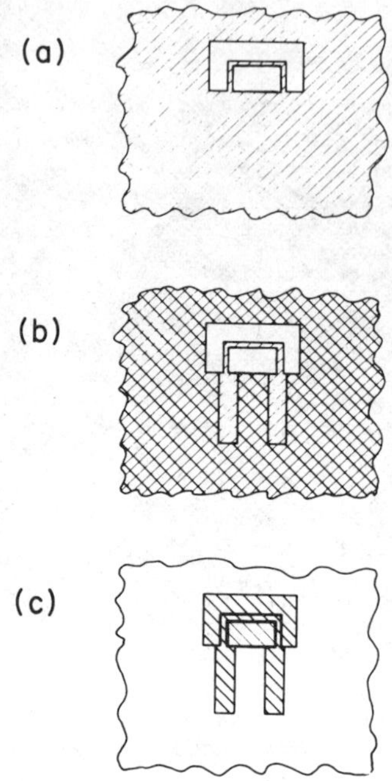

FIG. 7—*Fabrication steps in preparing test specimen.* (a) *Deposited film is patterned by sputter etching.* (b) *Photoresist mask (double crosshatch) covers film for electroplating; plating takes place only on singly shaded portion.* (c) *Photoresist pattern protects test structure while excess of original film is etched off.*

could not be caused by stresses up to 9000 psi (62 100 kPa); therefore, the graph shows that the failure stress was greater than this value. The expected improvement in bonding with increased evaporation temperature is apparent from these data. The upper portion of Fig. 6 shows two test patterns which have been detached from the substrate.

One room-temperature deposition specimen showed anomalously good adhesion while one specimen prepared at 200 °C (392 °F) was exceptionally poor. The measurements for these specimens fall well outside of the normal spread found for the other specimens and are plotted separately. Since all specimens for each deposition temperature were prepared identically, we can offer no explanation for the anomalous measurements on these particular specimens. During processing, however, we noted that this particular room-temperature specimen withstood much rougher handling without damage than is usually possible for wafers of copper evaporated on SiO_2 without substrate heating, and we believe that the measurements correctly reflect superior adhesion in this specimen. Likewise, the poor adhesion on the 200 °C (392 °F) specimen is consistent with the occasional fluctuations in bond quality that are a known occurrence in thin-film processing.

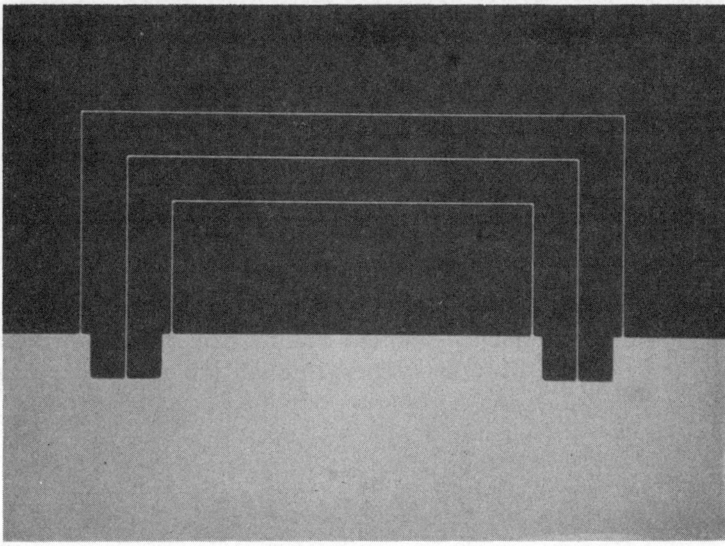

FIG. 8—*Typical 0.2-mil-wide interface achievable by current technology after etching as in Fig. 7*a.

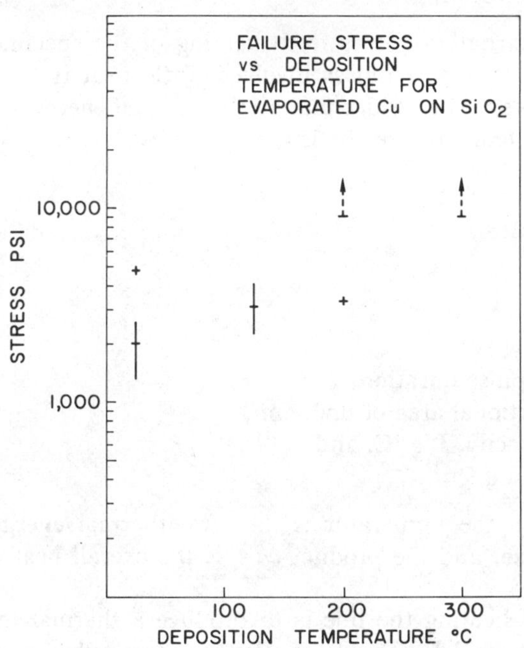

FIG. 9—*Failure stress versus deposition temperature for evaporated copper on SiO_2.*

The adhesive tape test was applied to the contact pads on several of the specimens after the aforementioned measurements were made. About half of the pads in the specimen deposited at 125 °C (257 °F) withstood the test, suggesting that films which "pass" the adhesive tape test have adhesion greater than 2200 to 4000 psi (15 180 to 27 600 kPa) while those which fail fall below this range.

Discussion

The electromagnetic adhesion test, like other methods [*8*], has advantages and limitations. The usefulness of the technique depends on the maximum stress which can be developed without unduly perturbing the interface. The stress is determined by the current through the specimen, magnetic field and interface width and is given in psi by

$$S = 14.5\frac{IB}{w} \tag{1}$$

where

I = current, A,
B = magnetic field, kOe, and
w = interface width, μm.

The major perturbation comes from heating of the specimen by the current pulse. For the short pulses involved, little heat is transferred to the substrate, and we may consider that all the thermal energy goes into heating of the line. The temperature rise is then given by

$$\Delta T = \frac{\tau I^2 \rho / A}{cdA} \tag{2}$$

where

ρ = resistivity, Ω,
τ = current pulse duration,
A = cross-sectional area of line, cm^2,
c = heat capacity, J/g °C, and
d = density in g/cm^3.

In this equation the numerator is the total thermal energy produced by the current pulse, and the product *cdA* is the overall heat capacity of the line.

The effect of heating the line is to produce a thermal mismatch which results in an interfacial shear stress. For the short pulses used in this work, the maximum value of this stress occurs at the end of the current pulse when all the resistive losses have been converted into thermal energy but no significant part of this energy has yet been transferred to the substrate. Under these conditions it is the expansion coefficient of the line itself rather than the difference in coefficients between line and substrate that is relevant, and the interfacial shear stress is given by [9]

$$S = \alpha \Delta T \frac{E_F}{1 - \nu_F} \tag{3}$$

where

α = temperature coefficient of expansion,
E_f = elastic modulus of film, and
ν_f = Poisson's ratio for film.

The thermal stress should be less than or at worst comparable to the typical intrinsic stresses in thin films [10] which are of the order of 10^9 dynes/cm^2 or 1.4×10^4 psi. Using values appropriate for copper in Eq 3 (α = 16×10^{-6} per °C, E_f = 17×10^6 psi, ν_f = 1/3), we see that the thermal mismatch stress is less than half the typical intrinsic stress if ΔT is limited to about 20°C (11°F). (Note that Eq 3 is not strictly applicable to our structure with modified cross-sectional shape, and that the parameters used are bulk values, which often are significantly different from the film properties. The calculation should, however, serve as a guide to the order of magnitude of the stress.)

It is apparent from Eqs 1 and 2 that a large magnetic field, narrow interface width, and large cross-sectional area are required. Fields of the order of 200 kOe are easily achievable in pulsed field magnets and are available in many laboratories. The literature also describes many systems where fields in excess of 400 kOe are produced [4,11,12], which would be quite suitable for the electromagnetic adhesion test. Specimens with a 5-μm interface and with a 10-μm-thick electroplated conductor—dimensions comparable to those found in some device structures—are readily prepared using current fabrication technology. Using a 400-kOe field and an 0.2-μs, 21-A current pulse, this test pattern can develop 24 000 psi (165 600 kPa) with a ΔT of 20°C (11°F). The same conditions can be achieved in a 200-kOe field using a 14-μm-thick conductor and a 42-A current pulse.

The maximum-temperature criterion is, of course, somewhat arbitrary. The thermal expansion stress can result in an increase or decrease of the interfacial shear stress depending on whether the film was initially in compression or tension. The thermal effects can be investigated empirically, however, by adjusting the current and magnetic field to produce the same normal stress while varying the thermally induced stress. Such a study would provide a better indication of the actual limitations on usefully achievable stress and would also give some insight into the interaction between shear stress and normal stress in producing bond failure.

A related consideration is the interfacial stress produced by the electroplated conductor. Such stress would arise from the intrinsic stress in the electroplated copper and would constitute a significant perturbation of the interface. Fortunately, copper plating formulations which produce near-zero stress deposits are available [*13*] and can provide a conductor which does not significantly disturb the interface.

The structures to which the electromagnetic method may be applied can now be considered. The technique is applicable to any deposit on a dielectric substrate provided the surface of the deposit will accept electroplating. A wide range of circuit and device structures is thus included. Furthermore, specimen preparation is essentially that used in fabricating actual devices so that the results provide practical information. Since essentially all the current can be made to flow in the plated copper conductor if the initial deposit is thin, almost equal stresses will be applied to the deposit-substrate interface and to the copper-deposit interface. Failure will then occur at the weaker bond, and the electroplated interface may thus be studied if it is the weaker one.

Moderately bonded materials such as metal on plastics can probably be tested by the electromagnetic method with negligible induced thermal stress. The thermally induced stresses become comparable to the intrinsic stresses for tests in the range 25 000 to 30 000 psi (172 500 to 207 000 kPa), and care would have to be taken in interpreting the results for stresses beyond this range. Extension of the useful stress range is possible if thicker conductors are electroplated. We note from Eq 2 that the temperature rise decreases as the fourth power of the thickness, since the latter quantity appears as a squared term in calculating the cross-sectional area, A, of the mushroom. Higher stresses can also be achieved by using advanced lithographic technology, which can achieve dimensions of 1 μm at the interface. In this case we note from Eqs 1 and 2 that, for a given stress, the required current decreases directly as the interface width and that ΔT decreases as the square of the current. Decreases in the pulse width τ are also possible but of more limited benefit since, first, ΔT depends only on the first power of τ, and, second, the skin depth becomes a more significant factor for shorter τ so that the apparent resistivity of the conductor pattern increases.

Conclusions

The interaction of an external magnetic field with an electric current in a patterned film structure has been shown to provide a means of measuring the adhesion of the film to its substrate. The test is particularly applicable to electronic microcircuit films since the preparation of the test structures uses conventional device fabrication technology. The adhesion of evaporated copper films on SiO_2 has been measured as a demonstration of the electromagnetic method. The principal limitation on this technique arises from heating of the conducting pattern by the test current. Consideration of present device fabrication technology and conveniently achievable magnetic fields has shown that stresses of the order of 24 000 psi (165 600 kPa) are practical and that further extensions are possible.

Acknowledgments

The assistance of B. J. Stoeber in the various aspects of this work is sincerely appreciated. Thanks also go to B. A. Marinello, who constructed the pulse field magnet power supply and to G. J. Morrisey, whose skillful machine work produced the magnet solenoid. The author is especially grateful to R. A. Laff and L. T. Romankiw for their encouragement and for many helpful discussions.

References

[*1*] Frederick, J. R. and Ludema, K. C., *Journal of Applied Physics*, Vol. 35, 1964, pp. 256-257.
[*2*] Murphy, N. F. and Swansey, E. F., *Plating*, Vol. 56, 1969, pp. 371-376.
[*3*] Jacobsson, R., *Thin Solid Films*, Vol. 34, 1976, pp. 191-199.
[*4*] Foner, S. and Kolm, H. H., *Review of Scientific Instruments*, Vol. 28, 1957, pp. 799-807.
[*5*] Montgomery, D. B., *Solenoid Magnet Design*, Wiley, New York, 1969, pp. 190-225.
[*6*] Romankiw, L. T., Krongelb, S., Castellani, E. E., Pfeiffer, A. T., Stoeber, B. J., and Olsen, J. D., *IEEE Transactions on Magnetics*, Institute of Electrical and Electronic Engineers, Vol. MAG-10, 1974, pp. 828-831.
[*7*] Koel, G. J., Postma, L., Gerkema, J. T., and Snijders, J. T., *Extended Abstracts of The Electrochemical Society Spring Meeting*, Washington, D.C., 2-7 May 1976, pp. 374-376.
[*8*] Mittal, K. L., *Electrocomponent Science and Technology*, Vol. 3, 1976, pp. 21-42.
[*9*] Hoffman, R. W. in *Measurement Techniques for Thin Films*, B. Schwartz and N. Schwartz, Eds., The Electrochemical Society, New York, 1967, p. 316.
[*10*] Hoffman, R. W. in *Physics of Thin Films*, Vol. 3, G. Hass and R. E. Thun, Eds., Academic Press, New York, 1966, pp. 211-273.
[*11*] Dworschak, G., Haberey, F., Hildebrand, P., Kneller, E., and Schreiber, D., *Review of Scientific Instruments*, Vol. 45, 1974, pp. 243-249.
[*12*] Knoepfel, H. and Luppi, R., *Journal of Physics E*, Vol. 5, 1972, pp. 1133-1141.
[*13*] Safranek, W. H., *The Properties of Electrodeposited Metals and Alloys*, American Elsevier, New York, Chapter 6, 1974, pp. 91-135.

DISCUSSION

J. J. Bikerman[1] (*written discussion*)—I have three questions. (1) Do you intend to determine, theoretically and experimentally, the effect of the surface roughness of the silicon dioxide substrate? (2) Do you intend to determine the effect of the stresses caused by the recrystallization of the copper coating? (3) Do you intend to determine the exact locus of failure?

Sol Krongelb (*author's closure*)—Our present program does not include extensive work on adhesion, so we have no immediate plans to investigate the questions which you raised. These points are important aspects of the adhesion problem, and I believe they can be effectively investigated using the electromagnetic method.

L. H. Sharpe[2] (*written discussion*)—What is the test sequence? Do you hold the current constant and increase the field in successive pulses until failure occurs, or what? Near failure, don't you start to accumulate damage which could influence your results significantly; that is, don't successive pulses lead to the accumulation of such damage as you attempt to locate the combination of current and field which leads to failure?

Sol Krongelb (*author's closure*)—In general, we prefer to use a large field and a small current to minimize the I^2R heating. We therefore keep the magnetic field constant and increase the stress by increasing the current.

We attempted to account for the possibility of accumulated damage by repeating the same stress condition several times before increasing the current further. Specimens which failed did so on the first or second pulse at the failure stress level, whereas the same specimen had previously withstood five or six pulses at a somewhat lower stress.

K. L. Mittal[3] (*written discussion*)—(1) Is the electromagnetic technique for measuring adhesion applicable to nonmetallic films? (2) Did you examine the site or locus of failure? If yes, what did you find?

[1]Department of Chemical Engineering, Case Western Reserve University, Cleveland, Ohio 44106.
[2]Bell Laboratories, Engineering and Development, Murray Hill, N.J. 07974.
[3]IBM Corporation, Hopewell Junction, N.Y. 12533.

Sol Krongelb (author's closure)—(1) A conductor must be present to carry the current in the electromagnetic method. However, the test specimen can be fabricated just as well if a multilayer dielectric structure is present under the conductor. If the adhesion between the various layers happens to be such that failure occurs between two of these dielectric layers, then the adhesion between these materials could be determined. The interface would, of course, have to be examined to determine where failure occurred. (2) We have looked at the failure interface only by optical microscopy to determine the interface width. What really should be done is to use techniques such as Auger spectroscopy to determine how the materials separate.

J. L. Vossen[1]

Measurements of Film-Substrate Bond Strength by Laser Spallation

REFERENCE: Vossen, J. L., **"Measurements of Film-Substrate Bond Strength by Laser Spallation,"** *Adhesion Measurement of Thin Films, Thick Films, and Bulk Coatings, ASTM STP 640,* K. L. Mittal, Ed., American Society for Testing and Materials, 1978, pp. 122–133.

ABSTRACT: Laser spallation is a technique for detaching thin films from substrates by impinging a pulsed, high-energy laser beam onto the back side of a substrate (made opaque in the case of transparent substrates). The explosive evaporation of absorbing material on the back side sends a recoil compressive shock wave through the substrate toward the film-substrate interface. The substrate then snaps back in tension. Threshold values for film detachment can yield information about the film-substrate bonds without contact with or disturbance of the film surface. Thus damage propagated to the interface from film surface disturbance is eliminated. In some cases it is possible to examine [for example, by Auger electron spectroscopy (AES)] the spalled dot and the region of the substrate from which it was spalled to determine the locus of failure and to obtain information about interface formation.

KEY WORDS: adhesion, interfaces, thin films, laser shock waves, laser damage

The "adhesion" of thin films to substrates has been one of the most elusive parameters in all of thin-film technology. The word "adhesion" is often used in ill-defined ways, but, in general, it may be considered as the sum of the attachment (+) and detachment (−) forces acting on film-substrate interfaces. The attachment forces may be van der Waal's, electrostatic, or chemical. The detachment forces built into the film are thermal and intrinsic stress. Superimposed on these are any externally applied forces required by the application. Most adhesion tests are really engineering tests of the durability of a thin film in a particular fracture mode. There have been several reviews of adhesion measurements [*1–3*].[2] Common to nearly all of these tests is the requirement of either attaching something to the surface of the film (with which to pull it off), or somehow damaging the surface (for example, by scratching or abrading). The only

[1] Member of technical staff, David Sarnoff Research Center, RCA Laboratories, Princeton, N.J. 08540.

[2] The italic numbers in brackets refer to the list of references appended to this paper.

exceptions to this are the ultracentrifuge and the ultrasonic vibration tests, but these work only for very poorly bonded films.

One can well question whether the very act of disturbing the film surface does not propagate damage to the film-substrate interface which could materially affect the interfacial bond strength.

We are developing a technique called "laser spallation" [*4,5*], with which a film can be detached from a substrate without any prior disturbance of the film surface. This is the main difference between this technique for adhesion measurement and most others. The technique involves impinging a high-energy, pulsed laser beam on the back side of the substrate (Fig. 1). In the case of transparent substrates, a thick, absorbing layer is first deposited on the back side of the substrate. The incident laser radiation is converted rapidly to thermal energy in the absorbing substrate or coating, and the explosive evaporation of the material sends a com-

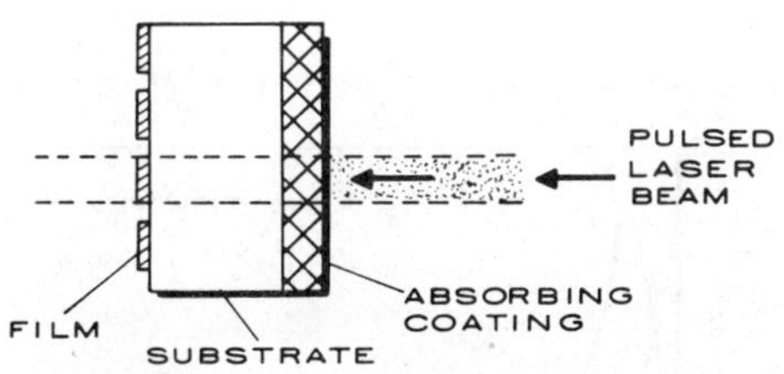

FIG. 1—*Schematic representation of laser spallation. The laser beam typically used is about ten times larger than the dot, so that a plane shock wave is approximated.*

pressive shock wave through the substrate toward the film-substrate interface and then the substrate snaps back in tension. It is the reflection of the compressive wave from the surface that gives rise to the tensile wave and leads to removal of the film. By increasing the incident laser power, one can find a threshold at which the film is torn loose from the substrate. This threshold can be related directly to the bond strength of the film to the substrate if corrections are applied for the decay in energy as the shock wave proceeds through the substrate, and for stress in the film.

Internal stresses in films produce shear stresses at the film-substrate interface [*6*]. These shear stresses depend directly on the internal stress in the film. Since a film in compressive stress will first be stress-relieved and then spalled off, compressive stress must be subtracted from the measured values. If the film is in tensile stress, the opposite is true.

Furthermore, it is necessary that the film be patterned into dots smaller in diameter than the incident laser beam. This may be done either by deposition through a mask or by photolithographic techniques after film deposition. If this is not done, the film will first tear and then spall off. In this case, the threshold is the sum of the film-substrate bond strength and the tensile strength of the film. The latter is a largely unknown quantity. In the case of dots, the entire dot is removed by the shock wave.

Behavior of Stress Waves in Solids

The propagation of stress waves in solids is exceedingly complicated [*7*] and, in practical situations, is studied empirically using quartz stress gages [*8*]. To put this measurement technique into better perspective, it is necessary to understand, at least on an empirical basis, those effects which could influence the measurements.

Stress waves generated by a pulsed laser vary with time. The stress builds rather rapidly to a maximum and decays somewhat more slowly (Fig. 2). The relatively long decay time leads to uncertainties of the order of 20 percent. The peak stress is inversely proportional to pulse width (Fig. 3). The damping at longer pulse widths is due to destructive interference

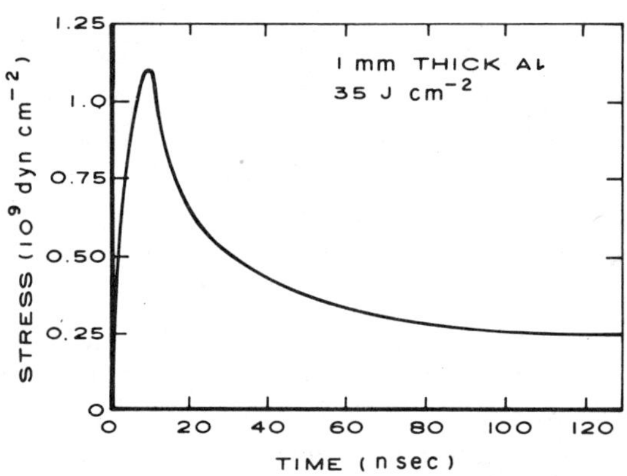

FIG. 2—*Typical stress history for laser pulse (pulse width = 30 ns; fluence = 35 J/cm^2); incident on 1-mm-thick 6061-T6 aluminum (after Ref* 3).

caused by reverberation of the initial wave (ringing) [*7*]. The peak stress is also a function of laser fluence or energy density (Fig. 4). The peak stress rises to a maximum as a function of fluence and then decreases due absorption by microplasmas generated in the recoiling vapor [*10*]. In addition, the stress wave is damped by the substrate in transit from the back side to the film-substrate interface (Fig. 5). This necessitates calibration of the system with stress gages for each type of substrate employed and for various thicknesses of a given substrate material.

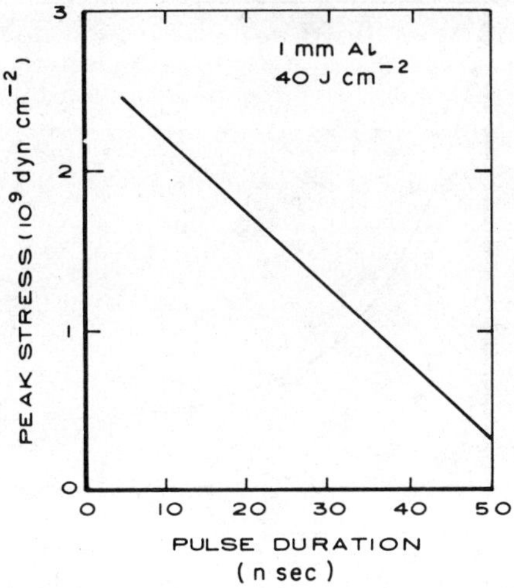

FIG. 3—*Peak stress versus pulse duration* (*after Ref* 3).

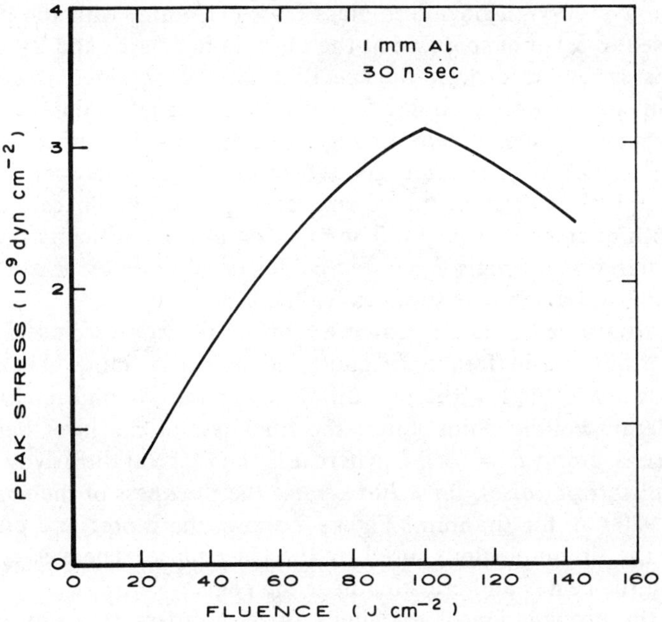

FIG. 4—*Peak stress versus laser fluence* (*after Ref* 3).

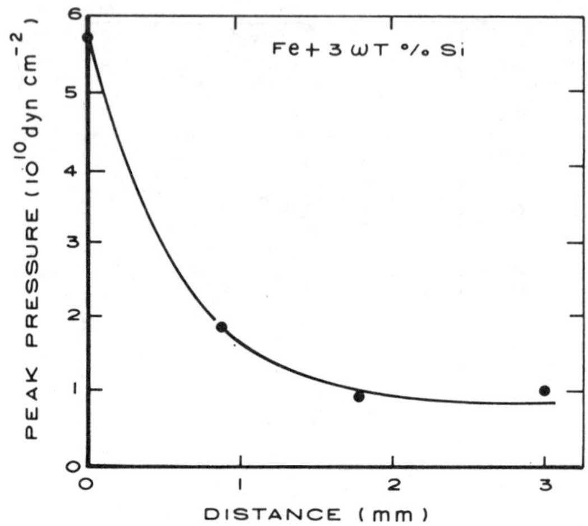

FIG. 5—*Damping of stress wave by substrate* (*after Ref* 4).

Experimental Techniques

We use a 1500-W neodymium-glass laser (1.06-μm) with the beam usually focused to a 1-mm spot size at the point of incidence and 0.1-mm dots. From this system, the peak fluence that can be obtained is about 1200 J/cm^2; this has proven adequate for all systems tested to date. The overall system is shown schematically in Fig. 6. Owing to the rather short pulse widths (~20 ns) required for good measurements, there are numerous instrumental difficulties in monitoring each pulse, and in calibrating the system with quartz stress gages. The most significant difficulty is related to spurious triggering signals generated by electrical noise associated with the pulsed, high-voltage power supplies for the laser.

When absorbing layers are required on the substrates, we use 5-μm-thick titanium films (radio frequency sputtered at 500 Å/min). Titanium was chosen because of its low thermal diffusivity, relatively high melting point, and moderate boiling point. Since the thickness of the layer heated by a laser pulse is simply $d = (at)^{1/2}$ (where a is the thermal diffusivity and t the time from start of pulse), for a 20-ns pulse the thickness of the heated layer is about 9200 Å for titanium. Figure 7 shows the center and edge of the crater in the titanium film caused by the laser pulse. There is evidence of melting in the center and cracking near the edge.

While the propagation of the shock wave is orders of magnitude faster than the propagation of the thermal wave [*10*], the thermal wave does arrive at the interface, but this causes melting—not detachment. Charac-

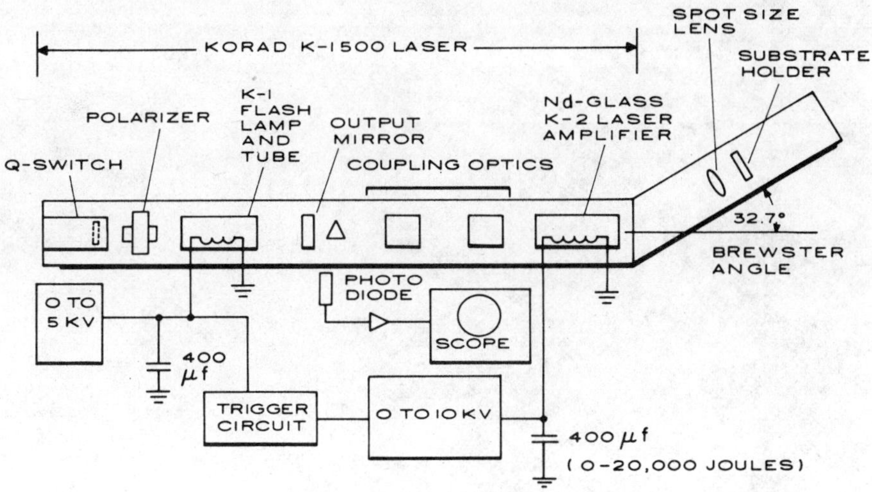

FIG. 6—*Schematic of laser spallation apparatus.*

teristic signs of melted material are generally found in subthreshold laser pulses. The thermal energy involved also limits the types of substrates that may be used with this technique, as will be discussed later.

Experimental Results

Up to now, we have been unable to assure ourselves that all of the pertinent corrections noted in the foregoing have been made in order to obtain quantitative measurements. However, certain qualitative results have been achieved which validate the measurement technique. The following are a few examples.

We find a direct correlation of metal film bond strength on silicon dioxide substrates with the free energy of oxide formation of the arriving metal vapor for both evaporated and sputtered metals (Fig. 8). This is in agreement with other workers [*12*]. It has been shown that heat treatment of substrates either during or after deposition increases the bond strength of chromium to SiO_2 by a factor of 2. The effect of film stress on the measurements has been examined using bias-sputtered molybdenum films on molybdenum substrates where the metal-metal bond should be extremely high. When molybdenum is sputtered at low substrate bias voltages it is in tensile stress, while at high bias voltages it is in compressive stress (Fig. 9). Molybdenum films deposited with highly tensile stress can be spalled, while those with low tensile or compressive stress cannot be spalled at energies within the capability of the apparatus.

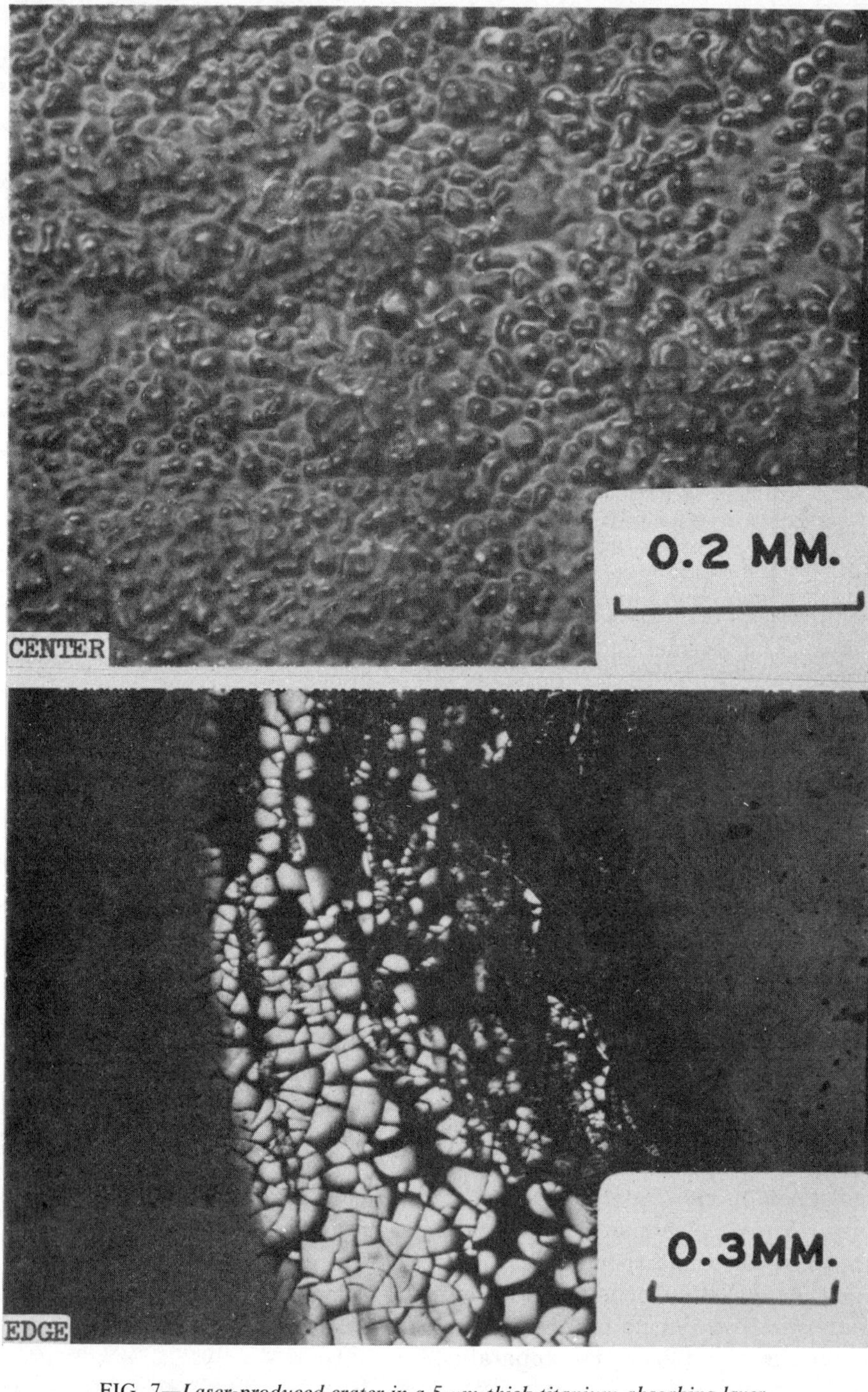

FIG. 7—*Laser-produced crater in a 5-μm-thick titanium absorbing layer.*

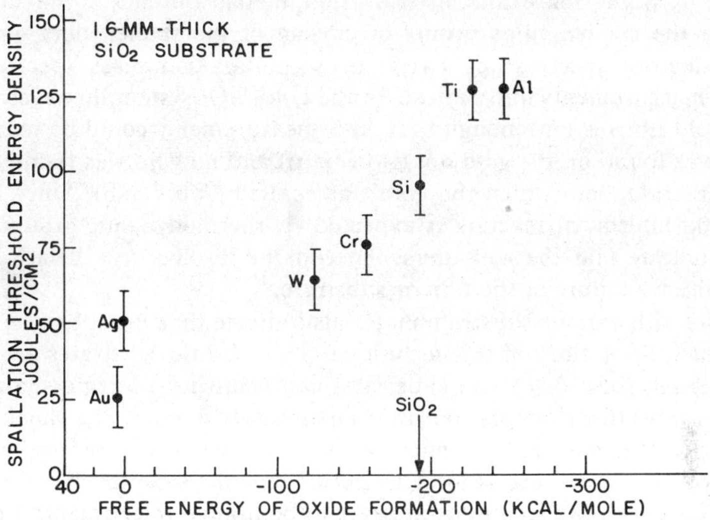

FIG. 8—*Spallation threshold for various metal films deposited on fused silica substrates. The energy density is that which is incident on the back side of the substrate, not the energy density at the interface.*

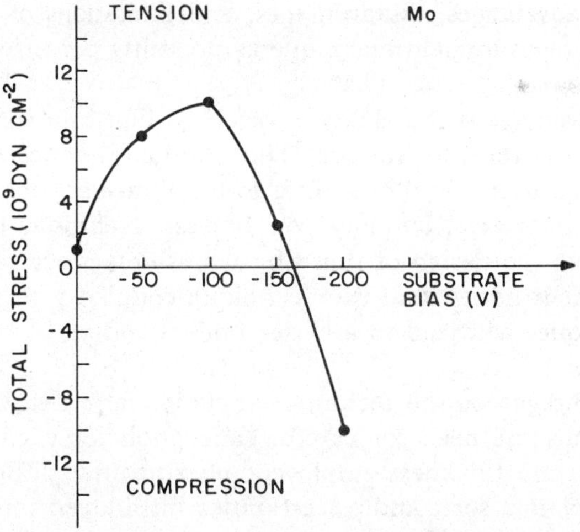

FIG. 9—*Stress in sputtered molybdenum films versus substrate bias.*

An interesting use of this technique is to spall a dot from a substrate and to examine the back (substrate side) of the spalled dot and the location on the substrate from which the dot was spalled by various surface analytic techniques. This leads to the possibility of gaining detailed information about the kinetics of interface formation. In systems in which the bond strength is high (for example, transition metals on SiO_2), the intrinsic stress in the spalled films results in curling of the spalled dots—making Auger electron spectroscopy (AES) or secondary ion mass spectrometry (SIMS) measurements impractical. In the gold-SiO_2 system, however, stress in the gold films is low enough that AES measurements could be made. No silicon was found on the gold dot (<0.2 *a*/o) and no gold was found on the SiO_2 substrate from which the gold was spalled (<0.4 *a*/o). Since in this system no interfacial reaction is expected on thermodynamic grounds, the results indicate that the measurement technique involves true detachment—not a cohesive failure of the film or substrate.

Results with various substrate materials indicate that the substrate is one of the main limitations of this technique. Thin, brittle substrates are prone to shattering; local defects in substrates can result in substrate shattering; soft substrates (for example, polymers) excessively dampen the shock wave; low-temperature substrates cannot be used because of decomposition; and soft metal substrates are subject to permanent deformation by the shock wave. In general, the technique appears to be limited to refractory films on refractory substrates.

Summary

The relative advantages, disadvantages, and limitations of laser spallation as an adhesion measurement technique as presently perceived may be summarized as follows.

The main advantage is the ability to detach a film in a controlled manner without disturbing the film surface. This eliminates concern about propagation of damage from the film surface to the film-substrate interface which may affect the interfacial bonding. We find the technique to be especially useful in guiding the design of thin film deposition processes for optimum film bonding. It is hoped that this technique coupled with various surface analytic techniques will lead to a better understanding of interface formation and failure.

The disadvantages of the technique include: large, expensive, complicated equipment; the need for careful calibration for each substrate material and substrate thickness employed; instrumental difficulties in monitoring pulses of this sort; and uncertainties introduced into the measurements by shock wave artifacts such as the shock wave tail, destructive interference due to ringing, possible variable damping effects from one ostensibly identical substrate to another, and the large thermal tail that follows the shock wave. Substrate material limitations related to shattering and decomposition suggest that the technique is mainly suitable for refractory substrates. Other uncertainties, as yet unresolved, include the possibility of shock-induced crystallization or other shock-induced microprocesses that may change interfacial structure, bonding, etc. in an interval shorter than the duration of the pulse.

References

[1] Mittal, K. L., *Electrocomponent Science and Technology*, Vol. 3, No. 1, Jan. 1976, pp. 21-42.

[2] Weaver, C., *Journal of Vacuum Science and Technology*, Vol. 12, No. 1, Jan. 1975, pp. 18-25.

[3] Chapman, B. N., *Journal of Vacuum Science and Technology*, Vol. 11, No. 1, Jan. 1974, pp. 106-113.

[4] Anderholm, N. C. and Goodman, A., U.S. Patent No. 3,605,486 (20 Sept. 1971).

[5] Anderholm, N. C., *Applied Physics Letters*, Vol. 16, No. 3, Feb. 1970, pp. 113-115.

[6] Hoffman, R. W. in *Physics of Thin Films*, Vol. 3, G. Hass and R. E. Thun, Eds., Academic Press, New York, 1966, pp. 211-274.

[7] Kolsky, H., *Stress Waves in Solids*, Dover Publications, New York, 1972.

[8] Graham, R. A., Neilson, F. W., and Benedick, W. B., *Journal of Applied Physics*, Vol. 36, No. 5, May 1965, pp. 1775-1783.

[9] Fox J. A., and Barr, D. N., *Applied Optics*, Vol. 12, No. 11, Nov. 1973, pp. 2547-2548.

[10] Cohen, M. I. in *Laser Handbook*, Vol. 2, F. T. Arecchi and E. O. Schulz-DuBois, Eds., North Holland Publishing Co., Amsterdam, 1972, pp. 1579-1647.

[11] Fairand, B. P., Clauer, A. H., Jung, R. G., and Wilcox, B. A., *Applied Physics Letters*, Vol. 25, No. 8, Oct. 1974, pp. 431-433.

[12] Bateson, S., *Vacuum*, Vol. 2, No. 4, Oct. 1952, pp. 365-376.

DISCUSSION

H. E. Ashton[1] *(written discussion)*—You stated that organic substrates cannot be used because of charring. Does this also apply to organic coatings on metal substrates?

J. L. Vossen (author's closure)—We have not investigated organic films on metal substrates to date. As I mentioned, I believe this technique to be useful mainly for refractory materials on refractory substrates. I suspect the problem with organic films on metal substrates would be related to the thermal tail following the shock wave. Results would be clouded if this thermal energy decomposed or evaporated the organic films, or both. Since the specimen is at atmospheric pressure, organic reactions with atmospheric oxygen, water vapor, etc. could lead to chemical conversion and evaporation.

[1] Division of Building Research, National Research Council of Canada, Ottawa, Ont., Canada.

J. J. Bikerman[2] (*written discussion*)—Is it correct to predict that very similar results would have been obtained if you used a hammer to hit the back side of the support instead of illuminating it with a laser light, assuming that the contact time of hammer and substrate could be reduced to 20 ns?

J. L. Vossen (*author's closure*)—To first order, yes, but numerous complications arise due to the thermal and optical effects noted.

W. D. Bascom[3] (*written discussion*)—Evidently it is assumed that there is no microdamage by the compressive wave. I find such an assumption difficult to believe especially in a brittle metal oxide interfacial layer.

J. L. Vossen (*author's closure*)—We do not assume that there is no microdamage. As mentioned, there can occur numerous effects such as shock-induced crystallization that presumably can cloud results. These effects are still under investigation.

L. H. Sharpe[4] (*written discussion*)—Since you saw large effects (for example, factor of 2) of annealing the films on spallation energy and since the dots probably don't move monolithically—that is, failure starts at one spot and propagates sequentially across the dot—why would you expect to find a good correlation between heat of formation of the oxide and spallation energy?

J. L. Vossen (*author's closure*)—If, as has been proposed in the literature, chemical bonds form when arriving metal vapor can reduce (chemically) an oxide substrate, then there should be some correlation between heat of formation of the oxide of the metal vapor and the interfacial bond strength. Annealing the chromium films on SiO_2 may have produced higher thresholds for spallation for one or more of several reasons (for example, increased chemical bonding or stress relief).

With regard to the movement of the dots, at subthreshold energies localized damage is observed. I feel that this is mainly due to the thermal tail because the damage appears characteristic of micromelting. However,

[2] Department of Chemical Engineering, Case Western Reserve University, Cleveland, Ohio 44106.

[3] Adhesion Section, Naval Research Laboratory, Washington, D.C. 20375.

[4] Engineering and Development, Bell Laboratories, Murray Hill, N.J. 07974.

one cannot rule out the possibility of small local delamination from these observations. Since one cannot really distinguish between the two unequivocally, we call the threshold the point at which the entire dot spalls. Therefore, this represents the maximum energy density for that dot.

J. Ahn,[1] *K. L. Mittal,*[2] *and R. H. MacQueen*[3]

Hardness and Adhesion of Filmed Structures as Determined by the Scratch Technique

REFERENCE: Ahn, J., Mittal, K. L., and MacQueen, R. H., **"Hardness and Adhesion of Filmed Structures as Determined by the Scratch Technique,"** *Adhesion Measurement of Thin Films, Thick Films, and Bulk Coatings, ASTM STP 640,* K. L. Mittal, Ed., American Society for Testing and Materials, 1978, pp. 134–157.

ABSTRACT: In the first part we discussed the evolution of the scratch technique for measuring adhesion of thin films and the associated difficulties involved in the interpretation of the results. Subsequently, results on scratch hardness and adhesion of multilayered structures as determined by a single point loaded scratch tester are presented. The controlled scratches produced by the tester were examined by a very sensitive surface profilemeter and by scanning electron microscopy (SEM) coupled with energy-dispersive X-ray spectroscopy (EDX) for their scratch topography, modes of film deformation, and the mechanisms for film material removal. It is found that (*a*) the mechanism of film failure depends on whether the film is ductile or brittle; (*b*) there may not always be a clear channel formation, that is, complete removal of the film; and (*c*) scratch depth increases linearly with increasing load, which provides a means to determine scratch hardness for various layers.

KEY WORDS: adhesion, thin films, scratch test, scribe test, scratch hardness, multilayers, scratch test history, threshold adhesion failure

The use of thin films has proliferated in a variety of electronic, engineering, optical, biomedical, nuclear, space, and other applications; and all signals indicate that thin films will be increasingly used in the future. The reason why thin films are so widely used is that materials in monolithic form are not suitable for diverse and special requirements, and thin films provide the requisite answer. Thin films are used for a variety of purposes: to provide resistance to abrasion, erosion, corrosion, galling, tarnish, wear, radiation damage, or high-temperature oxidation; to reduce friction or electrical resistance, provide lubrication, prevent sticking; and also to pro-

[1]Advisory engineer, International Business Machines Corporation, San Jose, Calif. 95193.
[2]Staff engineer, East Fishill Facility, International Business Machines Corporation, Hopewell Junction, N.Y. 12533. To whom all correspondence should be addressed.
[3]Senior associate engineer, International Business Machines Corporation, Poughkeepsie, N.Y. 12602.

vide special magnetic or dielectric properties. Mittal [*1*][4] has listed many reasons for the importance of adhesion of thin films. Irrespective of the intended use or function of thin films, the properties, structure, integrity, functional characteristics, and performance all depend, among other things, on adhesion between the thin film (coating) and the substrate.

In many applications, instead of a single thin film, multilayer structures are required, and the mechanical properties of a multilayered structure such as scratch resistance and adhesion of individual layers are of great interest, especially if such a device structure is to perform properly under load-bearing conditions.

In the present paper, we first discuss the evolution of the scratch technique for measuring adhesion of thin films and the associated difficulties involved in the interpretation of the results. Subsequently, our results on the scratch resistance and adhesion of multilayered structures as determined by a scratch tester are presented. The controlled scratches produced by the tester were examined for their scratch topography, modes of film deformation, and mechanisms for material removal. The methods of calculating "scratch hardness" of each layer in multilayered structures as well as adhesion value are presented. Our present work describes some refinements of previous works by (*a*) determining scratch depth within 10-nm resolution via a very sensitive surface profilemeter, and (*b*) examining the nature of the film damage associated with scratches by scanning electron microscopy (SEM).

Evolution of the Scratch Technique

Mittal [*1*] has recently reviewed techniques for the adhesion measurements of thin films, and he discusses the scratch test in detail. Essentially, in this test, a smoothly rounded chrome steel, tungsten carbide, or diamond tip (~ 0.003 to 0.005 cm radius) is drawn across the film surface and a vertical load, applied to the point, is gradually increased until the film is completely removed, resulting in a clear channel. The critical load at which the clear track is formed is taken as a measure of adhesion. Heavens [*2*] and Heavens and Collins [*3*] were the first to use this test to study quantitatively the adhesion of thin evaporated metallic films on glass and the effect of chromium interlayers. Weaver and Hill [*4*] studied the increase in adhesion of aluminum films to glass by the predeposition of films of chromium, taking the load necessary to strip the film completely as a measure of adhesion.

In 1960 Benjamin and Weaver [*5*] analyzed the scratch test in detail. Their analysis showed that the action of the point always involves plastic deformation of the substrate and that this deformation produces a shearing

[4]The italic numbers in brackets refer to the list of references appended to this paper.

force at the film substrate interface around the rim of the indentation produced by the point. Thus a simple relationship between the applied load and the shearing force was developed, so that adhesion could be calculated as a shearing force

$$F = \frac{AP}{\sqrt{R^2 - A^2}}$$

where

$$A = \sqrt{\frac{W}{\pi P}}$$

where

W = critical load,
R = radius of the tip,
F = shearing force per unit area due to the deformation of the substrate,
A = radius of the circle of contact, and
P = indentation hardness of the substrate material.

Based upon their studies of adhesion of evaporated silver films on various substrates and of various evaporated metal films on glass, Benjamin and Weaver [*5*] found the following: (*a*) In each case, the measured load became constant for film thicknesses exceeding a certain value; this value was close to 80 nm. For thinner films, a slightly smaller load was required and they surmised that this could be attributable to film structure and was no innate weakness of the method. (*b*) The load did not depend directly upon the mechanical properties of the film—hardness, elasticity, or tensile strength—as the load did not vary with the thickness of the film. (*c*) The load depended upon the nature of both the film and the substrate, without being directly attributable to the mechanical properties of either. From these observations Benjamin and Weaver concluded that the load is determined essentially by the properties of the interface, which implies that the scratch adhesion is a measure of "basic adhesion" as defined elsewhere in this publication [*6*].

It should be pointed up that although Benjamin and Weaver found no effect of film thickness, Chopra [*7*] found that the critical load (in the case of polycrystalline gold films deposited on glass) increased nearly linearly with film thickness above 200 nm, which means that the critical load is determined both by the adhesion at the interface and the mechanical strength of the film.

In 1970 Butler, Stoddart, and Stuart [*8*] using SEM and optical interference microscopy showed that the process of scratch formation was very complex and varied with the film material, indicating that it is not possible to deduce absolute values of adhesion using a simple general theoretical model as proposed by Benjamin and Weaver. Some of their comments are summarized in the following.

1. Their work has shown the unreliability of simple optical methods of detecting readherence caused by the sliding stylus. A film can become detached before the formation of a cleared track and conversely a film may be thinned to optical translucency without being removed.
2. The form of the track depends on the elastic and plastic deformation of the film and substrate; that is, it depends primarily on their hardnesses.
3. There is no preferential failure (that is, yielding of a permanent nature) at the film-substrate interface. Factors such as the size and surface finish of the stylus as well as the thickness of the film can determine whether failure occurs first in the film or in the substrate or at the interface.
4. Film detachment often occurs at lighter loads than required for track clearance, which appears to depend on film tearing, film pile-up in front of the stylus, dust, imperfections, etc.

Although Weaver [*9*] has recently offered explanation for all the observable details in the micrograph of the scratch as taken in the work of Butler, Stoddart, and Stuart, it should be borne in mind that there are too many variables affecting scratch test results.

Hamersky [*10*] in his studies of adhesion of aluminum films deposited on glass slides concluded that the accuracy of the scratch test result is dependent on the point radius and only indirectly on its hardness. The method gives results which are reproducible using a point radius larger than 0.2 mm. Berendsohn [*11*] has used the critical load as the criterion for comparing adhesion of a variety of films on an assortment of substrates.

So far most of the work had been done on glass substrates because of the difficulties in observing a clear channel on opaque substrates. Greene et al [*12*] have described a technique which is not limited by the optical transmission of the substrate, and thus extends the utility of the scratch test to all kinds of substrates. Also in the work reported above rigid substrates were used, and recently Goldstein and Bertone [*13*] have discussed the use of the scratch test to evaluate metal film adhesion onto flexible substrates.

So far the discussion of the scratch test has centered around the concept of critical load, but Oroshnik and Croll [*14*], during their detailed studies of the scratch test, found that a complete removal of a film (for example, aluminum) by a stylus was an infrequent event. At stylus loadings below (and often above) the so-called "critical load" there was only partial removal of the film, but not necessarily in relative proportion to loading. This led them to develop the concept of "Threshold Adhesion Failure" as an operational criterion for adhesion measurement. It is defined as follows:

> Threshold Adhesion Failure occurs if, within the boundaries of a scratch and over its 1-cm path, removal of the film from its substrate can be detected by transmitted light with a microscope (×40 magnification) at even one spot, no matter how small.

For details of the Oroshnik and Croll concept and the results obtained using it, the reader should consult Ref *14*. Among other things, they found the following: (*a*) No two styli of any one material (diamond or tungsten carbide) of the same nominal tip radius of curvature yield the same Threshold Adhesion Failure load values. (*b*) Any one stylus exhibits its own individual scribing and testing characteristics. (*c*) Each material (for example, gold as compared with aluminum) shows a different scratch test behavior.

It is obvious from the discussion of the evolution of the scratch technique that (*a*) there are too many variables affecting scratch adhesion values, which means that the basic adhesion[5] cannot be directly determined from the scratch values; (*b*) in many cases, there is no clear channel formed, so one can only determine the minimum scratch adhesion values; and (*c*) if separation occurs either at the interface or in the interphase, the scratch test should be a measure or practical adhesion,[5] and it should be interesting to compare scratch test results with those obtained using other adhesion measurement techniques.

Experimental

A schematic diagram of the scratch tester as used in our investigations is shown in Fig. 1. The zero load condition was achieved by balancing the lever on an air-bearing pivot on load W'. Positioning of the probe was facilitated by the vertical movement of the probe and the x-y stage on which the test specimen was attached. Use of the air-bearing pivot enabled us to zero-balance the tester to within ±100 mg. Typically, specimens were scratch tested with increasing load W, such as 0.2, 0.6, 0.8, 1.0 g. The maximum load used depends on the nature of the test specimen and the probe radius. Typical values reported in this paper range from 0.2 to 20 g with a diamond probe of 2.54×10^{-3} cm radius. Scratches were formed by placing the loaded probe on specimen surface and moving the stage by hand. This manual movement was approximately 0.05 cm/s for all scratches made.

The scratches were examined using Reichart MeF_2 Metallograph interference contrast microscopy and SEM coupled with energy-dispersive X-ray

[5] The term "basic adhesion" is used to signify the summation of all interfacial intermolecular interactions; whereas the term "practical adhesion" is used to represent the forces or the work required for the disruption—either at the interface or in the interfacial region—of the adhering system. For details see Ref *6*.

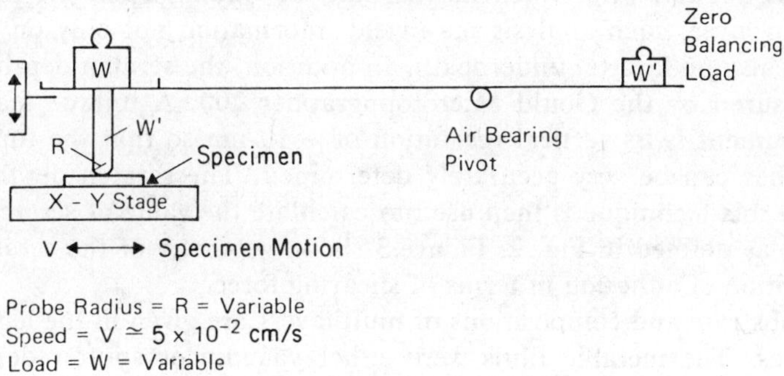

FIG.1—*Schematic diagram of scratch adhesion tester.*

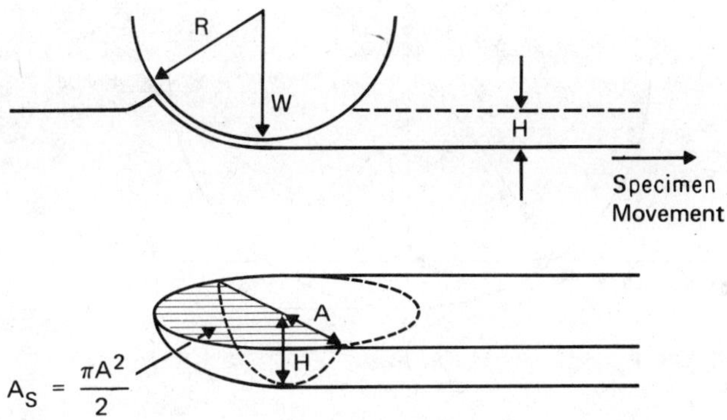

$$S_H \equiv \frac{W}{A_S} = \frac{W}{\pi RH} \quad \text{if} \quad A^2 = 2RH - H^2 \simeq 2RH \text{ for } R >> H$$

For $R = 25.4 \mu m$

$$S_H = 12.53 \frac{W}{H}$$

where S_H is in Kg/mm^2,

W is gm, and H in μm

For each layer

$$W = KH \quad (K = \text{const.})$$

Hence, $S_H = 12.53 K$

FIG. 2—*Definition of scratch hardness.*

spectroscopy (EDX). The limitation of the in-depth resolution of the EDX analysis is about 1 μm, which means that, depending upon the thickness of the top layer, such analysis may yield information not only on the top layer but also on layer(s) underneath. In addition, the scratch depth profile was measured by the Gould Microtopographer 200. A unique feature of this instrument is its vertical resolution of ~10 nm so that the topograph of scratches can be very accurately determined. The scratch depth determined by this technique is then used to calculate the value of scratch hardness, S_H, as defined in Fig. 2. Figure 3 shows geometry of the scratch and the definition of adhesion in terms of shearing force.

The substrate and compositions of multilayers are given in the legends to the figures. The metallic films were either vacuum or sputter deposited, and thin organic films were deposited by the gas phase polymerization.

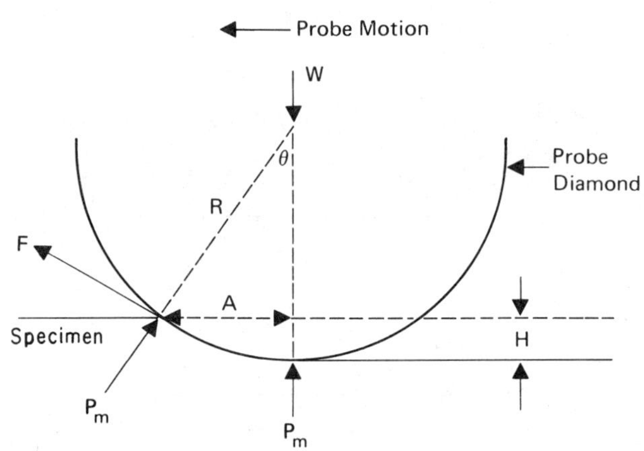

$$\text{Hardness} = P_m \equiv \frac{W}{\pi A^2}$$

$$\text{Adhesive Strength} = F \equiv P_m \tan\theta$$

$$= \frac{W}{\pi A^2} \frac{A}{\sqrt{R^2 - A^2}}$$

$$F \simeq \frac{W}{\pi A R} \quad \text{if } R >> A$$

W = Load
A = Contact Radius
H = Contact Depth

FIG. 3—*Definition of adhesion strength.*

Results and Discussion

Scratch Hardness

Examination of scratches by optical microscopy or SEM generally reveals the mechanism of film failure. Figure 4*a* depicts the location of maximum stresses in tension or in shear. For example, a brittle film fails in tension, forming the characteristic tensile cracks shown in Fig. 4*b*. Figures 5-7 show scratches of a ductile type, while Fig. 8 (load 9.0 g) and Fig. 9 show the scratches with predominantly tensile cracks. A ductile scratch signifies that the film material is pushed aside by the moving probe but is still attached to the specimen substrate, whereas in brittle scratch the film material fragments, and these fragments may or may not stay on the substrate. EDX comparisons between areas in and out of the scratch were made. For ductile scratches (Figs. 5-7), increases in substrate element counts and

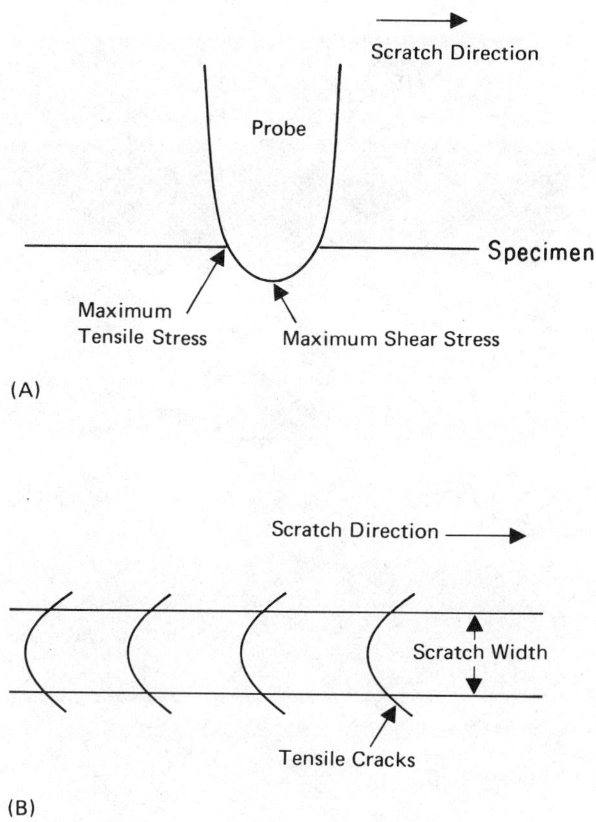

FIG. 4—(a) *Location of maximum shear and tensile stresses in region of contact.* (b) *Brittle materials fail where tensile stress is maximum, that is, the trailing edge of the probe as shown.*

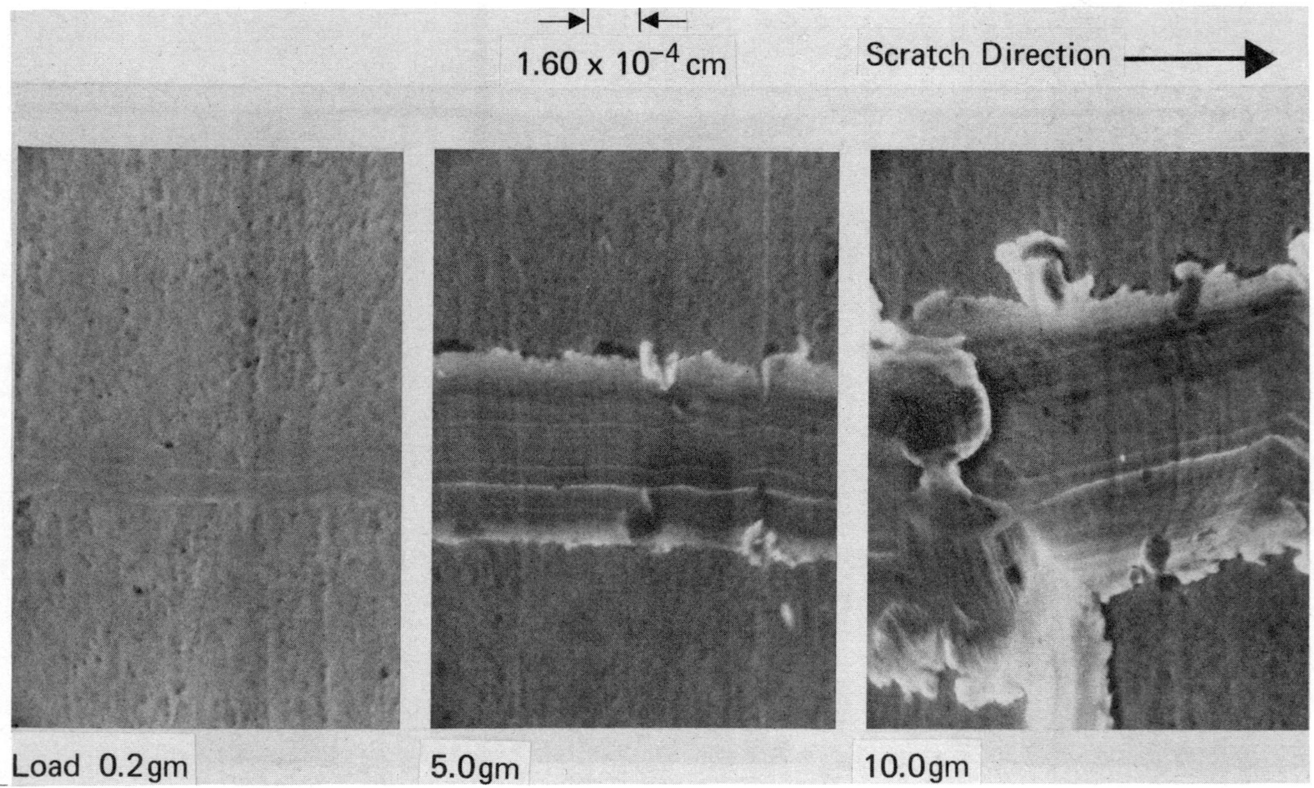

FIG. 5—*SEM micrographs of scratch formed by diamond probe* (R = *2.54* $\times$ *10*$^{-3}$ *cm*) *on specimen surface: gold* (*100 nm*)/*nickel alloy* (*1000 nm*)/*aluminum alloy substrate.*

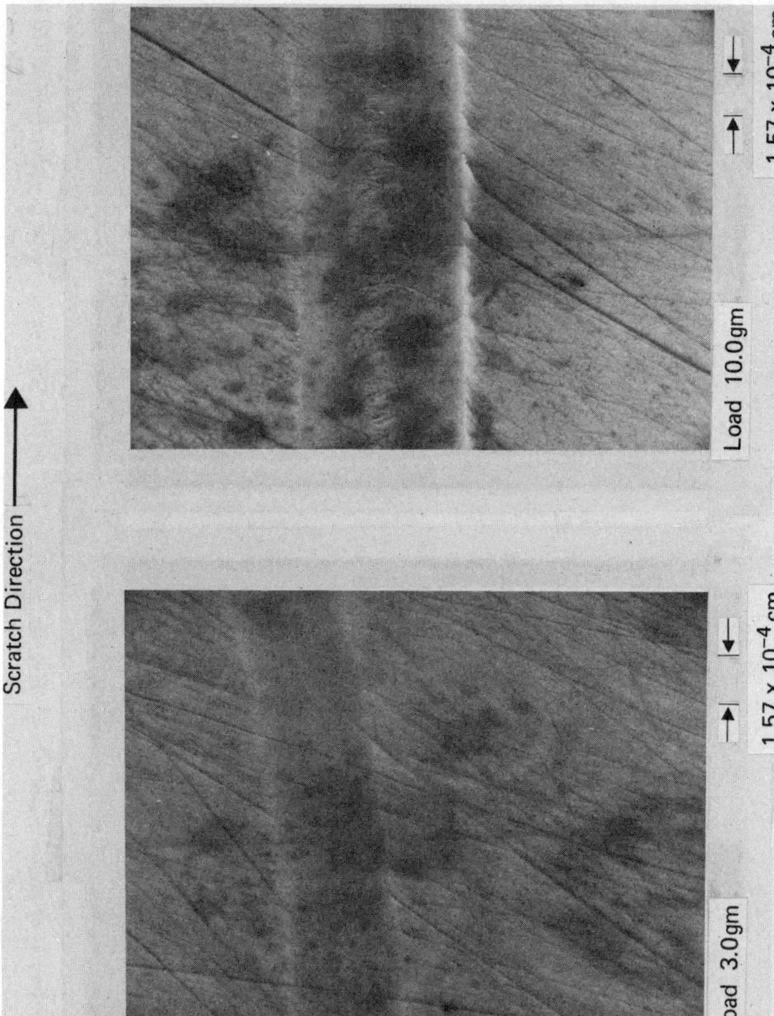

FIG. 6—*SEM micrographs of scratch formed by diamond probe* (R = 2.54 × 10^{-3} *cm*) *on specimen surface: rhodium (240 nm)/cobalt alloy (150 nm)/nickel alloy (52 μm)/aluminum alloy substrate.*

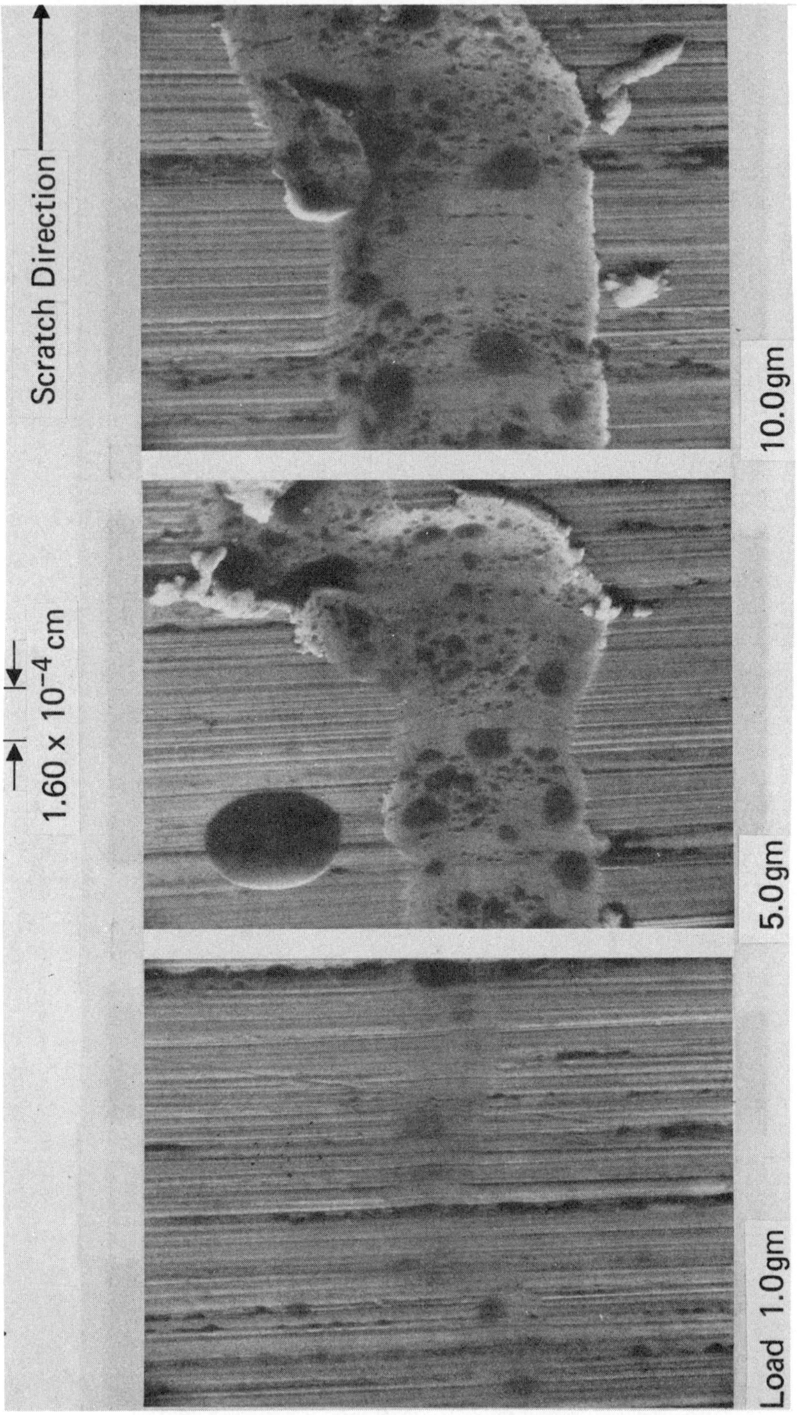

FIG. 7—*SEM micrographs of scratch formed by diamond probe* (R = 2.54 × 10^{-3} cm) *on specimen surface: rhodium (100 nm)/aluminum alloy substrate.*

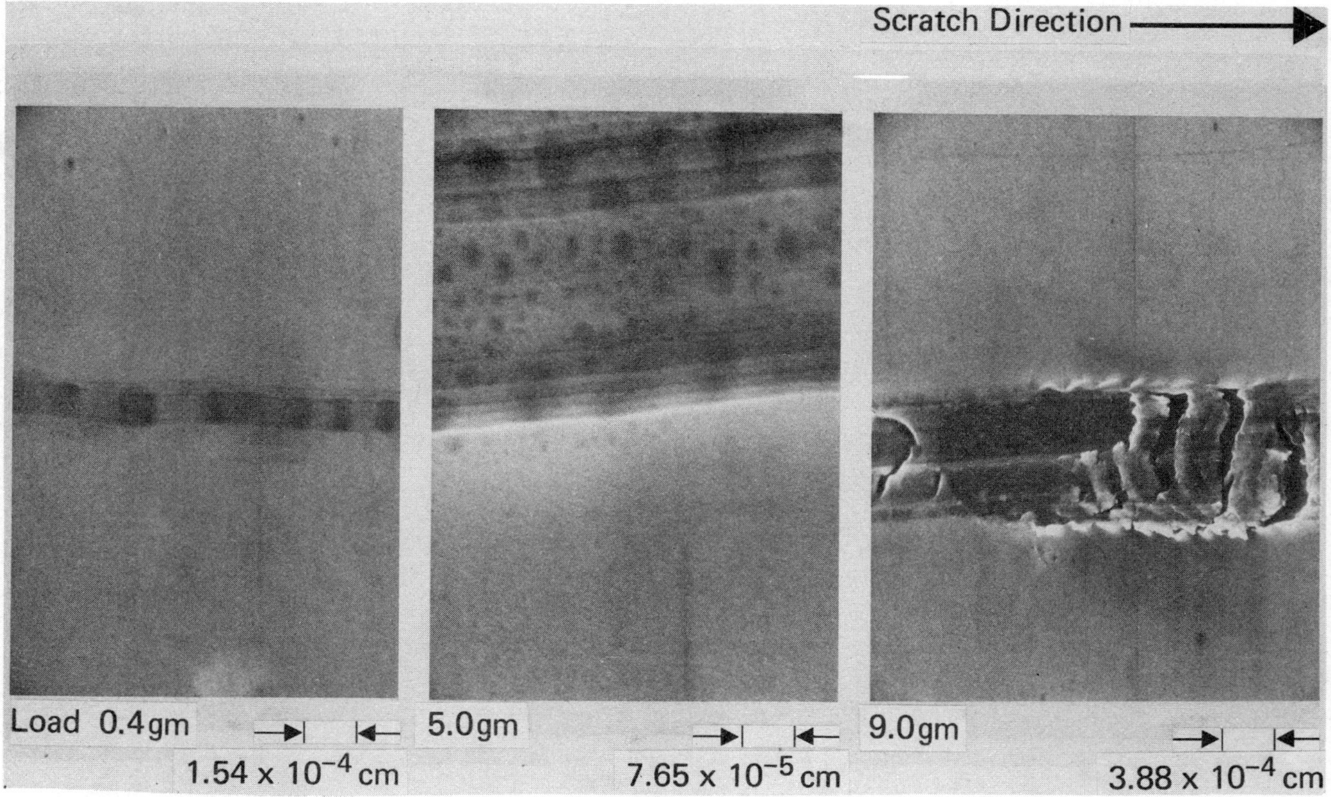

FIG. 8—*SEM micrographs of scratch formed by diamond probe* (R = 2.54 × 10^{-3} *cm*) *on specimen surface: iron alloy (36 nm)/aluminum alloy substrate.*

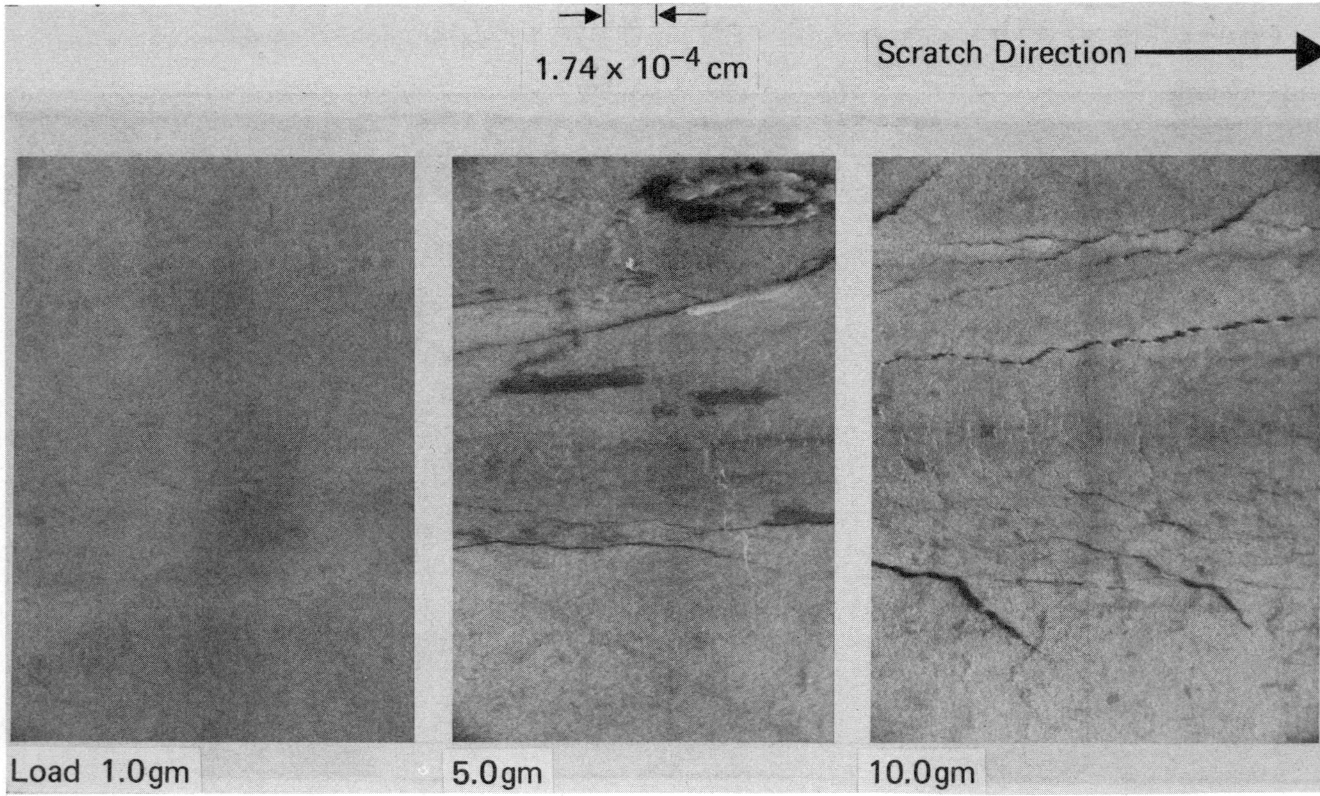

FIG. 9—*SEM micrographs of scratch formed by diamond probe* (R = *2.54* × *10*$^{-3}$ *cm*) *on specimen surface: iron alloy (36 nm)/chromium (690 nm)/aluminum alloy substrate.*

decreases in surface element counts were observed, indicating gradual thinning of the surface material as load is increased. For brittle cracks (for example, Fig. 8, load 9.0 g), only substrate elements were detected, indicating a complete removal of surface material. At lower loads, film thinning is observed as shown in Fig. 8. These results certainly bring out the effect of the mechanical properties, for example, the hardness, of the film material.

Figure 10 shows the Gould Microtopographer traces of a specimen at increasing loads. From such traces the scratch depth, *H*, can be determined to within 0.03 μm and Fig. 11 is constructed. From the breaks in Fig. 11 the scratch hardnesses were calculated and plotted in Fig. 12. The S_H value obtained for Region I (Figs. 11 and 12) extends into the surface depth of ~0.4 μm, which is an order of magnitude greater than the thickness of the top metal layer. The low value (64 kg/mm^2) of this region suggests other possible mechanisms. Interdiffusion of the layers is important, and such a process will culminate in an interlayer with mechanical properties different from those of the original layers. The break between Regions II and III occurs at a depth very close to the chromium-substrate interface, reach-

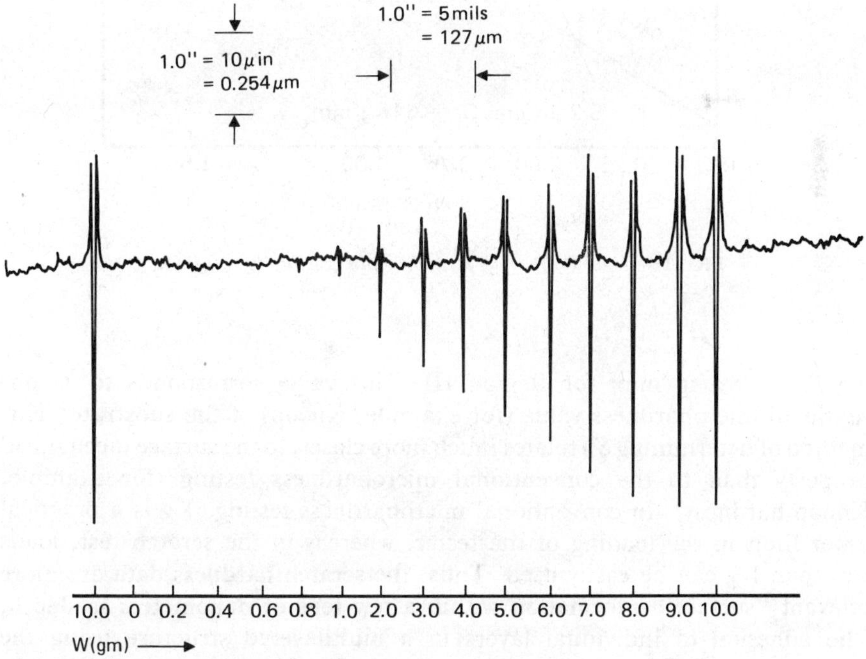

FIG. 10—*Gould Microtopographer trace of scratches formed by diamond probe* (R = 2.54×10^{-3} cm) *on specimen surface: iron alloy (50 nm)/chrominum(675 nm)/aluminum alloy substrate.*

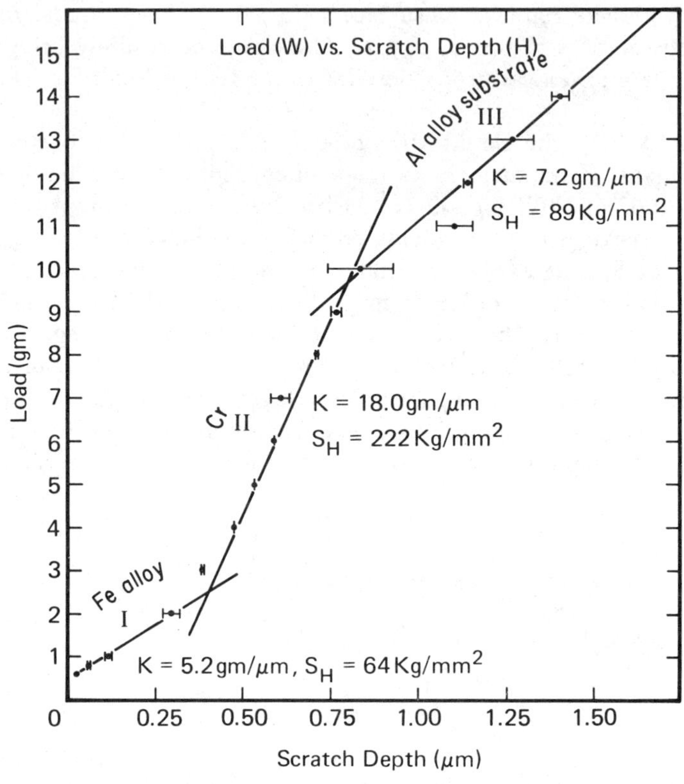

FIG. 11—*Load* (W) *versus scratch depth* (H) *for Fig. 10 film.*

ing $S_H = 89$ kg/mm² for Region III. This value corresponds to a conventional microhardness value (for example, Knoop) of the substrate. This method of determining S_H relates much more closely to the surface mechanical property than to the conventional microhardness testing (for example, Knoop hardness). In conventional microhardness testing, 1 g is a practical lower limit in the loading of the tester, whereas in the scratch test, loads less than 1 g can be easily used. Thus, the scratch hardness data are more relevant "surface" information within a few tens of nanometres in depth. The adhesion of individual layers in a multilayered structure using the scratch technique has not been investigated, and the determination of S_H values in such structures may have important implications for their adhesion behavior.

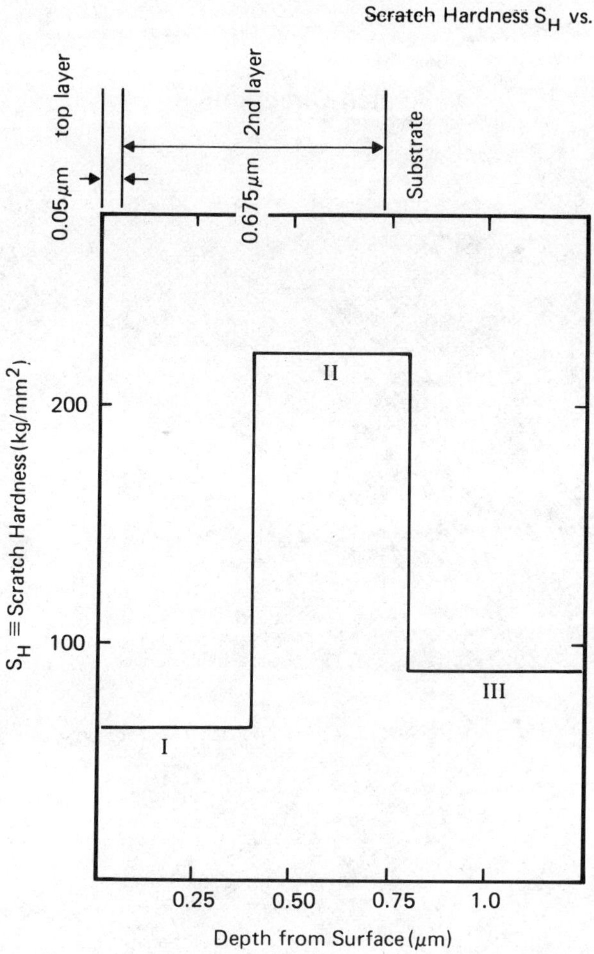

FIG. 12—*Scratch hardness* S_H *versus distance from surface.*

Adhesion

A qualitative measure of film adhesion can be easily made with the scratch tester. Examples are given for "good" (Fig. 13) and "bad" (Fig. 14) adhesion of an organic film. This great disparity in their behavior could be attributed to the way they were deposited; the higher rate of deposition gives poorer adhesion. These results show that, if the practical adhesion of a film is very poor, then it can be easily discerned using the scratch technique.

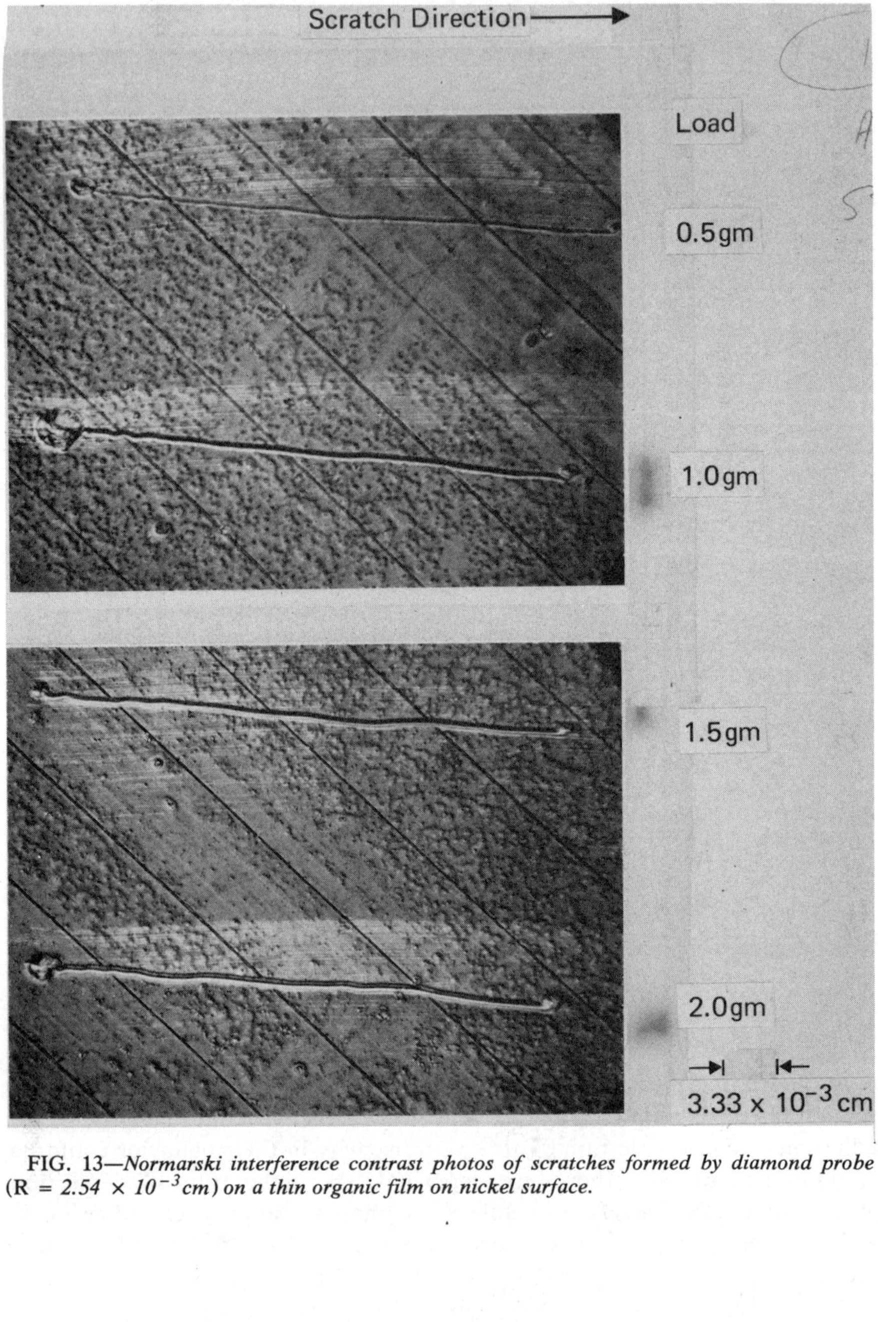

FIG. 13—*Normarski interference contrast photos of scratches formed by diamond probe* (R = *2.54* × *10^{-3} cm*) *on a thin organic film on nickel surface.*

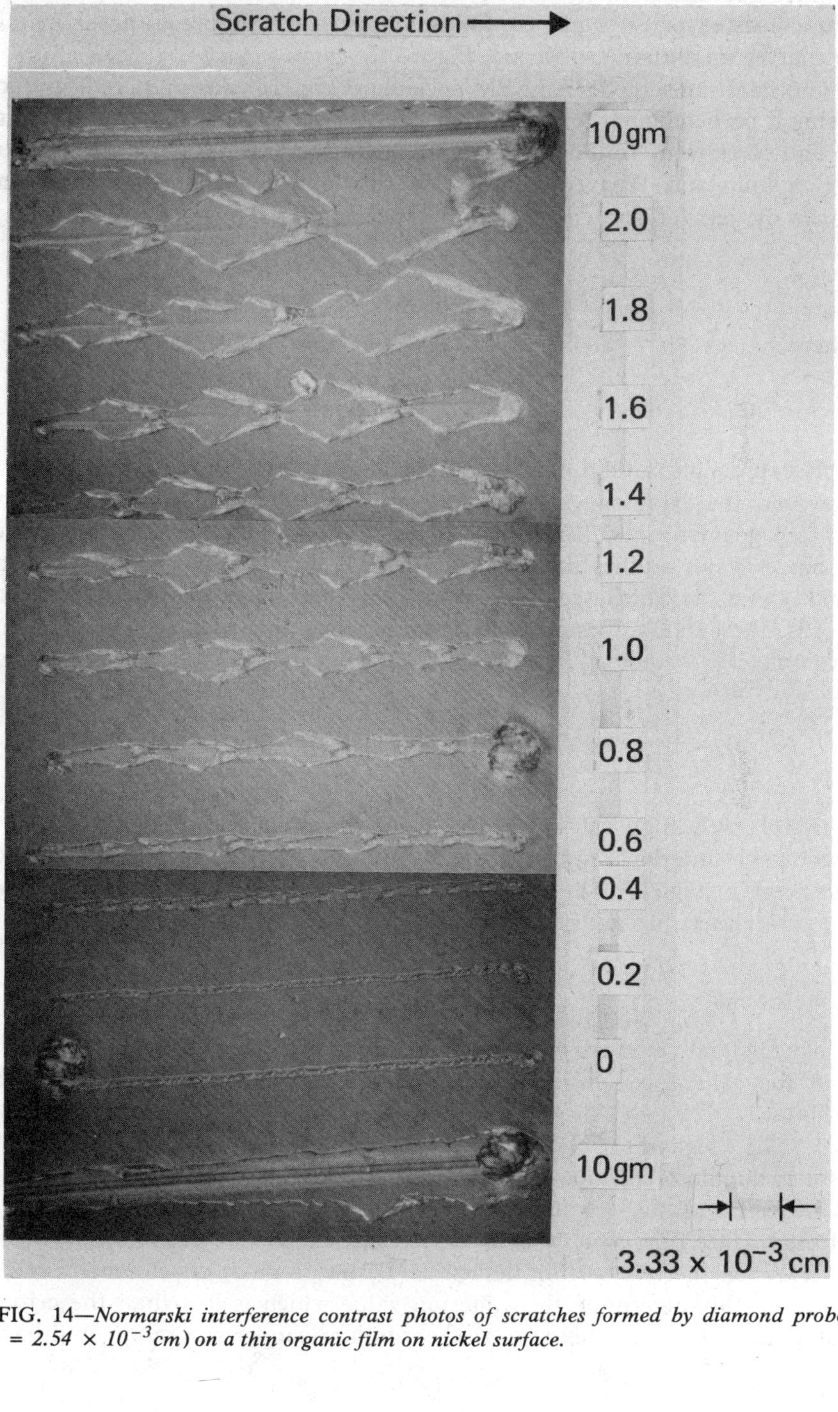

FIG. 14—*Normarski interference contrast photos of scratches formed by diamond probe* (R = *2.54* × *10*$^{-3}$*cm) on a thin organic film on nickel surface.*

More quantitative adhesion measurements were carried out on a specimen that consisted of ~ 10-μm chromium-copper-chromium multilayer metallic coating on alumina substrate. Figure 15 shows some SEM views of areas in and near the scratch. An SEM specimen (Fig. 16) was prepared by fracturing it perpendicular to the scratch after the scratch was made. Both Figs. 15 and 16 show no film detachment at the inside or the edge of the scratch.

Benjamin and Weaver had derived the following expression (as mentioned earlier) for scratch adhesion as shearing force, F

$$F = \frac{W}{\pi A^2} \frac{A}{\sqrt{R^2 - A^2}}$$

where

$$A = \sqrt{\frac{W}{\pi P}}$$

This expression is valid only if there is a clear track formation. If $R \gg A$, however, the expression reduces to $F \simeq W/\pi AR$, and, by measuring the scratch geometry and load, one can calculate the scratch adhesion value.

Since in our studies there was no sign of film removal, all that can be said is that the calculated value of F is less than the critical value. In other words, the actual adhesion value is larger than the calculated one. In this example, $W = 1660$ g, $R = 4.32 \times 10^{-2}$ cm, and $A = 9.0 \times 10^{-3}$; thus

$$F \simeq \frac{W}{\pi AR} = 13.64 \text{ kg/mm}^2 \qquad (1 \text{ kg/mm}^2 = \sim 10^7 \text{ Pa})$$

This relatively high value of scratch adhesion may be attributable to the mechanical interlocking of the metal layers to the rough substrate surface as shown in Fig. 17. The effect of the substrate rugosity on the validity of the scratch test has not been studied.

Conclusions

The general comments on the scratch technique have been made earlier, and the salient conclusions from our investigations are summarized as follows.

1. The mechanism of film removal or failure depends on whether the film is ductile or brittle. For ductile films, a gradual thinning of the film is observed as load is increased. For brittle films, a complete removal of the surface film is observed.

2. When there is no film detachment, that is, no clear channel formed, one can only determine the adhesion value which is less than the critical value as is done for the case of a chromium-copper-chromium multilayer

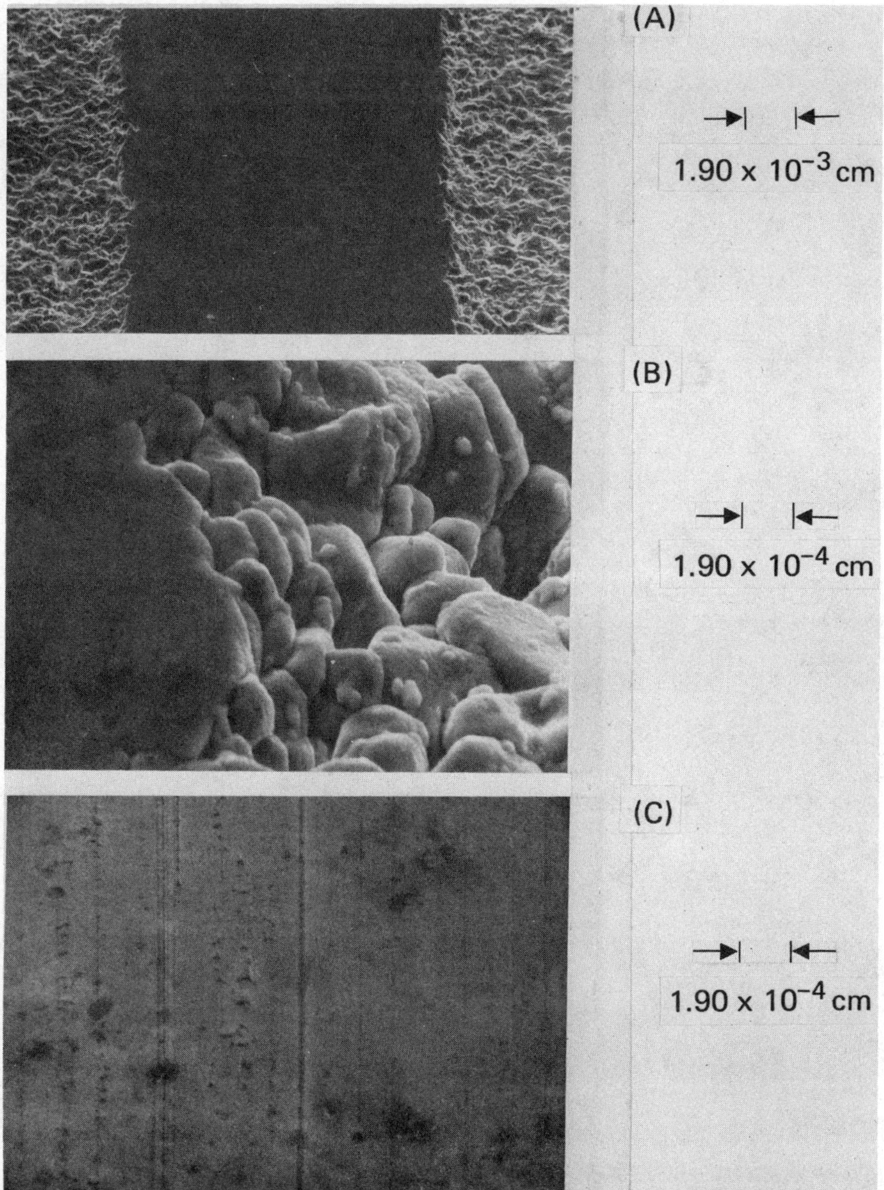

FIG. 15—*SEM micrographs of scratched area formed by diamond probe (R = 4.32 × 10^{-2} cm) on specimen surface: chromium-copper-chromium (~10 μm)/alumina substrate,* (a) *scratch and surrounding areas;* (b) *edge of scratch;* (c) *midregion of scratch.*

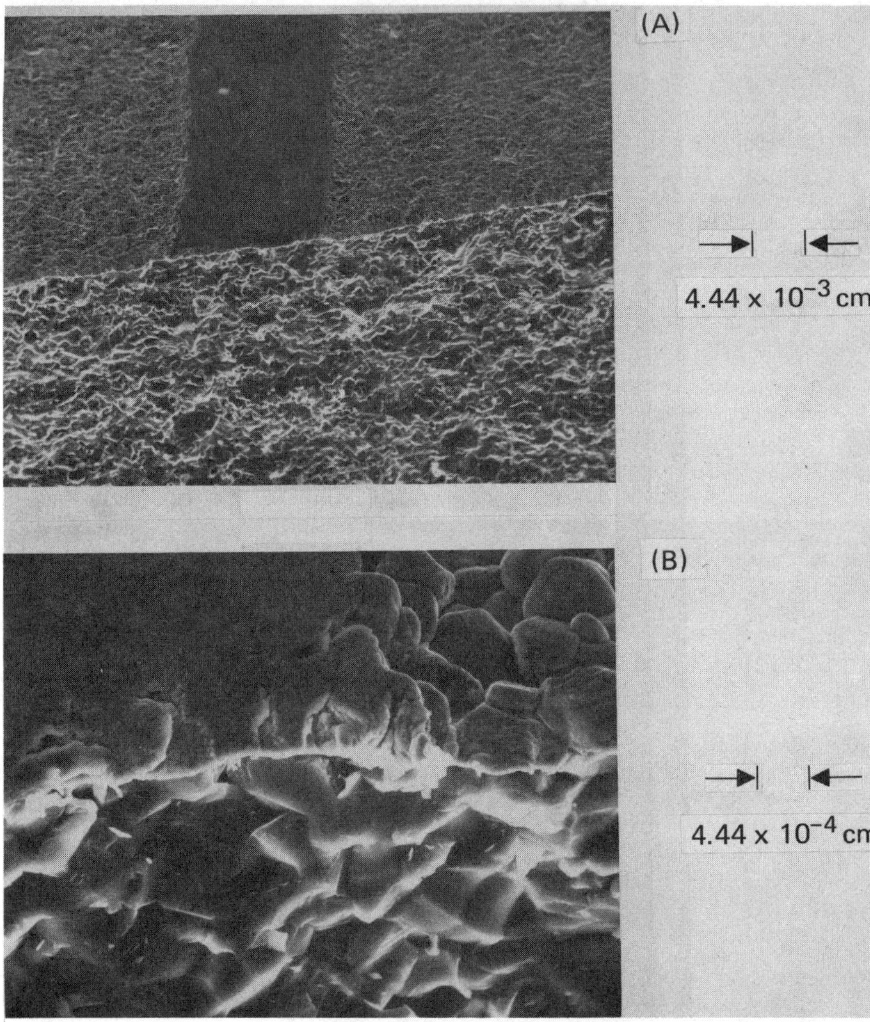

FIG. 16—*SEM micrographs of scratched area formed by diamond probe* (R = *4.32* $\times$ *10^{-2} cm*) *on specimen surface: chromium-copper-chromium* (*~10 μm*)*/alumina substrate.* (a) *lower part is the transverse section perpendicular to the scratch;* (b) *larger magnification.*

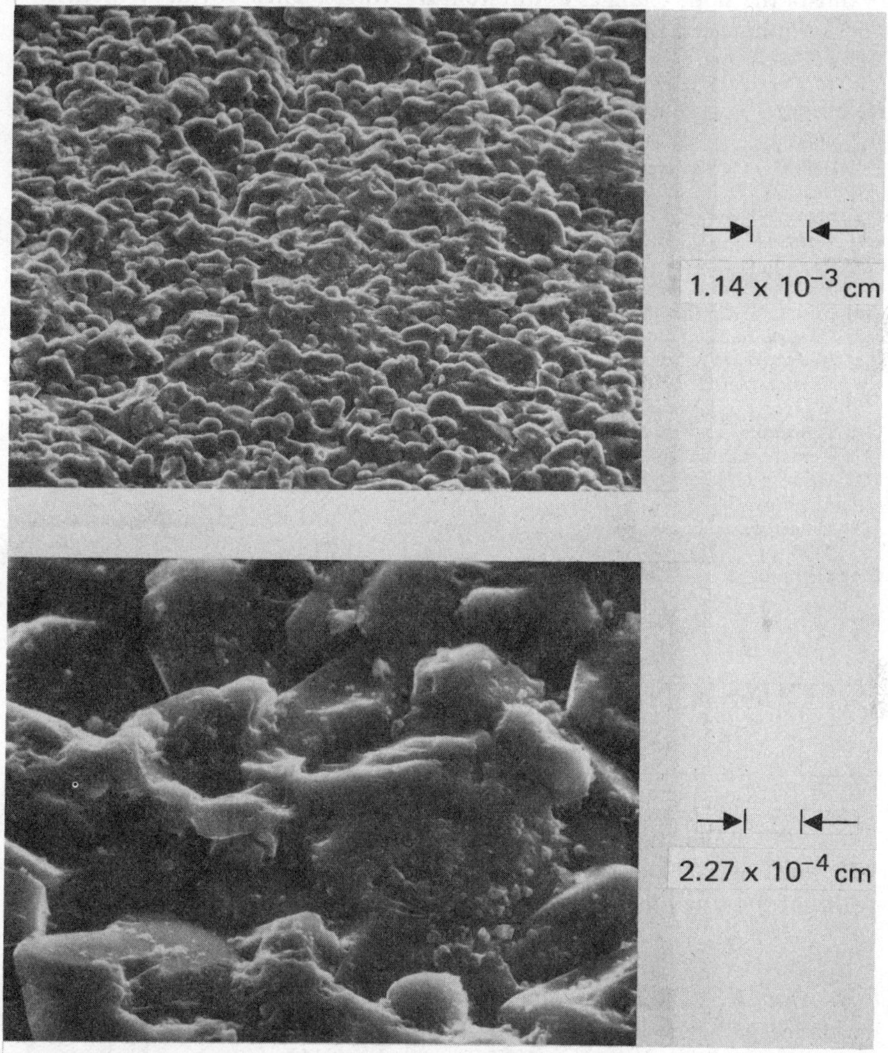

FIG. 17—*SEM micrographs of alumina surface showing rough topography. The surface is coated with 20 nm carbon.*

metallic coating deposited on alumina substrate. For this system, the adhesion value was calculated to be 13.64 kg/mm^2.

3. Scratch depth increases linearly with increasing load, and the numerical value of the slope changes at different interfaces. This provides a method for determining scratch hardness for various layers.

References

[*1*] Mittal, K. L., *Electrocomponent Science and Technology*, Vol. 3, 1976, pp. 21–42.
[*2*] Heavens, O. S., *Journal of the Physics of Radium*, Vol. 11, 1950, pp. 355–360.
[*3*] Heavens, O. S. and Collins, L. E., *Journal of the Physics of Radium*, Vol. 13, 1952, pp. 658–660.
[*4*] Weaver, C. and Hill, R. M., *Philosophical Magazine*, Vol. 3, 1958, pp. 1402–1410.
[*5*] Benjamin, P. and Weaver, C., *Proceedings of the Royal Society* (London), Vol. 254A, 1960, pp. 163–176.
[*6*] Mittal, K. L., this publication, pp. 5–17.
[*7*] Chopra, K. L., *Thin Film Phenomena*, McGraw-Hill, New York, 1969, pp. 313–323.
[*8*] Butler, D. W., Stoddart, C. J. H., and Stuart, P. R., *Journal of Physics (D): Applied Physics*, Vol. 3, 1970, pp. 887–883.
[*9*] Weaver, C., *Journal of Vacuum Science and Technology*, Vol. 12, 1975, pp. 18–25.
[*10*] Hamersky, J., *Thin Solid Films*, Vol. 3, 1969, pp. 263–267.
[*11*] Berendsohn, D., *Journal of Testing and Evaluation*, Vol. 1, 1973, pp. 139–143.
[*12*] Greene, J. E., Woodhouse, J., and Pestes, M., *Review of Scientific Instruments*, Vol. 45, 1974, pp. 747–749.
[*13*] Goldstein, L. F. and Bertone, J. J., *Journal of Vacuum Science and Technology*, Vol. 12, 1975, pp. 1423–1426.
[*14*] Oroshnik, J. and Croll, W. K., this publication, pp. 158–183.

DISCUSSION

J. J. Bikerman[1] (*written discussion*)—Scratch hardness was defined about 130 years ago as the ratio of the load on the stylus to the square of the width of the track left on the film.

J. Ahn, K. L. Mittal, and R. H. MacQueen (*authors' closure*)—In accordance with the symbols used in our paper, scratch hardness (defined 130 years ago, as pointed up by Bikerman) would be

$$S_H = \frac{W}{4A^2}$$

[1] Department of Chemical Engineering, Case Western Reserve University, Cleveland, Ohio 44106.

whereas we have described

$$S_H \equiv \frac{W}{\pi A^2/2} \equiv \frac{W}{1.57A^2}$$

For $R \gg H \; A^2 \simeq 2RH$

$$S_H = \frac{W}{\pi RH}$$

The vertical resolution (see Fig. 10 of paper) of the microtopographer is far better than the horizontal resolution; therefore, S_H can be determined more precisely.

H. E. Ashton[2] (*written discussion*)—(1) You are not measuring "hardness" with the scratch technique, at least with organic coatings, because one can get higher scratch readings by making the film softer. In Canada, we use the term "scratch resistance" because that is all that can be said that is measured. (2) Reproducibility cannot be determined in one laboratory. Both the American Society for Testing and Materials and the International Standards Organization restrict that term to several laboratories; repeatability is within one laboratory.

J. Ahn, K. L. Mittal, and R. H. MacQueen (*authors' closure*)—(1) It is a matter of semantics. Both "scratch hardness" and "scratch resistance" should be acceptable terms. A. P. Gerk[3] from England has also used the term "scratch hardness." (2) We fully agree with your comment; however, we fail to see its relevance to our paper.

[2]National Research Council of Canada, Division of Building Research, Ottawa, Ont., Canada.

[3]Gerk, A. P., *Journal of Physics (D): Applied Physics,* Vol. 9, 1976, pp. L179–L181.

J. Oroshnik[1] *and W. K. Croll*[2]

Threshold Adhesion Failure: An Approach to Aluminum Thin-Film Adhesion Measurement Using the Stylus Method

REFERENCE: Oroshnik, J. and Croll, W. K., **"Threshold Adhesion Failure: An Approach to Aluminum Thin-Film Adhesion Measurement Using the Stylus Method,"** *Adhesion Measurement of Thin Films, Thick Films, and Bulk Coatings, ASTM STP 640,* K. L. Mittal, Ed., American Society for Testing and Materials, 1978, pp. 158–183.

ABSTRACT: A detailed study of the scratch test was undertaken to determine its sensitivity, reproducibility, and soundness as a means of measuring the adhesion of aluminum films on fused quartz substrates. The complete-removal failure criterion was found to be quite subjective and often ambiguous, finally leading to the threshold adhesion failure (TAF) concept. Using the TAF approach, it was found that the scratch test is sensitive enough to discriminate between failure and nonfailure within 0.1 g of stylus loading at both low- and high-load levels. Also, the reproducibility of the mean TAF loads is excellent. In addition to the expected TAF load dependence on stylus tip radius of curvature and film thickness, it was also found that any one stylus has its own scribing characteristics which are reflected in the TAF load. The mean TAF loads are, therefore, relative measurements.

The occurrence of crescent fractures in the substrate at high-stylus loads puts an upper bound on acceptable testing loads. TAF loads are almost always well below the crescent fracture load level, but critical loads determined by the complete removal criterion (being so much higher) may have doubtful meaningfulness.

It appears that the TAF approach is measuring the lower limits, while the critical load approach is measuring the upper limits of film adherence. In either case, one obtains an estimate of the quality of film adherence, but not an unequivocal measurement of the adhesive forces.

KEY WORDS: adhesion, scratch tests, thin films, measurements, failure

[1] Physicist, U.S. Naval Weapons Engineering Support Activity, Washington Navy Yard, Washington, D.C. 20374; presently on assignment to the Naval Research Laboratory, Washington, D.C. 20375.

[2] Cherryfield Stoneware, Cherryfield, Me. 04622.

The remarkable growth of thin-film and semiconductor devices was not only made possible by the parallel development of thin film technology, but placed severe demands upon it. The sophistication of electronic semiconductor devices, for example, requires that the variety of thin films present in a device simultaneously possess the electrical and physical characteristics necessary for device operation as well as the capacity to survive the device fabrication processes. Strongly adherent metal films on various substrates is one of the most pressing requirements of thin-film devices, but a reliable quantitative measurement of the adhesion is still elusive. The scratch test [*1,2*][3] appeared to hold some promise as a quantitative test; therefore, a detailed examination of it was undertaken. The goals of this study were to determine how quantitative or how qualitative the scratch test measurements are, to determine the sensitivity and reproducibility of the measurements, and, perhaps, to gain some idea of the scope of applicability of this test.

Failure Criterion

As originally conceived, the criterion for adhesion failure is, ". . . the film was stripped cleanly from the substrate, leaving a clear channel" [*1,2*]. The force applied to the stylus ". . . which just removed the film cleanly . . ." [*2*] is called the "critical load," and is taken as a measure of the adhesion. However, it was noticed early in this study that complete removal of a film, leaving a clear track over a specified length, was a rather infrequent event. Repeatable complete-removal scratches on any particular film sample were difficult to obtain. Also, at loads below but close to the critical load, there was partial film removal in varying amounts, though not necessarily in proportion to the applied load.

Communications with earlier workers in the field regarding the complete-removal criterion revealed that determination of complete removal (and therefore of the critical load) is often a matter of weighted judgement. It was further pointed out that it was only under the most favorable experimental circumstances that complete removal was "generally" achieved. The subjectivity of the complete-removal criterion as reported by earlier workers, taken together with the authors' own experience, left one with doubts regarding the quantitative reliability of the scratch test as a measure of thin-film adhesion.

In the course of these exploratory measurements, it was always observed that at stylus loads well below the "critical load" there was some film detachment[4]—not long streaks, but in spots. These small areas of removed film were always in the central portion of the track, and the largest of them

[3] The italic numbers in brackets refer to the list of references appended to this paper.

[4] This observation and other behavioral phenomena discussed later in this paper were simultaneously and independently observed by Butler et al [*4,5*] and Chapman [*6*].

never extended beyond the boundaries (that is, the width) of the track. The breaks occurred at some minimum load, and did not appear to be associated with any visibly identifiable dust particle, film imperfection, or other surface asperity, although these possibilities are not set aside. The experimental evidence heavily favored the idea that the low-load breaks were associated with localized low film-to-substrate adhesion regardless of cause. The sum total of the foregoing considerations finally led to the concept of "Threshold Adhesion Failure." The operational definition adopted throughout this study is as follows:

> Threshold Adhesion Failure occurs if, within the boundaries of a scratch and over its 1-cm path, removal of the film from its substrate can be detected by transmitted light with a microscope (×40 magnification) at even one spot, no matter how small.

The question remained whether or not the adhesion of a thin film could be characterized through measurements based on the Threshold Adhesion Failure (TAF) concept. The main thrust of this effort was addressed to this question.

Equipment

A scratch test instrument must necessarily provide the means to apply a known force to a stylus, a controlled movement of the stylus across the film to make the scratch, and provisions for examining the scratch. The instrument constructed to conduct these experiments is shown schematically in Fig. 1*a*; Fig. 1*b* is a photograph of the completed instrument. Referring to the numbered parts of Fig. 1*a*, the instrument was built around a commercial triple-beam laboratory balance ③ that provided the applied force to the stylus and also held the substrate bearing the thin film ①. The original weighing platform was replaced by a suitable stage. A massive *x*-*y* movement ⑤-⑥ supported a stylus in its holder ②, carried in turn by its vertical adjustment mechanism ④, the fiber optic light source ⑦, and a viewing microscope ⑧. The motorized *x*-movement ⑤ in the plane of the page drives the stylus from left to right to make the scratch, and the manual *y*-movement ⑥ (perpendicular to the plane of the page) indexes the stylus for the next scratch. The microscope is positioned just to the right of the stylus holder so that the scratch is observable as the *x*-movement is manually returned to its zero position, and while the stylus is no longer in contact with the film; that is, while the balance is unloaded.

In order to have the stylus contact the film with a known force, the riders on the balance beams are first adjusted to read that desired force (in grams) while the balance is unloaded. The stylus is then brought down into contact with the specimen and its vertical position is adjusted until the

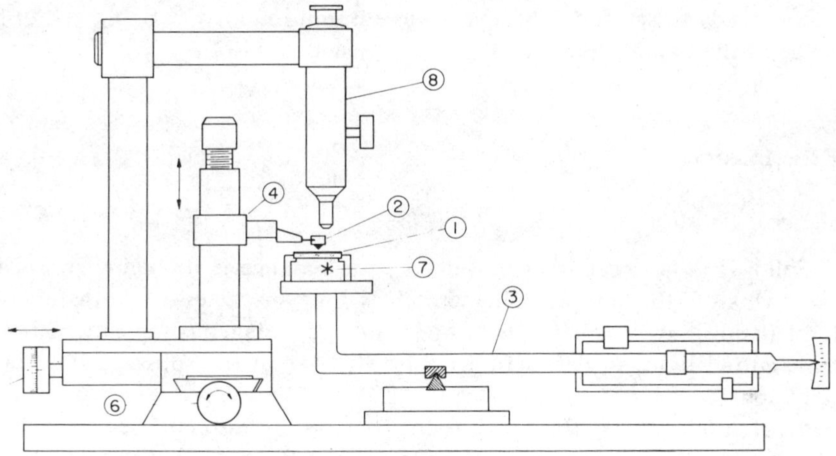

FIG. 1*a*—*Scratch test instrument*: (1) *Specimen under test*; (2) *stylus* (*in holder*); (3) *triple-beam balance*; (4) *vertical adjustment for stylus*; (5) x-*movement drive for scratching*; (6) y-*movement drive for indexing*; (7) *fiber optic light source*; (8) *viewing microscope.*

FIG. 1*b*—*Scratch test instrument.*

beam pointer indicates that the balance is in its balance position. The stylus is now ready to scratch. The stylus force is continuously variable within the range of the balance (0 to 1600 g) with a least count of 0.1 g.

Experimental

General Conditions

Following the scratching movement, the balance is unloaded by raising the stylus off the sample. The scratch is then inspected by both reflected light (using any suitable microscope lamp advantageously positioned) and transmitted light, to detect film failure, by way of the substage fiber optic light source.

Scratch length was fixed at 1.0 cm, plus an additional 1 mm at the beginning of the scratch to allow any transient effects of the initial stylus-to-film relative motion to dissipate. Scratching speed was fixed at 1 cm/min.

The measurements were made on vacuum-evaporated aluminum films (99.999 percent purity and approximately 0.5 μm thick) deposited on transparent fused quartz plates (75 by 25 by 1.5 mm). Prior to the aluminum deposition, the quartz plates were cleaned and heat-treated in air to remove adsorbed water [*3*]. In all cases, fully aged aluminum films were examined, ranging in age from 14 days to 9 months.

Diamond and tungsten carbide styli were used. Tungsten carbide styli with a 50-μm nominal tip radius of curvature were used for some of the measurements, but the predominant number of measurements was made with diamond styli whose tip radius of curvature ranged between 15 and 120 μm.

Measurement Procedure

The procedure for making a scratch has already been described in the section on apparatus description. Each test scratch is made at a predetermined load level that is held constant throughout that scratch. At first, one has to estimate (or probe for) the stylus force level at which threshold adhesion failures are expected. The stylus is then incrementally loaded until a TAF is observed at say, W_0. In the next test the load is reduced by one increment to W_1 ($= W_0 - \Delta W$), and then by successive decrements of ΔW until at load W_k there is no TAF. The succeeding test will be at load $W_{k+1} = W_k + \Delta W$. The experimenter continues this procedure until the number of tests made is believed to be representative of the specimen. Data of this kind are conveniently treated statistically by the "up-and-down" or "staircase" method wherein the stimulus (load) level of each successive test is determined by the result of the immediately preceding test [*8*]. The statistical treatment applied to the data set yields the mean failure load and the sample standard deviation.

Sensitivity

Sensitivity is defined here as the smallest differential stylus loading (in grams) that can distinguish between a TAF failure and nonfailure. In the experiments described in this paper, the highest sensitivity obtainable was 0.1 g, which happens to be the resolution limit of the balance.

Reproducibility

Measurements were conducted with the objective of determining both the sensitivity and reproducibility of the scratch test as used in the TAF mode. The combinations of nominal film thickness and nominal stylus tip radius of curvature used for these determinations were as follows:

	Nominal Film Thickness μm	
Nominal Tip Radius of Curvature, μm	0.5	1.0
18		*S, R*
50	*S**, *R*	
120	*S, R*	

NOTE—*S* is sensitivity and *R* reproducibility. (*) Sensitivity measurement for tungsten carbide and diamond. All other measurements are with diamond styli only.

The reproducibility measurements for the combination [0.5-μm films; 46-μm stylus] were made by interdigitating four sets of 20 scratches each so that four independent measurements of the mean TAF level may be obtained within a single well-defined patch of aluminum film. Successive scratches in any one set were spaced 0.010 in. (0.25 mm) apart. The interdigitation was arranged so that the track centers of all four sets together were 0.0025 in. (0.06 mm) apart.

For the combination [0.5-μm films; 120-μm styli], a comprehensive set of measurements to determine TAF load reproducibility was not carried out as it was for the 46-μm stylus. However, three pairs of runs were made which serve to indicate the load reproducibility using 120-μm styli.

An extensive series of scratch tests was made on a 1.03-μm-thick aluminum film with 18, 45, and 120-μm styli. A negligible number of failures of the threshold type were observed with the 45- and 120-μm styli, but significant success was achieved with the 18-μm stylus.

Loads in the range 30 to 120 g were used for scratching with the 18-μm stylus. Progressive thinning of the film with increasing load took place, and translucency of the film was detectable at a stylus loading of approximately 40 g. However, within the scratches made with the 18-μm stylus, many tiny spots appeared to be significantly brighter than the surrounding translucent region and to have a distinct white-light diffraction pattern associated with them. These spots strongly resembled previously observed threshold adhesion failures, although they were decidedly smaller. Since the diffraction pattern was present, it is believed that these tiny brighter spots are transparent holes in the film and represent failure points. That the occurrence or nonoccurrence of these spots was quite distinct with rather small differential stylus loadings is taken as further evidence of the correctness of this conclusion. On the basis that the tiny bright spots with associated diffraction patterns were threshold adhesion failures, a number of measurements were made to determine sensitivity and reproducibility.

Stylus Tip Radius of Curvature

An exhaustive examination was conducted of styli tips by the scanning electron microscope (SEM), the optical microscope, and optical interferometry for purposes of determining tip radius of curvature. Since only the first few tenths of a micrometre of the stylus tip make contact with the film, it is the curvature and surface character of that small region that is important in the scratch test behavior. Detailed information about the surface features of the tip of the stylus that does the actual scratching was obtained with the SEM.

A modified shadowgraph technique was found most convenient to measure the overall tip radius of curvature. A photomicrograph of the tip profile is made with a magnification somewhat in excess of $\times 100$. A reticle-type comparator is then used to match semicircles of known radii to the photograph of the tip profile. Matching can be done within 0.5 mm so that with $\times 130$ magnification a determination of the nominal tip radius within approximately ± 4 μm is obtained.

Interference fringes (mercury green line, $\lambda = 0.5461$ μm) providing a contour map of the stylus tip and a possible means of determining the tip radius of curvature were generated by resting a microscope cover glass on the stylus tip. A number of different diamond styli were examined in this way.

Individual Stylus Behavior—One experiment was performed to determine whether or not styli showed individual scribing and testing behavior. Using two different 45-μm diamond styli, a pair of interdigitated test scratches was made on the same film to determine their respective TAF load levels.

Results

Sensitivity

[*0.5-μm films*; *50-μm styli*]—Figure 2 is the plot of a typical set of scratch test measurements made on a fully aged, 0.566-μm-thick aluminum film with a 50-μm tip radius tungsten carbide stylus. Scratches that showed TAF are plotted with a spade (▽) and those that did not are plotted with a circle. A similar measurement set using a tungsten carbide stylus is shown in Fig. 3. Equally sensitive results are obtainable with diamond

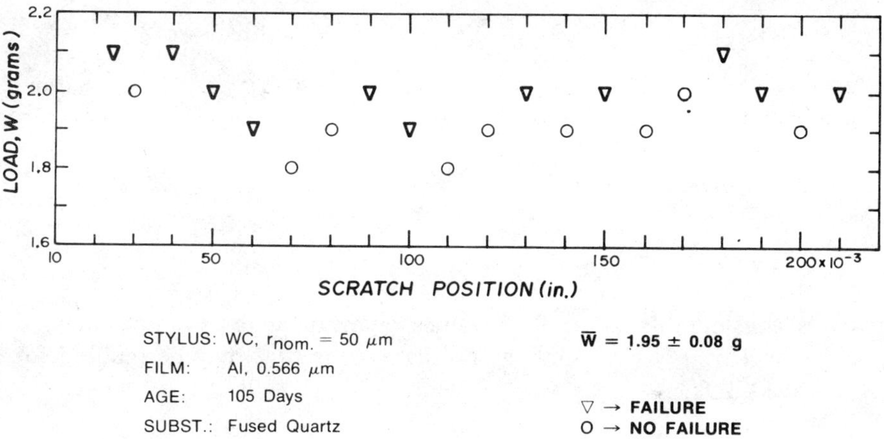

FIG. 2—*Test data from a typical set of scratch test measurements.*

styli as illustrated in Figs. 4 and 8. It is amply demonstrated here that the scratch test in the TAF mode can discriminate between differential loads as small as 0.1 g.

[*0.5-μm film*; *120-μm styli*]—Two diamond styli (113 and 120-μm tip radius) were used to test five different films, each being close to 0.5 μm thick. The results are given in Table 1.

It is immediately apparent that the TAF loads using the larger styli lie in the range of several hundreds of grams compared with a few tens of grams for 45-μm diamond styli, and 2 to 3 g for 50-μm tungsten carbide styli. Even so, substantially good sensitivities are obtainable. From the data on Specimen L1, it will be seen that sensitivities of 0.1 g can be achieved with styli in the range of 120 μm.

[*1.0-μm film*; *18-μm stylus*]—The results for this combination are shown in Table 2, from which it can be seen that the test can distinguish between failure and nonfailure for a differential load of 2.0 g. Although this sensi-

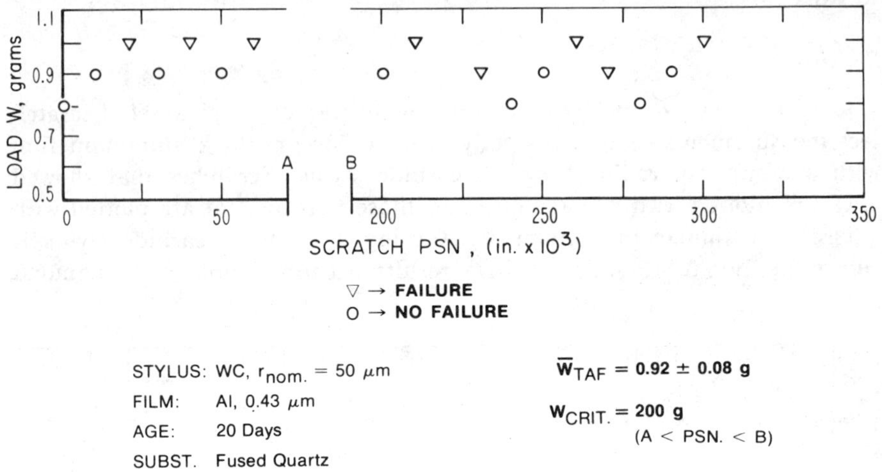

FIG. 3—*TAF versus complete removal.*

tivity is considerably lower than values determined for 0.5-μm films, it does not necessarily represent the ultimate value that may be realized for 1-μm-thick films.

TABLE 1—*Threshold adhesion failure loads for aged, highly adherent, 0.5-μm-thick aluminum films on quartz substrates.*

Specimen	Stylus	Tip Radius, μm	Increment, g	Mean Failure Load, g	Sample Standard Deviation, g
L-1	D#1	113	0.1	79.04	0.25
				82.88	0.15
B3-S2	D#1	113	0.2	345.5[b]	0.06
			0.5	343.8[b]	0.6
				341.1[a]	1.4
				348.9[a]	1.6
B1-S3	D#1	113	1.0	424.1	1.2
				424.3	2.9
B4-S1	D#3	120	0.5	231.8	4.2
			2.0	244.4	7.2
B4-S3	D#3	120	5.0	220.4	20.7

[a] Data taken in neighboring regions of film.
[b] Data taken in adjacent regions.

TABLE 2—*Results of scratch tests on a 1.0-μm-thick aluminum film with an 18-μm diamond stylus.*

Run No.	Load Increment, g	Mean Failure Load, g	Sample Standard Deviation, g
1	5.0	59.2	10.0
2	2.0	57.9	4.4
3	2.0	55.7	2.0
4	2.0	59.0	4.8
Grand mean[a]		57.6	1.8

[a] Interdigitated Runs 1-3 only.

Reproducibility

[*0.5-μm film*; *46-μm stylus*]—All the measurements for this group of four interdigitated sets are shown in Fig. 4. Each set of 20 scratches is connected by lines to indicate the order in which each scratch measurement was made. (The lines do not indicate a physical process.) For each set of 20 scratches a mean TAF load is calculated, and a summary of these calculations appears in Table 3.

The sample standard deviation (0.34) is 2.8 percent of the grand mean of the TAF load for the four measurement sets. If one assumes a normal distribution for these data, then the relative standard deviation of the mean for the entire population is 8.2 percent—within a 95 percent confidence limit. This later figure of 8.2 percent may be taken as a measure of the reproducibility.

[*0.5-μm films*; *120-μm styli*]—As mentioned earlier, a comprehensive set of measurements to determine reproducibility was not made as it was for the preceding group. However, the measurements indicate that for 120-μm styli the TAF load reproducibility is comparable to that for the 45-μm styli. For example, when examining the measurements on Specimen B3-S2 in Table 1, one notices the closeness of the mean failure loads for the pairs marked *a* and *b*. Also, Table 4 shows the results for an interdigitated pair of runs, which essentially duplicate each other.

[*1.03-μm films*; *18-μm stylus*]—The reproducibility for this combination is calculated from the measurements summarized in Table 2. The mean TAF load for each run was determined from at least 20 scratches.

Runs 1 through 3 were interdigitated. If the differences in load increment are ignored and the data are treated together, the grand mean failure load is 57.6 ± 1.8 g. If the data are assumed to be normally distributed, it can be asserted with 95 percent confidence that the relative standard deviation for the entire population is less than 14 percent.

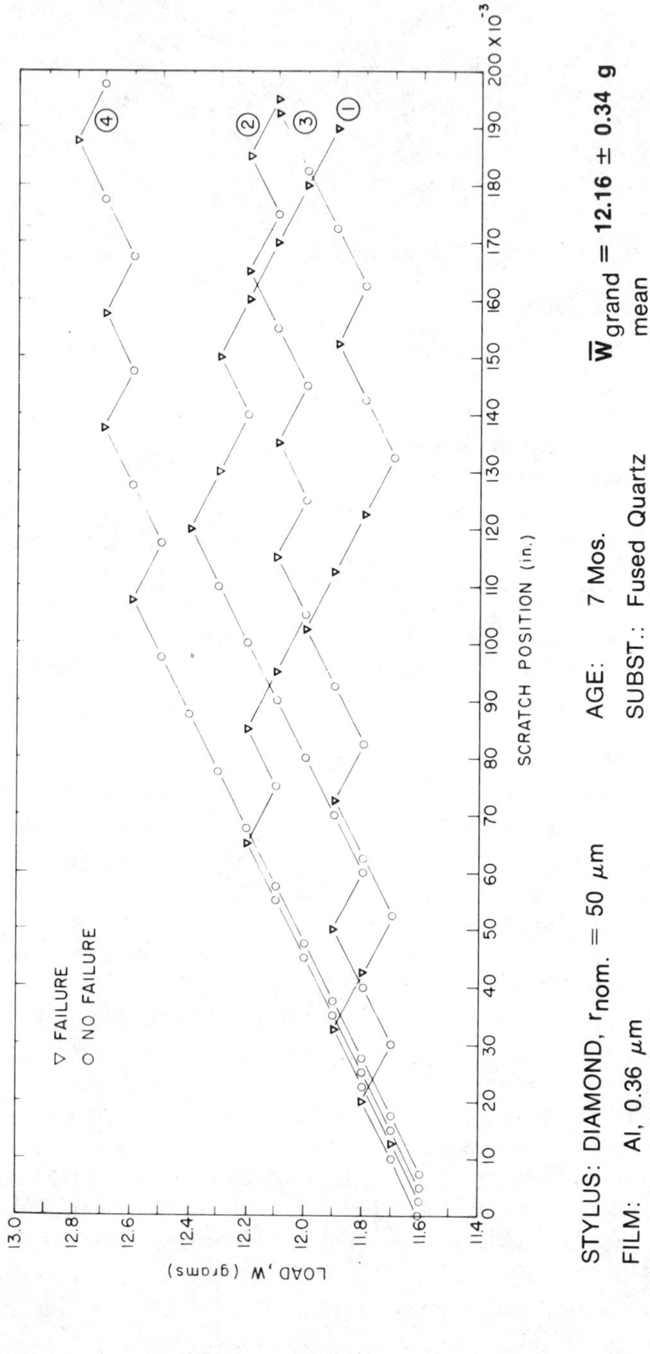

FIG. 4—*TAF reproducibility test.*

TABLE 3—*Analysis of scratch test reproducibility data.*

Set No.	Mean Failure Load, g	Sample Standard Deviation, g
1	12.05	±0.65
2	12.10	±0.05
3	11.86	±0.14
4	12.65	±0.09
Grand mean	12.16	±0.34

TABLE 4—*TAF reproducibility data. Stylus—diamond;* r_{nom} = *120 μm. Film—aluminum;* τ = *0.48 μm.*

Data Run No.[a]	Load Increment, g	Mean Failure Load, g	Sample Standard Deviation, g
1	2.0	244.4	7.2
2	2.0	244.8	5.3

[a] Both runs are interdigitated.

Stylus Tip Radius of Curvature

The shadowgraph technique described earlier proved useful for determining the overall tip radius, and for identifying any one stylus. Nevertheless, it was evident from the shadowgraph measurements that the very tip of the stylus was far from an ideal spherical surface.

That the stylus tip is not an ideal spherical surface is clearly shown by interference techniques. Interferograms of two different diamond styli are shown in Fig. 5. It is immediately obvious that the tips are neither spherically nor radially symmetric. Neither stylus can be assigned or said to possess a single unambiguous tip radius to characterize it.

The SEM was invaluable for examining the topography and surface features of the styli tips. The tip of a tungsten carbide stylus is shown in Fig. 6*a*. With higher magnification, the individual grains of tungsten carbide are clearly evident, the smaller ones being about 0.6 μm in size. The tip of a stylus actually used in these experiments is shown in Fig. 6*b*. Aluminum from the scratched films can be seen on the surface of the tip area. From this photograph it also appears probable that a single granule was responsible for all the cutting. Further enlargements of this area show that the two clearly exposed edges of this granule are about 1.2 μm long and that the radius of the included tip is about 0.18 μm. Relative to a metal film thick-

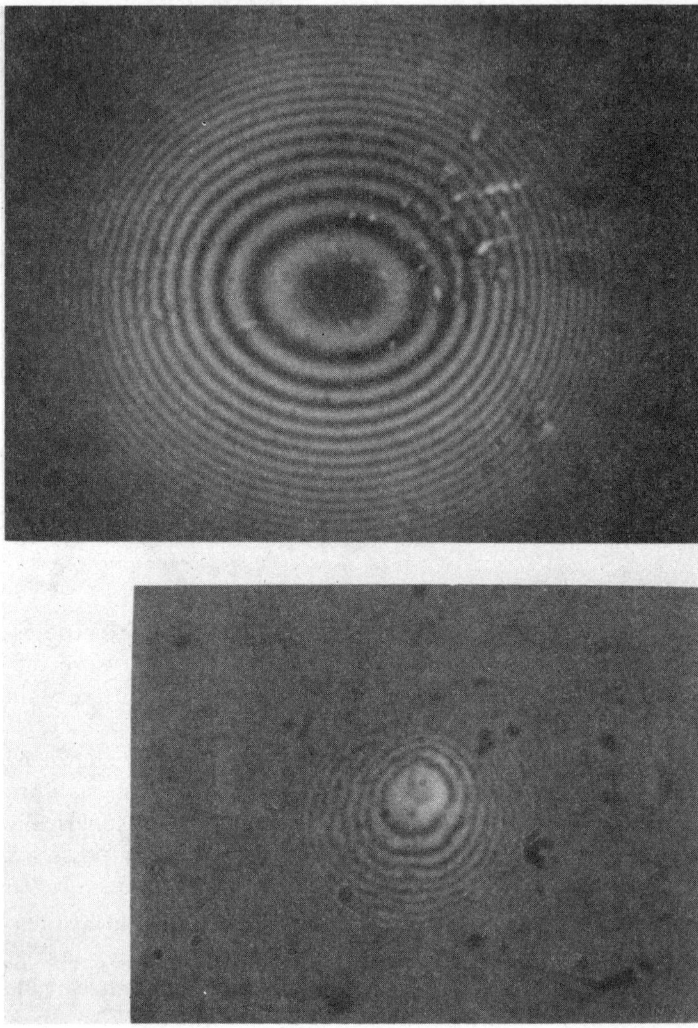

(A) $r_{nom.}$ = 45 μm

(B) $r_{nom.}$ = 120 μm

FIG. 5—*Interferograms of two diamond stylus tips photographed with the mercury green line* (λ = *0.5461 μm*). *Magnification* ×966.

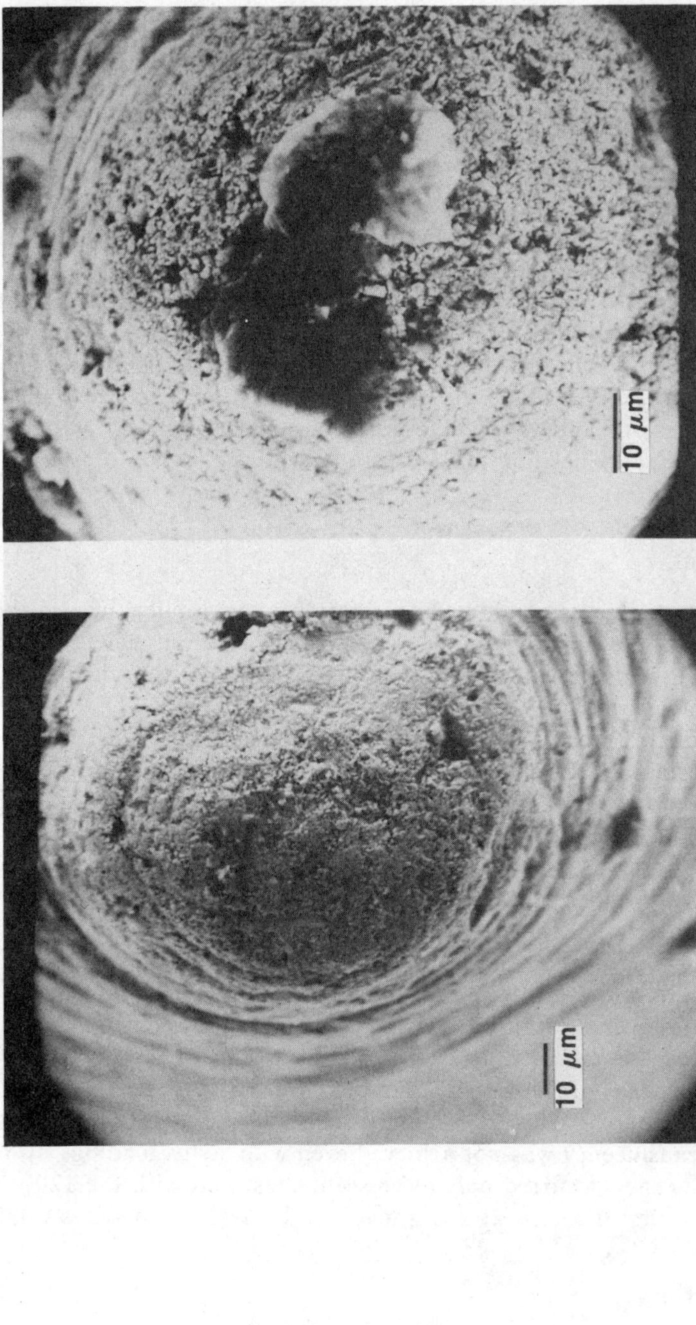

(A) BEFORE $r_{nom.} = 50\ \mu m$ (B) AFTER

FIG. 6—*Tip of a typical tungsten carbide stylus before and after scratching. Magnification* (a) ×592; (b) ×960.

ness of 0.5 μm, this granule represents a fairly sharp cutting edge. Parenthetically, plastically deformed aluminum was never found to adhere to a diamond tip.

Diamond tip surfaces are far smoother and more regular, as will be seen from Fig. 7. One must bear in mind that the magnification in Fig. 7 are slightly more than four times that of Fig. 6*b*. The 15-μm tip is quite polished but pitted, while the 75-μm tip has a "lemon peel" type of surface.

Individual Stylus Behavior—During the course of this work, no two diamond styli were found to be identical in tip contour or surface quality. The consequence of this is most important since it leads one to expect that each stylus will show a scribing and testing behavior peculiar to itself. This is indeed the case and is confirmed by testing of the same film areas with different styli. An example of this is shown in Fig. 8, where two interdigitated runs were made by two different 45-μm diamond styli. It is quite clear that although each test set results in a different TAF mean load with a narrow spread, the two styli yield decidedly different TAF loads. This means that the scratch test in the TAF mode will give indications of only relative values of thin-film adhesion, and those determinations must be made with the same stylus in the same position in the instrument.

Crescent Fracture

During the experiments conducted to determine the applicability of the TAF approach to 1-μm films, one phenomenon was observed which cannot be lightly treated. This was the appearance of "crescents" within the boundaries of the scratch track.

The crescents are fractures in the fuzed quartz beginning at the quartz-aluminum interface. The crescents were circular arcs, and open in the scratching direction. They appeared at relatively high stylus loadings; in the case of 18 and 45-μm styli, at 80 and 160-g loadings, respectively. The crescents were observed when the tracks were already thinned down to translucency, the latter becoming noticeable at 40 and 80 g for the 18 and 45-μm styli, respectively. They were clearly distinguishable because of their distinctly higher translucency than the rest of the track.

At no time was there any general removal of the film, leaving a clear substrate. The crescents, each one having a finite width across the track, were always covered over with a layer of aluminum, however thin, except for an occasional clear break directly on a crescent site.

Although thinning of the 1-μm film occurred with the 120-μm stylus, overall track translucency was not achieved even with stylus loadings up to 1100 g. Translucency occurred only at crescent sites, and with the 120-μm stylus they were first detected at 260-g loading. Table 5 summarizes mini-

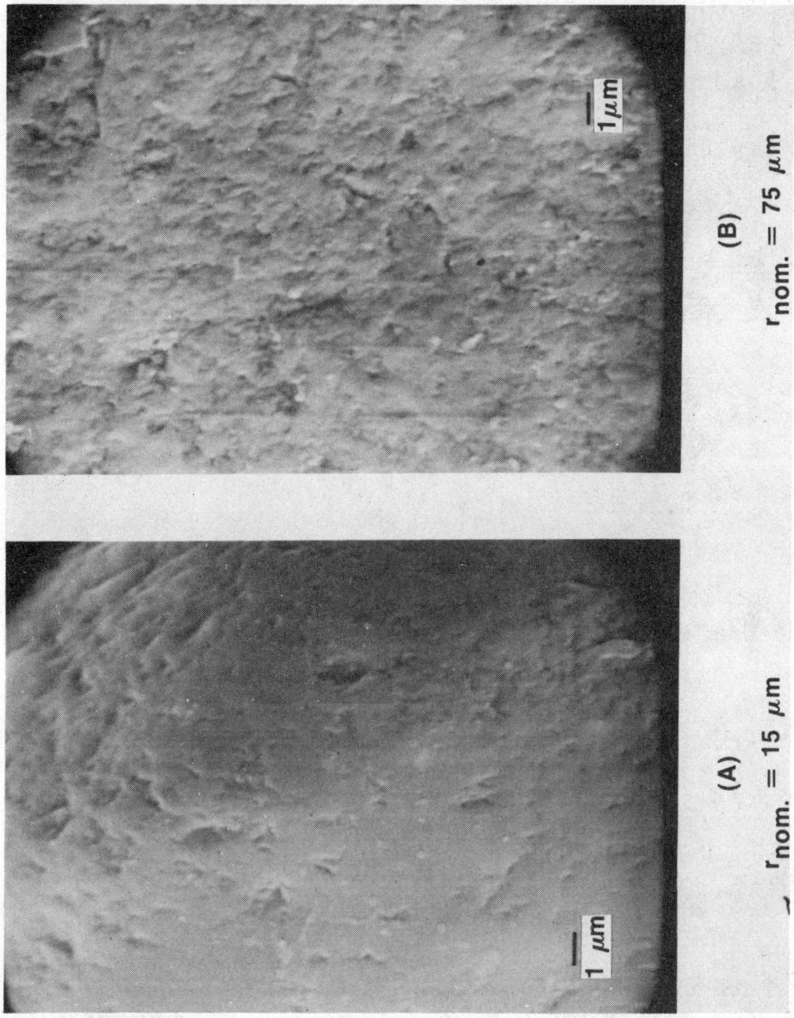

FIG. 7—*Varying surface quality of different diamond styli. Magnification* (a) ×3920; (b) ×3780.

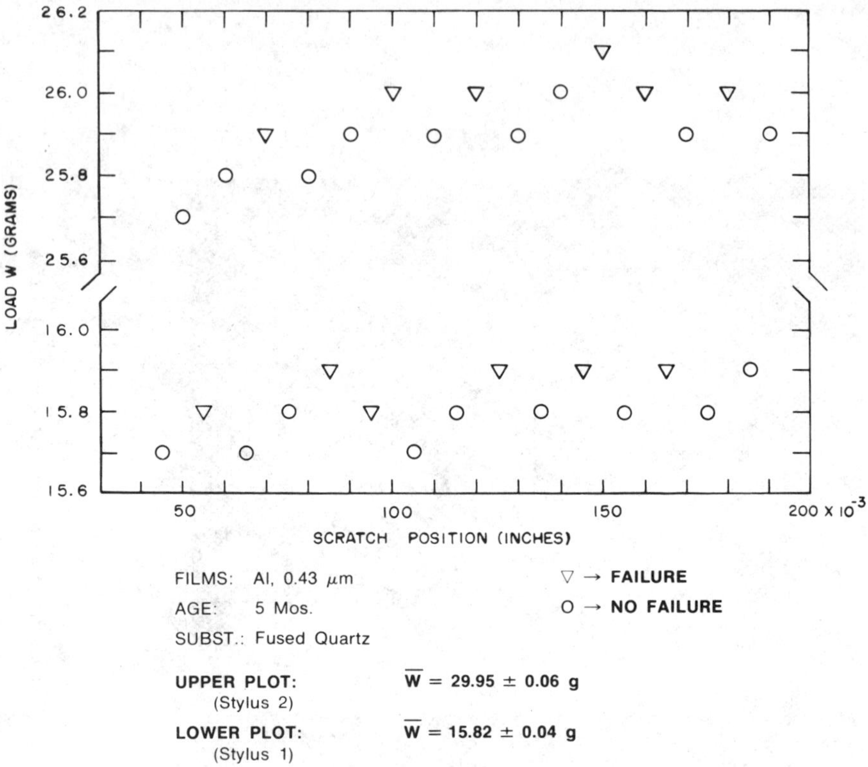

FIG. 8—*TAF test data using two different diamond styli, each having* r_{nom} = *45 μm.*

mum loads at which translucency and crescent fracture were observed in the 1-μm-thick films. Parenthetically, thinning of the aluminum film to translucency was observed only in the experiments on 1-μm-thick films.

Discussion

Sensitivity

The scratch test in the TAF mode is measuring, as its designation infers, the lower limit of the effective adherence of a thin film. It is detecting highly localized regions of low film-to-substrate adherence with exceptional sensitivity. The sensitivity of the test is such that the observer can obtain a quantitative estimate of the range of load levels over which the TAF's occur. With a constant load increment, the sample standard deviation of the TAF mean for any one determination is an indication of the scatter of the TAF load levels. A perusal of the data presented here will reveal that standard

TABLE 5—*Crescent fracture and translucency: minimum stylus loads; 1-μm-thick aluminum on fused quartz, diamond stylus.*

	Lowest Observed Load	
r_{nom}, μm	Crescent Fracture, g	Translucency, g
18	80	40 to 50
45	160	80
120	260	(none at 1100)[a]

[a] See text.

deviations run the gamut from very much smaller to much larger than the load increment.

Stylus Tip Radius of Curvature

Measurements of stylus tip radius of curvature serve only to identify a particular stylus, and possibly to characterize it in a very limited way.

When examining the complete-removal criterion early in this study, an attempt was made to verify the expression for the shear force between film and substrate, as derived by Benjamin and Weaver [*2*]. The shear force, F (g/cm^2), may be expressed as

$$F = \frac{P}{\left[\pi P \dfrac{r^2}{W_c} - 1\right]^{1/2}} \quad (1)$$

where

r = stylus tip radius (cm),
W_c = critical load (g), and
P = indentation hardness of the nonmetallic substrate (g/cm^2).

From this

$$W_c = r^2 \left[\frac{1}{\pi P} \left(\frac{P^2}{F^2} + 1 \right) \right]^{-1} \quad (2)$$

and when measurements are made on the same film and substrate, then

$$W_c = kr^2 \quad (3)$$

The relationship Eq 3 was not verifiable with 45-μm diamond and 50-μm tungsten carbide styli from interdigitated runs on the aluminum-quartz system.

The derivation of Eq 1 is based on a smooth, spherical stylus tip in contact with a flat plate. In the light of the findings reported here, it is not at all surprising that Eq 1 was not verifiable. In the range of a few tens of micrometres to 120 μm for tip radii, the diamond and tungsten carbide styli simply cannot fulfill the physical requirements demanded by Eq 1. Steel (phonograph) styli were not examined, and, unfortunately, the literature [*2,4,16*] does not contain any reports on steel tip radius of curvature and surface topography in detail comparable to the work reported here for diamond and tungsten carbide; thus, regrettably, a comparison cannot be made.

Individual Stylus Behavior—If diamond styli were ideal right circular cones with truly spherical tips having homogeneous and identical surfaces, it would be reasonable to expect substantially reproducible scratching behavior and TAF load levels. Rather, the evidence presented here leads one to expect the interaction between stylus and film to be quantitatively different from stylus to stylus with the same film. The measurements of Fig. 8 are clear, corroborative evidence.

Stylus/Film Thickness Combinations

The data gathered from these experiments suggest the probable existence of a range, or neighborhood, of combinations of film thicknesses and styli tip radii conducive to meaningful TAF determinations. Although the combination [1.0-μm film; 18-μm stylus] yielded good TAF measurements, it is felt that the upper limit of film thickness is 1.0 μm. A more realistic upper bound to film thickness is more likely about 0.8 μm.

Film thicknesses significantly thinner than 0.5 μm (that is, about 0.1 to 0.2 μm) have not been explored, but would probably respond well to the TAF approach. It seems probable that for films considerably thinner than 0.5 μm, styli with large tip radii (for example, 75 μm) would be in order so as to avoid penetration of the film. For films significantly thicker than 0.5 μm, experience has already shown that TAF measurements are more easily obtained with smaller styli tip radii, for example, 25 μm or less.

Crescent Fracture

This type of surface fracture in glass is well known and has been amply reported on [*9–11,13*]. Very recently, Sliney [*12*] studied these cracks, observing them directly as they were actually forming and growing. The same type of fracture appeared in the work by Goldstein and Bertone [*14*] when scratch testing metal films deposited on tetrafluoroethylene resin substrates. An example of crescent fractures in a bare quartz slide caused by a tungsten carbide stylus is shown in Fig. 9, as observed with dark-field illumination. It is apparent that the cracks propagated well beyond the scratch boundaries—an indication of the lateral extent of the stresses involved in their formation.

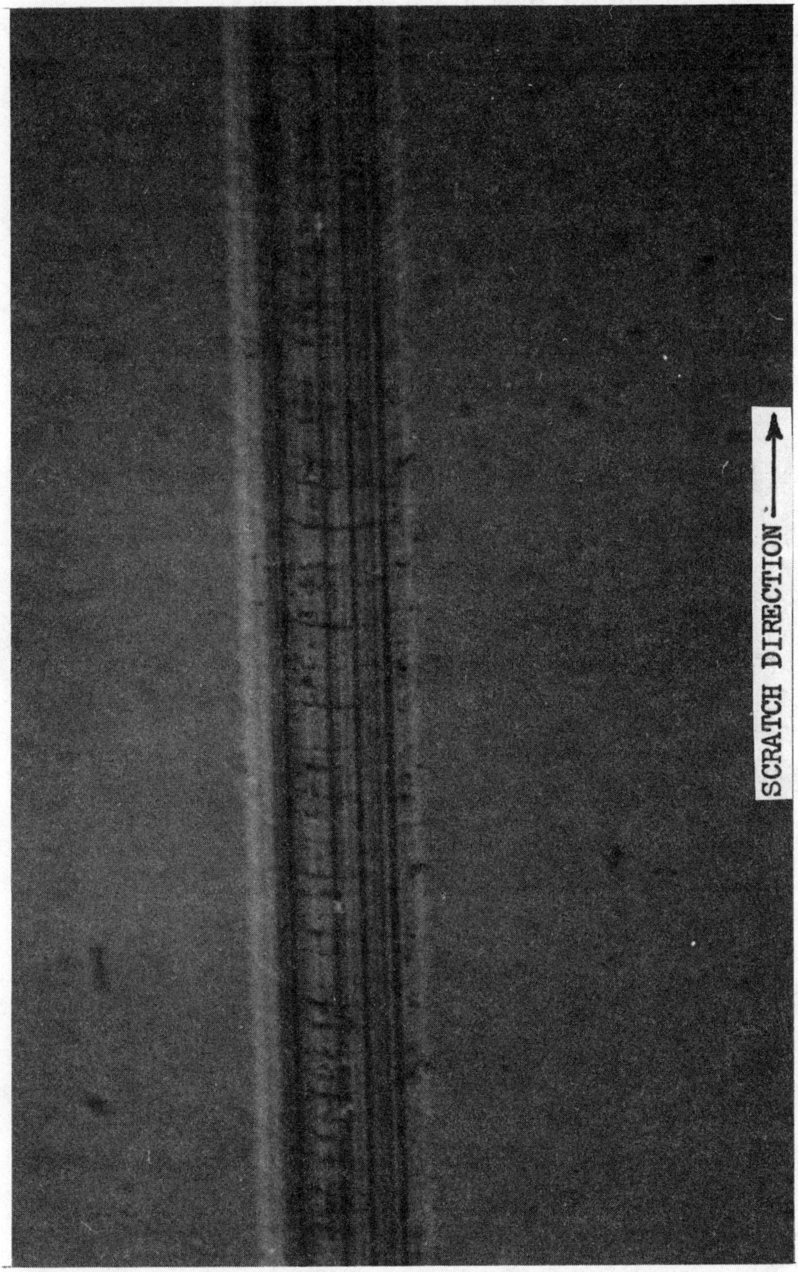

FIG. 9—*Crescent-shaped cracks produced in a fused quartz slide by the passage of a tungsten carbide stylus* (r_{rom} = 50 *μm*) *at a 100-g load. Magnification* ×640.

When a normally loaded ball or stylus is drawn across a glass surface, the friction between the two generates tangential stresses in the glass surface, parallel to the traverse direction. It has been shown [*15*] that a compressive stress is created ahead of the contact and a corresponding tensile stress trails the contact. This stress-pair is critical to this type of crack formation. Also of special importance is the demonstrated fact that crack initiation occurs *within the area of contact.*

The significant circumstance here is that the crescent fractures are generated in the quartz in the presence of an *intervening layer* of aluminum. The stylus is not in contact with the quartz. It is unlikely that the stylus is smearing the aluminum over the crack after it has formed, because the crack initiates within the contact region and the aluminum is on-site at that moment. The tangential forces generated by the stylus as it ploughs through the aluminum film are transmitted across the interface to the quartz. The salient fact is that the substrate structure fractured before any aluminum was removed.

In order to get some idea of the stylus loads necessary to produce a detectable track in quartz, the measurements summarized in Table 6 were made, and observed with dark-field illumination. With this information and that of Fig. 9, let us refer back to Fig. 3.

Complete removal was obtained at a 200 g load in Region AB. Now Fig. 9 tells us that crescent fractures are produced in abundance at a 100-g load, while Table 6 shows that a 50-μm tungsten carbide stylus begins to make plastically deformed tracks at a 4 to 5-g load on bare quartz. Therefore, with a 50-μm tungsten carbide stylus, crescent fracture sets in at some load lying between 5 and 100 g. Consequently, the 200-g load (of Fig. 3), which is far in excess of the minimum load for crescent fracture onset, is most likely thoroughly fracturing the quartz surface, leaving the meaningfulness of the critical load as an adhesion measurement quite doubtful. If the substrate destructs, then the stylus test fails.

It is in order at this point to reexamine the data of Tables 1 and 2 vis-a-vis the information in Tables 5 and 6. It is assumed that the Table 5 data are essentially applicable to 0.5-μm films. Referring to Specimen L-1 of Table

TABLE 6—*Visible track on fused quartz: minimum stylus load.*

Stylus	r_{nom}, μm	Visible Track Load (M = ×165), g
Tungsten carbide	50	4
Diamond	18	< 30
Diamond	45	15
Diamond	46	12
Diamond	120	140

1, the mean failure loads are well below the crescent fracture load of 260 g, and are therefore credible TAF determinations with good sensitivity. The mean failure loads for B3-S2 and B1-S3 are well above the crescent fracture load. Although the data are consistent and indicate good reproducibility, they are suspect because of possible substrate surface damage. On the other hand, the mean failure loads for B4-S1, B4-S3, and Table 4 lie quite comfortably between the crescent fracture level of 260 g and the deformation level of 140 g, and thereby are considered fully acceptable.

Friction

In a rather extended and sophisticated paper, Weaver [*16*] reexamines his original theory [*2*] and develops it further. In this recent paper, the interaction of an indenter with a glass substrate (in its elastic and elasto-plastic range) is developed in great detail. Weaver concludes that ultimately for a thin film to be separated from its substrate by the action of a moving stylus there should be enough friction between stylus and film so that a shearing force appears at the film-substrate interface sufficient in magnitude to break the adhesive bonds. Interaction other than friction between the moving stylus and the metal film is not considered in Weaver's paper.

Friction is not the only force of interaction between a moving stylus and a thin film. A cogent discussion of the possible interaction forces between stylus and film appear in an earlier paper by Butler et al [*4*]. The treatment is necessarily qualitative because very little is known with any certainty about the elastic constants and other mechanical properties of thin films.

The coefficient of friction between diamond and metals is known to be very small [*17*]. The additional forces that come into play with a moving stylus are a result of stylus loads sufficient to permanently deform the metal film, and very likely the substrate as well. Ploughing forces, shear forces within the body of the film just under the stylus tip, lateral shear forces at the conical sides of the stylus, the forces involved in the leading surface of the stylus "cutting" through the metal and pushing it aside to form its groove, forces required to extrude some of the metal, and frictional forces may, for simplicity, be lumped together and described as a drag force.

How large can the drag force be? Let us refer to the crescent fractures observed with 1-μm films. Tension tests on (5-mm-diameter) fused quartz rods at 20°C (68°F) exhibit a tensile strength of 7000 psi (48 300 kPa). If this value is applicable to the quartz substrates, then it means that a net tangential force in the neighborhood of 7000 psi (48 300 kPa) is appearing at the aluminum-quartz interface and is causing a crescent fracture in the quartz. Reiterating, this magnitude tangential force is being transmitted through an intervening layer of aluminum to the quartz; the stylus never penetrates the aluminum film completely. Moreover, it appears that the shear forces necessary to move (detach) an aluminum film over a quartz surface may exceed 7000 psi (48 300 kPa).

Summary

The scratch test as used in the TAF mode has been shown to be very sensitive, capable of good reproducibility, and more than adequate for testing aluminum thin films. Mean TAF loads are relative values that give some indication of the quality of film adherence.

For aluminum films, it appears that there exists a range of film-thickness/stylus-tip-radius combinations that will lead to meaningful TAF level determinations. Films up to about 0.8 μm in thickness may be tested. Diamond styli did not present any working problems.

Each stylus exhibits its own individual scribing and testing characteristics. To be able to compare TAF mean loads of one thin film sample with another (or against a laboratory "standard") requires that all the measurements be taken with the same stylus in the same position in the instrument.

Stylus loadings should not exceed those levels at which crescent fracture of the substrate will occur, if the TAF determination is to be valid.

Conclusions

The authors are not convinced that the scratch test in any form is an unambiguous measurement of adhesion. Neither the TAF technique nor the complete-removal criterion provide access to the determination of the absolute values of the adhesive forces. The TAF method is an approach to determining the lower bounds or limits of film *adherence*, while the complete-removal scratch test attempts determination of what is perhaps the upper limit of film *adherence*. In either case, the measurement is an indication of the *quality* of film adherence and is in no way an unequivocal measurement of adhesive forces.

In advance of testing any thin film, the substrate/stylus system of choice deserves careful scrutiny, specifically to determine the load level at which crescent fracture or other permanent structural damage (exceeding the plastic range) to the substrate occurs. Confidence in the TAF level is preserved provided it is below the minimum load for crescent fracture initiation. For the complete-removal test, this consideration is particularly important because of the much higher stylus loads involved. It is quite clear that more detailed study of the mechanical interaction between a translating stylus and substrate—especially in the presence of an interposing thin film—is definitely in order.

Acknowledgment

The authors are indebted to Leonard Smith, whose skills contributed immeasurably towards the fulfillment of this work.

This work was originally done at the U.S. National Bureau of Standards.

References

[*1*] Heavens, O. S., *Journal of Physics and Radium*, Vol. 11, No. 7, July 1950, pp. 355-359.
[*2*] Benjamin, P. and Weaver, C., *Proceedings of the Royal Society*, London, Vol. 254A, 1960, pp. 163-176.
[*3*] Hair, M. L. and Filbert, A. M., *Research/Development*, Vol. 20, No. 10, 1969, pp. 34-38.
[*4*] Butler, D. W., Stoddart, C. T. H., and Stuart, P. R., *Journal of Physics* (*D*), Vol. 3, 1970, pp. 877-883.
[*5*] Butler, D. W., Stoddart, C. T. H., and Stuart, P. R. in *Aspects of Adhesion*, Vol. 6, D. J. Alner, Ed., University of London Press, London, England, 1971.
[*6*] Chapman, B. N. in *Aspects of Adhesion*, Vol. 6, D. J. Alner, Ed., University of London Press, London, England, 1971.
[*7*] Collins, L. E., Perkins, J. G., and Stroud, P. T., *Thin Solid Films*, Vol. 4, No. 1, 1969, pp. 41-45.
[*8*] Natrella, M. G., *Experimental Statistics*, NBS Handbook 91, National Bureau of Standards, Washington, D.C., 1 Aug. 1963, pp. 10-22 to 10-23.
[*9*] Bowden, F. P. and Tabor, D., *The Friction and Lubrication of Solids*, Oxford at the Clarendon Press, Oxford, England, 1950, p. 169.
[*10*] Preston, F. W., *Transactions of the Optical Society*, Vol. 23, No. 3, 1921-1922, p. 10.
[*11*] Hamilton, B. and Rawson, H., *Journal of Physics* (*D*): *British Journal of Applied Physics*, Vol. 3, No. 9, Sept. 1970, pp. 140-144.
[*12*] Sliney, H. E., "Dynamics of Solid Lubrication as Observed by Optical Microscopy," NASA Technical Memorandum TMX-71880, National Aeronautics and Space Administration, Washington, D.C., Sept. 1976 and American Society of Lubrication Engineers, Paper No. 76-LC-1B-4, 5 Oct. 1976.
[*13*] Bowden, F. P. and Tabor, D., *The Friction and Lubrication of Solids, II*, 1964, pp. 19-20.
[*14*] Goldstein, L. F. and Bertone, T. J., *Journal of Vacuum Science and Technology*, Vol. 12, No. 6, Nov./Dec. 1975, p. 1423.
[*15*] Seely, F. B. and Smith, J. O., *Advanced Mechanics of Materials*, Wiley, New York, 1952.
[*16*] Weaver, C., *Journal of Vacuum Science and Technology*, Vol. 12, No. 1, Jan./Feb. 1975, p. 18.
[*17*] Mittal, K. L., *Electrocomponent Science and Technology*, Vol. 3, 1976, pp. 21-42.
[*18*] Chapman, B. N., *Journal of Vacuum Science and Technology*, Vol. 2, No. 1, Jan./Feb. 1974, p. 106.
[*19*] Oroshnik, J. and Croll, W. K., "Thin Film Adhesion Testing by the Scratch Method," American Vacuum Society, New Mexico Section, Surface Science Symposium, Albuquerque, N. Mex., 22 April 1970.
[*20*] Oroshnik, J. and Croll, W. K., "Metalization Evaluation," NBS Technical Note Nos. 520, 527, 555, 560, 571, 592, National Bureau of Standards, Washington, D.C., July 1969-Dec. 1970.

DISCUSSION

K. L. Mittal[1] (*written discussion*)—What are the main advantages of your TAF method over the conventional scratch test? As you are well aware of the criticisms of the scratch test, do you think that the TAF technique should mollify the critics of the conventional scratch test, or that the TAF technique is equally vulnerable to criticism?

J. Oroshnik and W. K. Croll (*authors' closure*)—The advantages of the TAF approach over the critical load method are as follows:

1. The determination of adhesion failure does not require any considered judgement. A failure in the scratch track is either present or not.
2. Its sensitivity, that is to say, its capacity to discriminate between failure and nonfailure for small load increments exceeds that of the critical load method by far.
3. A credible, measurable estimate (TAF mean load) of the quality of film adherence is available.
4. Reproducibility of the TAF mean loads is excellent.
5. Of great importance is the fact that TAF measurements are obtainable at load levels well below those at which catastrophic destruction of the substrate occurs, for example, crescent fracture.

In response to the latter part of your question, those experimenters who do not care for the scratch test, or cannot use the test to their advantage, are not expected to accept this test or any other scratch test. The TAF scratch test is an alternative, or parallel to the critical load approach.

No test is without its critics, and the TAF approach, I am sure, will receive its share of assorted attacks.

H. E. Ashton[2] (*written discussion*)—(1) Did I understand you to say that you could distinguish between adhesion levels of 0.5 g at the 240 to 250 g level when the standard deviation was 1.4 g? If so, this is incorrect, because for about 20 degrees of freedom the least significant difference is approximately 3 × 1.4 g. (I do not recall the gram value exactly.) (2) You were establishing the precision and not the reproducibility of the method. The

[1] IBM Corporation, Hopewell Junction, N.Y. 12533.
[2] National Research Council of Canada, Division of Building Research, Ottawa, Ont., Canada.

latter (as defined by the American Society for Testing and Materials and the International Standards Organization) is the agreement *between* laboratories, while repeatability is the agreement within one laboratory.

J. Oroshnik and W. K. Croll (authors' closure)—I believe you are referring to the measurements on specimen B3-S2 (mean load, 341.1 g) listed in Table 1. The maximum sensitivity (the smallest differential loading that will distinguish between film adhesion failure and nonfailure) is a function of the total system, that is, the instrumentation plus the film-substrate combination. Any sensitivity (ΔW) less than this is chosen by the experimenter and remains constant throughout a group of measurements. The standard deviation (and here we are addressing the *sample* standard deviation as distinguished from the population standard deviation) is a measure of the characteristic variability of performance. In our case the performance is the TAF mean load, $\overline{W}$. The sensitivity (ΔW) is independent of the sample standard deviation, since it is chosen by the operator and is a parameter of the test group.

If the increments ΔW are too large, then $\overline{W}$ will not be established with any reasonable precision. On the other hand, if the ΔW are too small it will take too long to find the load level at which failures occur about 50 percent of the time, namely $\overline{W}$. To run the test at optimal efficiency, Δw should be chosen so that $(2/3)s < \Delta W < (3/2)s$ (*see* Dixon, W. J., "The Up-and-Down Method for Small Samples," *Journal of the American Statistical Association,* Vol. 60, pp. 967–978, Dec. 1965).

The value of ΔW will affect both $|s|$ and $\overline{W}$. Suppose we have just completed one set of measurements with a sensitivity of ΔW_1, and obtained a TAF mean load of $\overline{W}_1 \pm s_1$. If we were now to perform a second set of measurements on an identical film using a sensitivity of ΔW_2 ($> \Delta W_1$) we would obtain a TAF mean load of $\overline{W}_2 \pm s_2$, where $\overline{W}_2 \approx \overline{W}_1$, but $s_2 > s_1$. The sample standard deviation can distinguish between the effects of different ΔW applications. Also, the adherence uniformity of a film will be reflected in $|s|$, the characteristic variability of performance.

Roland Faure[1]

Adhesion of Granular Thin Films

REFERENCE: Faure, Roland, "**Adhesion of Granular Thin Films,**" *Adhesion Measurement of Thin Films, Thick Films, and Bulk Coatings, ASTM STP 640,* K. L. Mittal, Ed., American Society for Testing and Materials, 1978, pp. 184–197.

ABSTRACT: With a view to measuring the adhesion of the microscopic metallic grains which constitute granular thin films, we have perfected a dynamic method. Very high accelerations (10^7 m/s^2) are obtained by means of ultrasonic vibration amplifiers.

Investigation of the adhesion of granular silver films deposited on carbon provided very interesting results: the presence of a coalescence phenomenon shows that there are interactions between grains. The adhesion varies according to the size of the grains, and is very low (150 N/m^2) compared with that of thick films (5.0×10^6 N/m^2).

Finally, our measurement of the adhesion of thick films revealed that the break occurs in the silver very near the silver/carbon interface, and subsequent studies of the growth of silver/carbon show that the grains form on one or several monolayers of silver.

KEY WORDS: adhesion, thin films, granulation, adhesive face

An electron microscope investigation of vacuum-evaporated metallic thin films indicates that in the first stages the films consist of a two-dimensional arrangement of grains. These grains are of a size such that their physical properties are intermediate between those of an isolated atom and those of a bulk crystal. We thus observe in these films properties significantly different from those of the bulk metal. The study of the adhesion of the particles which constitute most very thin metallic films is therefore of fundamental interest and can reveal new phenomena as yet unknown on this type of film.

Granular Thin Films

When investigating adhesive forces, the granular thin films can be ranged in two categories according to the method used in their preparation. The first method of preparation is classical; the thin film is obtained by vacuum

[1] Ingenieur au Centre National de la Recherche Scientifique Centre d'Etudes des Couches Minces, Laboratoire associe, C.N.R.S., Universite d'Aix-Marseille III, 13397 Marseille Cedex 4, France.

evaporation of a metal and condensation of the metallic vapor on a substrate. In this case, two growth mechanisms for granular films are suggested [*1*][2]:

1. Formation of nuclei made up of some atoms on the substrate sites, and then growth of these nuclei. This constitutes the Volmer-Weber mechanism.

2. Formation of one or several atomic monolayers and then on this system formation and growth of three-dimensional grains. This constitutes the Stranski-Krastanov mechanism.

The grain-substrate interface or grain-monolayer evolves throughout the evaporation; it is connected with the growth and eventual displacement of the grain.

The second method of preparation is less classical and has been developed by Kimoto et al [*2,3*]. When a metal is heated in an inert gas atmosphere, the metallic vapor is cooled by the gas and a smoke containing small particles is formed. The particles are then collected by placing a substrate above the smoke. With many metals these latter are single crystal, less than a micron in size. In this case the particle-substrate interface is formed at once upon contact.

The only means of measuring the adhesion of the grains which constitute the very thin metallic films, and which have a diameter of the order of a few nanometers, is to use a dynamic method. A force opposite to the adhesive force which maintains the grain on the substrate is due to the inertia of each grain. To obtain this force of inertia the continuous-acceleration method can be used; for example, an ultracentrifuge technique [*4,5*], or sinusoidal acceleration with ultrasonic vibrations [*6*].

Grain Adhesion

Each grain in a granular thin film may be considered as undergoing, in every case, two types of interactions (Fig. 1):

1. A force, $F_{\perp}$, normal to the supporting plane is the effective adhesion between the cluster and the substrate.

2. Other forces, $F_{\parallel}$, parallel to the substrate originate from the surrounding grains and appear during the coalescence itself.

When we attempt to measure experimentally the adhesion of a grain by applying an external force F_E opposite to $F_{\perp}$, two major possibilities exist: The grain is ejected from the substrate, if $F_E \geq F_{\perp}$ max; in this case F_E is a true measure of the adhesive force. Or, if $F_E \leq F_{\perp}$, we can induce the migration of the grain; in this instance, F_E acts to reduce the adhesion to a sufficient extent that the parallel forces $F_{\parallel}$ can come into play.

Since in all our experiments we find that the ejection of the grains is

[2] The italic numbers in brackets refer to the list of references appended to this paper.

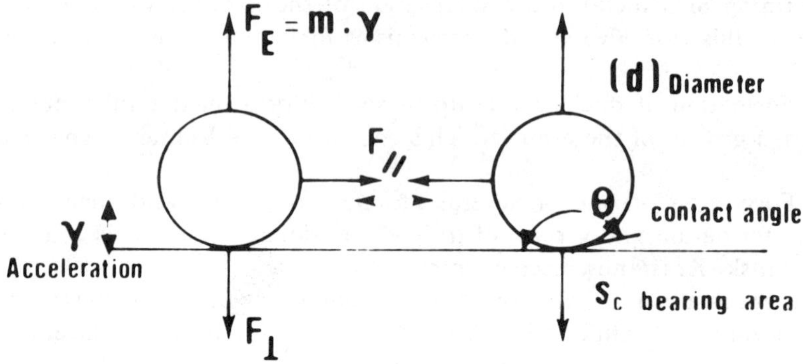

FIG. 1—*Grains and the two types of interaction:* $F_{\perp}$ = *true adhesion;* $F_{||}$ = *attractive or repulsive forces intervening in the coalescence.*

always preceded by dynamic-coalescence phenomena, we have chosen to call "adhesive force" the force per unit area which causes the disappearence of a grain, whether by actual ejection or by coalescence.

Therefore, due to the forces between grains, the mechanical equilibrium of a thin granular film is metastable, and determination of the adhesive forces necessarily disturbs this equilibrium. Desrousseaux [*7,8*] has shown that forces of electromagnetic origin could play an important role in the adhesion and coalescence of grains in a film. With intergrain forces he finds that the sign (attractive or repulsive) and amplitude depend on numerous factors related to the materials present and the film morphology.

The first difficulty encountered in determining the adhesion of thin film grains is therefore intrinsic to the physical system studied. The second difficulty concerns the measurement of microscopic physical quantities. Indeed, to determine the adhesive force of a grain when an acceleration γ is applied to the system, the volume V as well as the area of contact with the substrate S_c of this grain must be known. This gives

$$F_E = m \times \gamma$$

$$\text{adhesive force} = \frac{F_E}{S_c} = \left(\frac{V}{S_c}\right) \times \bar{\omega} \times \gamma \qquad \left|\frac{\text{newton}}{\text{m}^2}\right|$$

$$\bar{\omega} = \text{density (mass volume)}$$

The dimensions of the larger grains (those with a diameter greater than 100 Å) can be measured by observing them with either an electron microscope, a scanning microscope, or even by using shadowing techniques; then the volume and bearing area of the grains can be calculated. With small grains, on the other hand, only the apparent diameter (*d*), visible with an electron microscope, can be measured, and to determine the ratio (V/S_c) we were led to make the following two hypotheses:

1. There is no connection between the shape of the grains and their dimensions.

2. The contact angle defined by Young's equation is independent of the dimension of the grains.

By means of these two hypotheses, where only the diameter (*d*) of a grain is known, the ratio (V/S_c) can be calculated; then, knowing the acceleration γ which induces its disappearance, the value of the adhesive force can be computed. Finally, a measurement of the adhesive force of a grain can be significant only if it is effected with a very clean substrate isolated from every external disturbance. Therefore the substrate and thin film must be prepared in a vacuum, in an ultra-high vacuum in certain cases, and acceleration must be applied in this same vacuum. In our experiment, therefore, we prepared two films simultaneously. One of them acted as a check sample or reference film and did not undergo any constraint, and the other one, the test film, was subjected to ultrasonic vibrations. Their structures were then examined with an electron microscope.

Experimentation

Apparatus

In the films that we studied the mass of the grains was very small and it was necessary to submit them to very strong acceleration (of the order of 10^6 to 10^7 m/s^2) to get a sufficient force of inertia to eject them. The two systems for attaining such accelerations and measuring them easily are the ultracentrifuge technique and the ultrasonic resonators.

We used an ultrasonic resonator system for mechanical simplicity and ease of use in an ultrahigh vacuum [all its elements can be baked up to 200°C (392°F)].

To measure the adhesion of organic coatings, Moses and Witt [6] in 1949 used a setup composed of an electromagnetic transducer producing longitudinal mechanical vibrations at a frequency of 25 kHz in a cylindrical bar of aluminium alloy. The acceleration obtained was 10^5 m/s^2.

At present, electromagnetic transducers have been replaced by transducers using piezoelectric ceramics. Studies in physical acoustics have given rise to resonators of variable cross sections from one end to the other, thus amplifying the displacement produced by these piezoelectric ceramics. The theory and design of these resonators (they are known as "amplitude transformers") are reminiscent of those of horns used in classical acoustics. The amplitude factor and figure of merit (which measures the ability of a resonator to withstand high amplitudes) are both dependent on the shape of the resonator. The "exponential" transformer was studied by Masson [*9*], who obtained 10^6 m/s^2, and various types, particularly the "Fourier" one, were studied by Eisner [*10*], who obtained 10^7 m/s^2.

The acceleration value is limited to the amplitude of the motion at the free end of the resonator and to the frequency. Thus we tried to achieve greater acceleration and decrease the volume of the resonator by increasing the working frequency up to 100 kHz [*11*]. With the devices we have used, γ was worth 2×10^7 m/s^2 on a working surface of 10 mm^2. The characteristics of the different transformers used in our experiments are summarized

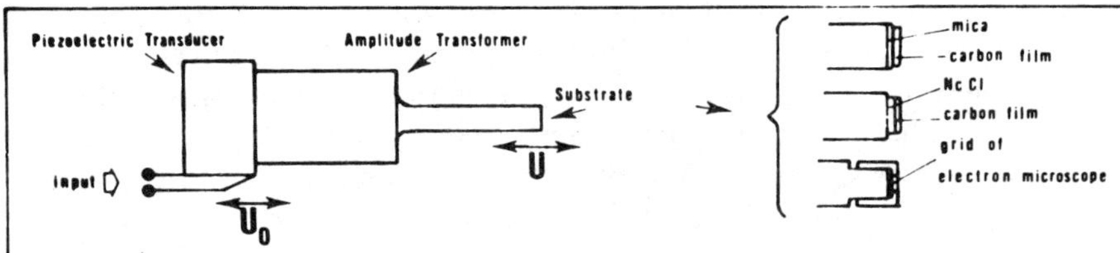

Type (Material: TITANIUM)	F KHZ	$M = \frac{U}{U_0}$	Φ*	U µm	γ m/s²	Working area mm²
STEPPED 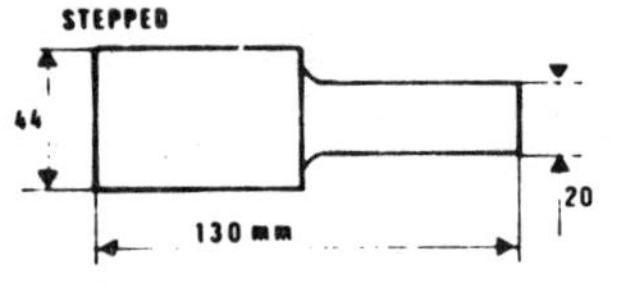	20	5	0.8	56	$8\ 10^5$	314
STEPPED 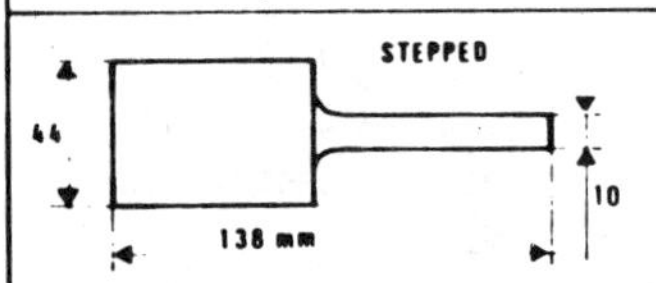	20	20	0.8	170	$2.6\ 10^6$	78
DOUBLE STEPPED 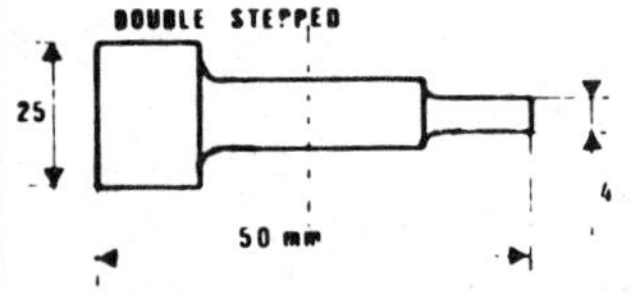	110	50	0.8 0.8	6	$2.9\ 10^6$	12
STEPPED + FOURIER 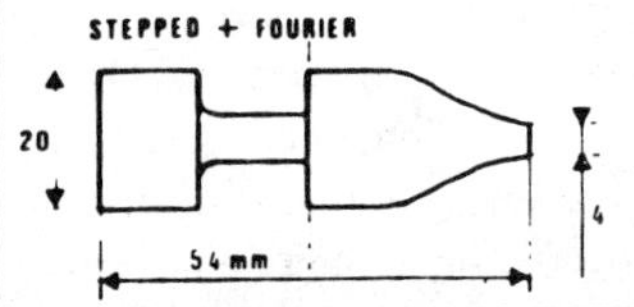	100	160	0.8 1.6	35	$14\ 10^6$	12

* Figure of merit $\Phi = \frac{\text{maximum particle velocity}}{\text{(maximum strain) X (speed of soud)}}$, $\Phi = 1$ for a uniform bar

FIG. 2—*Schematic diagram and characteristics of amplitude transformers used.*

in Fig. 2. Figure 3 shows a device we made, working at a frequency of 100 kHz and also an apparatus available commercially[3] which is normally used in biology to produce cavitation in liquids. The working frequency of this apparatus is 20 kHz. The substrate on which the film is condensed is fixed to the free end of the amplitude transformer; its mass must be small as possible and should allow observation of the thin film with an electron microscope. We used cleaved mica or cleaved sodium chlorine coated with a film of carbon, and the support grids of an electron microscope also coated with a carbon film. The cleaved substrates (mica, sodium chlorine) are fixed with Cyanoacrylate, and the support grids are fixed mechanically.

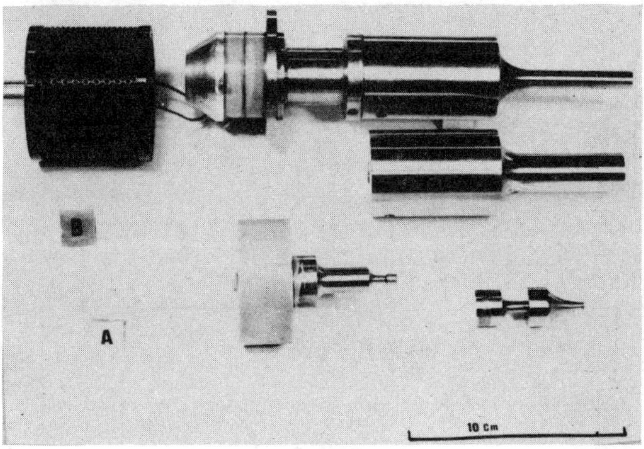

FIG. 3—*Measurement sets:* (a) *constructed by authors running on 100 kHz;* (b) *commercial apparatus used in biology running on 20 kHz.*

The whole measuring apparatus and the surface control device are shown in Fig. 4. For this device, we used the method of the vibrating capacitor of Kelvin-Zisman. This method can be set up very simply, and it enables us to observe the variation in the work function of a given surface. By measuring the work function of the substrate, uncoated as yet, and then after very small successive evaporations of metal, we can know the coverage rate of the metal on the substrate. Information about the structure and surface state of thin films can be obtained very rapidly *in situ* with this device.

[3] Branson Sonic Power Company, Eagle Road, Danbury, Conn. 06810.

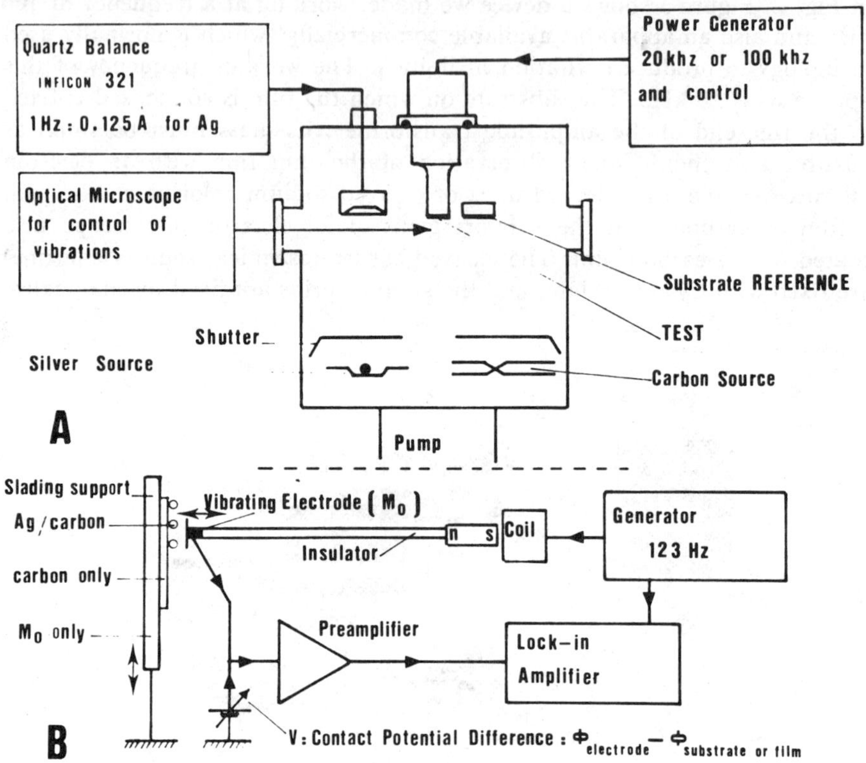

FIG. 4—*Apparatus under vacuum* (a) *and surface control device* (b).

Study of Silver on Carbon

Results

Very strong accelerations have been used on granular, lacunar, and continuous films of silver deposited on carbon films [*11*]. Checking reference films and test ones, after numerous observations, we came to the following conclusions:

1. The coalescence phenomenon begins to appear at a relatively low acceleration (approximately 6.0×10^5 m/s^2), Fig. 5.

2. A change in the morphology of the grains: This phenomenon is particularly visible with the very large grains which constitute the lacunar films, Fig. 6. Indeed these grains are shaped like flat islands before vibration. Subsequently the stresses to which they are submitted modify their form, in the first instance, and they assume a spherical shape which corresponds to better stability.

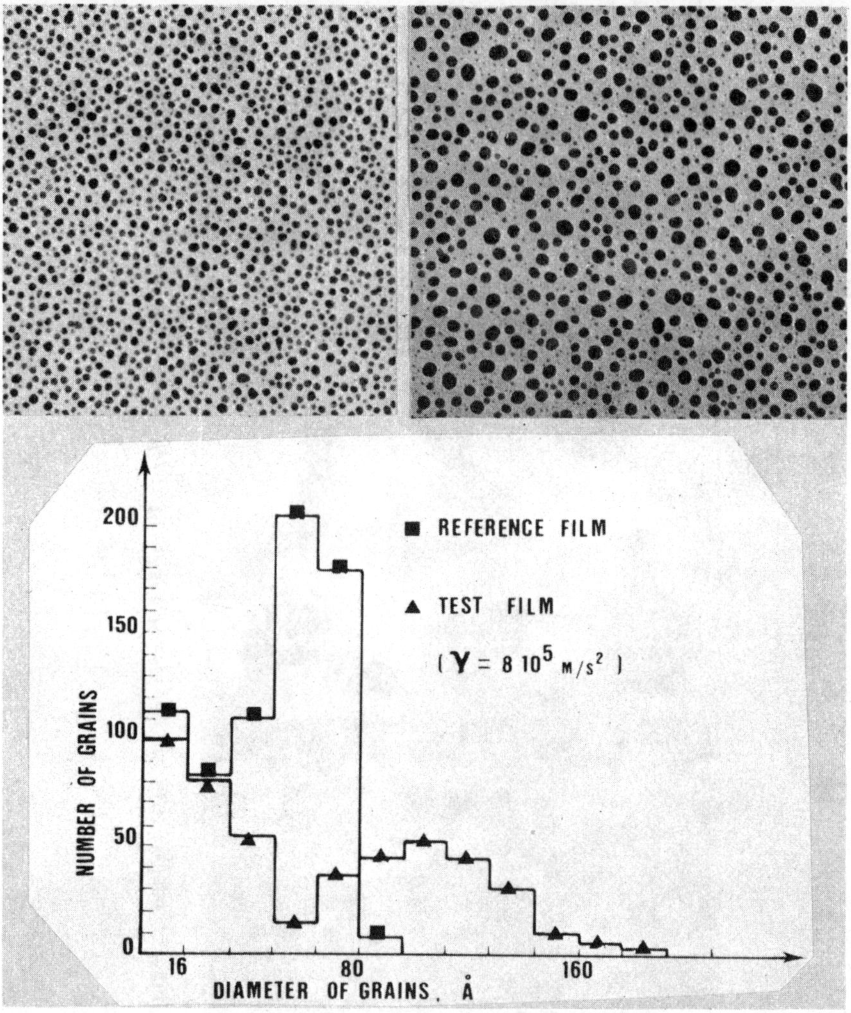

FIG. 5—*Coalescence of a granular film* (×*120 000*): (top left) *reference film, bulk thickness* = *25* Å; (top right) *test film,* $\gamma = 8.0 \times 10^5$ m/s^2; ■ *reference histogram*; ▲ *test histogram.*

3. By comparing the various histograms from the microphotographs, we were able to determine the adhesive force, depending on the various diameters of the grains, Fig. 7. Our calculations were based on the following assumption: A grain of diameter (d) is ejected under the influence of acceleration γ, when the value of its class (d) in the test histogram has been modified more than 50 percent by comparison with the value of the same class in the reference histogram.

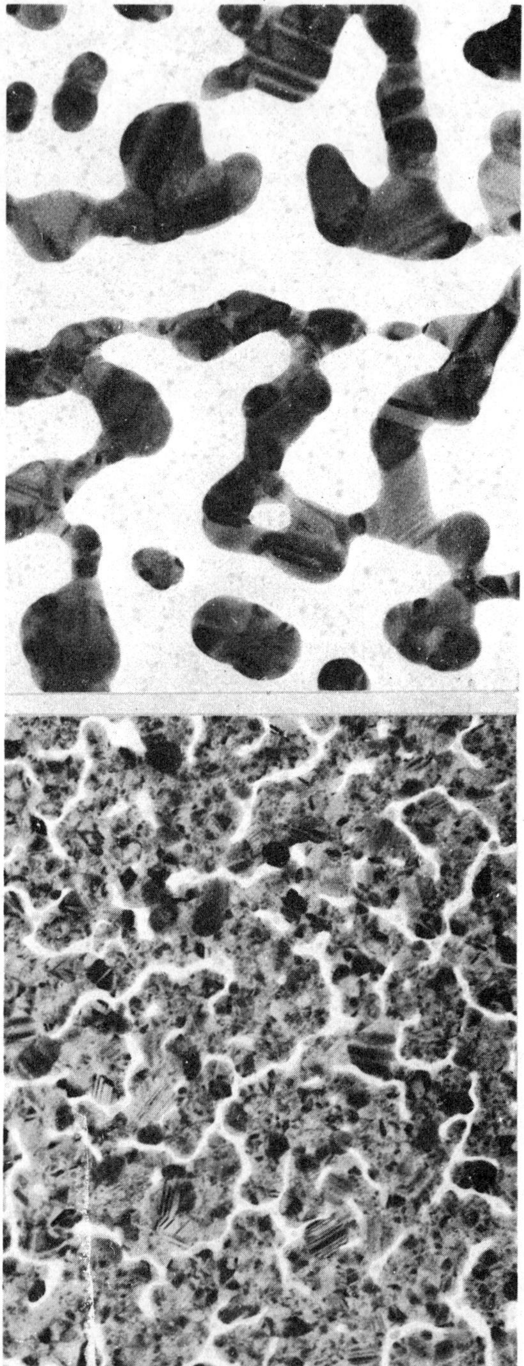

FIG. 6—*Effect of vibration on a lacunar film* ($\times$ *160 000*)*:* (left) *reference film, bulk thickness* = *130 Å;* (right) *test film,* $\gamma = 2.0 \times 10^6$ m/s^2.

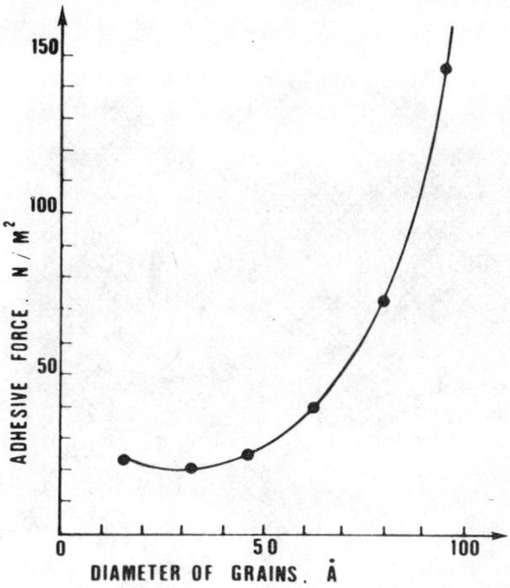

FIG. 7—*Adhesive force as function of diameter of grain. This curve has been determined by comparing histograms of microphotographs obtained for different accelerations.*

Because of the coalescence phenomenon, it is obvious that a grain may leave a certain class and join the grains of a superior class. We notice the following:

1. There is a minimum for a diameter close to 35 Å.
2. The very small grains ($d < 25$ Å) are very adhesive.
3. The adhesive force increases strongly from a diameter of 50 Å onward.
4. The values of adhesive force encountered are very low compared with the known values for continuous films (100 N/m^2 for a grain and 5.0×10^6 N/m^2 for a film 0.4 μm thick). In this connection it should be noted that the definition of "adhesive force" is not the same in both cases.

This set of results with granular and lacunar films prompted us to carry out measurements with thick continuous films (0.2 to 0.5 μm) [*12*]. The adhesive force of the films is so great that the force of inertia produced by the vibrations and bulk of the film alone is not enough to cause ejection. Therefore this force of inertia was increased by sticking small metallic overloads on the silver film. Thus a mean value of 5.6×10^6 N/m^2 was obtained for these films. The value found is rather easily reproducible and does not vary with the thickness of the film within the range of thickness used.

After the tearing of the silver film we observed the carbon substrate with an electron microscope. Silver grains were present at some very rare points, Fig. 8. The presence of these grains led us to suppose that there was a very thin silver film on the carbon substrate, invisible with an electron microscope. Electron spectroscopy chemical analysis (ESCA) measurements as

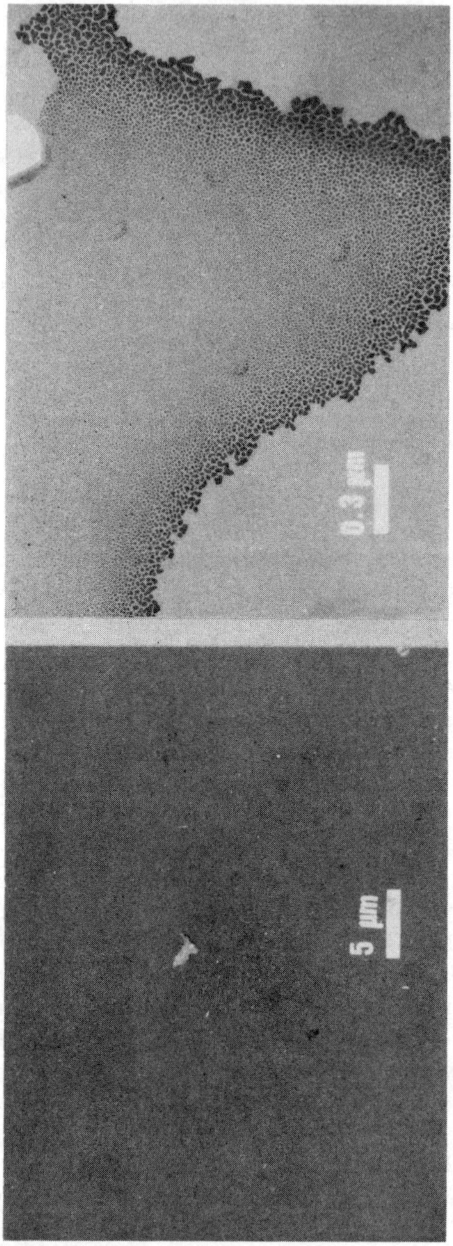

FIG. 8—*Carbon substrate after tearing of silver film: (left) general appearance (right) particular case where grains can be distinguished.*

well as zinc evaporations confirmed our assumption of this very thin silver film (estimated at 10 or 20 Å). The fracture therefore occurs in the silver film in a zone of least mechanical resistance.

Discussion

With a view to precise definition of the grain-substrate interface and consequently the thick film substrate, we have studied the growth of silver on carbon by measuring, in terms of bulk thickness of the silver film, the work function variations of the film-substrate system [*13*]. In Fig. 9 it is shown that with a bulk thickness of 3 Å the value of the work function of bulk silver is already obtained. These measurements, as well as electron microscope observations associated with gravimetric measurements using a quartz balance, have enabled us to show that the grains form on one or several monolayers not perceptible with an electron microscope. These latter monolayers are very adhesive, and the adhesive force is measured at the silver grain/silver monolayer interface. When the grains grow larger

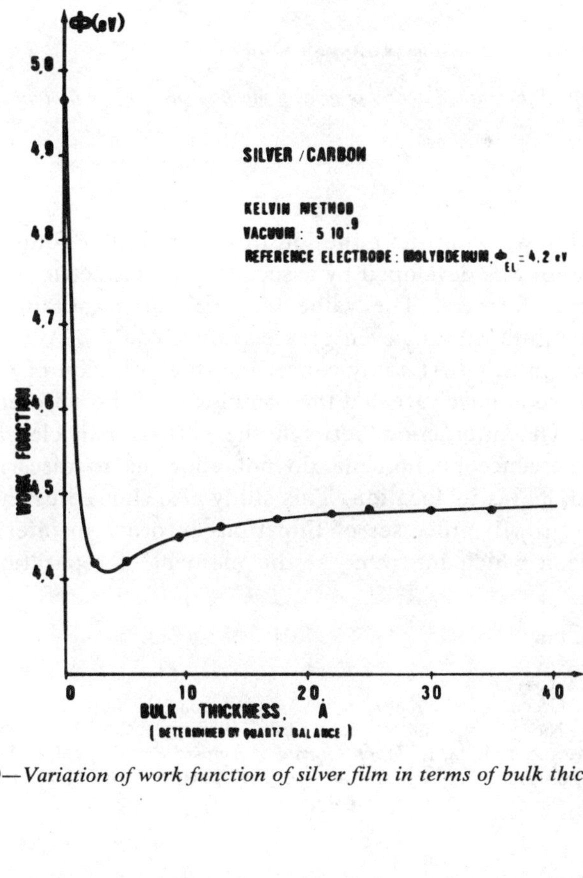

FIG. 9—*Variation of work function of silver film in terms of bulk thickness.*

and merge into a continuous film, a zone of least mechanical resistance is formed where the structure varies between that of the monolayer and that of bulk metal. The fracture occurs in this zone just above the monolayer, Fig. 10.

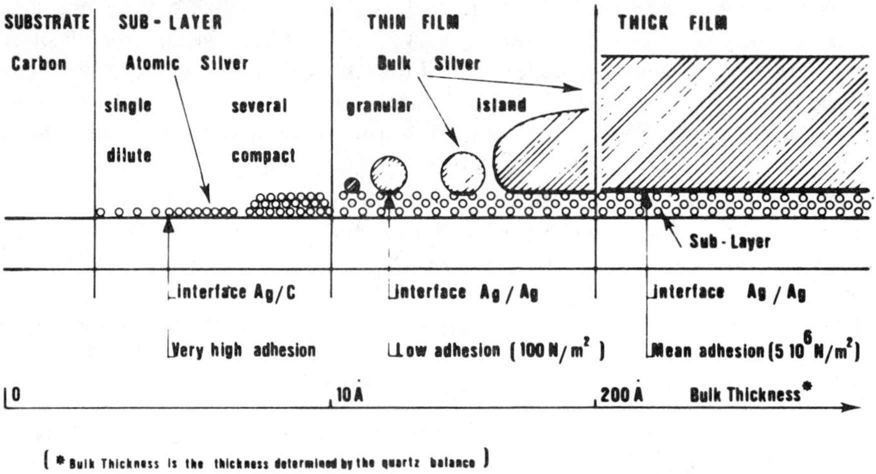

FIG. 10—*Schematic diagram of growth mechanism of silver film on carbon.*

Conclusion

A system for studying the adhesion of thin granular films easily carried out in a vacuum was developed by associating piezoelectric transducers and amplitude transformers. The value of acceleration obtained (10^7 m/s^2) does not constitute a limit; even greater values could be obtained. The results obtained in this first study concerning the adhesion of small particles of silver on carbon have revealed the complexity of the different intervening phenomena. The interaction between the various particles and resulting dynamic coalescence phenomena do not allow us to discern the way in which a grain leaves its location. This study also showed us the necessity of knowing the growth processes of thin films in order to interpret correctly the phenomena which intervene at the moment of separation of the two media.

References

[1] Bauer, E., Green, A. K., Kunz, K. M., and Poppa, H., *Basic Problems in Thin Film Physics*, R. Niedermayer and H. Mayer, Eds., 1965, pp. 135–152.

[2] Kimoto, K. and Nishida, I., *Japan Journal of Applied Physics*, Vol. 6, 1967, p. 1047.

[3] Kasukake, S. and Uyeda, R., *Journal of Crystal Growth*, Vols. 24/25, 1974, p. 315.

[4] Polke, R., *Colloque,* Adhesion et Physico-Chimie des Surfaces Solides, Mulhouse, 8–10 Oct. 1969, pp. 51-54.
[5] Krupp, H., *Advances in Colloid and Interface Science,* Vol. 1, 1967, p. 111.
[6] Moses, S. T. and Witt, R. K., *Industrial and Engineering Chemistry,* Vol. 41, 1949, p. 2334.
[7] Desrousseaux, G., *Thin Solid Films,* Vol. 22, 1974, pp. 317-321.
[8] Desrousseaux, G., *Thin Solid Films,* Vol. 32, 1976, pp. 255–258.
[9] Masson, W. P., *Physical Acoustical and the Properties of Solids,* D. Van Nostrand, New York, 1958.
[10] Eisner, E., *Journal of the Acoustical Society of America,* Vol. 35, 1963, p. 1367.
[11] Faure, R. et al, *Thin Solid Films,* Vol. 9, 1972, pp. 329-339.
[12] Robrieux, B., Fauer, R., and Desrousseaux, G., *Thin Solid Films,* Vol. 31, 1976, pp. 311-319.
[13] Faure, R., Desrousseaux, G., and Trompette, J., C. R. Academy of Science, Series B, t. 282, Paris, 24 May 1976, p. 491.

DISCUSSION

A. J. Berejka[1] *(written discussion)*—Could your analysis and methods for studying the adhesion of granular thin films be applied to or used in studying the adhesion of metal oxides to their respective metals? In some instances the oxide can be considered a granular thin film.

Roland Faure (author's closure)—It is possible to study the adhesion of metal oxides to their respective metals, with the following conditions:

1. The metal oxide is formed with particles or grains.
2. The grains are sufficiently large enough for visibility with a scanning microscope or an optical microscope.
3. The adhesion is low, less than 500 N/m^2.

If the grains are ejected from the substrate, it is necessary to control the nature of the new surface (ESCA, work function, etc.), in order to determine if the fracture is in the metal, in the oxide, or in the metal-oxide interface.

[1] Raychem Corporation, Menlo Park, Calif. 94205.

Kaizo Kuwahara,[1] *Hidenori Hirota,*[1] *and Nobuo Umemoto*[1]

Adhesion Measurement on Thin Evaporated Films

REFERENCE: Kuwahara, Kaizo, Hirota, Hidenori, and Umemoto, Nobuo, **"Adhesion Measurement on Thin Evaporated Films,"** *Adhesion Measurement of Thin Films, Thick Films, and Bulk Coatings, ASTM STP 640,* K. L. Mittal, Ed., American Society for Testing and Materials, 1978, pp. 198–207.

ABSTRACT: Methods to evaluate the adhesion of thin evaporated films to their substrate were devised. One was to pull down and the other to twist off a rod whose bottom was cemented with epoxy to the film. The critical force or torque to strip off the film from the substrate was taken as a measure of the adhesion. It was important to maintain adequate temperature and humidity [16°C (61°F) and below 60 percent] during the cementing and measurement, to assure the high adhesion of the cement and the reproducibility of the result.

The adhesion of thin films of aluminum, silver, and copper to substrates of mild steel and glass was studied, along with the adhesion of epoxy and other cements. It was found that the critical force and torque were proportional not to the 3rd power as expected but rather to the 2.5 ~ 2.9th power of the rod diameter. In the case of thin films, the dispersion of the measured values was very large. The dispersion and also the ratio of tensile to shear adhesive strength increased rapidly with the decrease in adhesion. This is contrary to the effects of the small dispersion, the independence of the dispersion, and the ratio on the adhesion in the case of cement itself. Also mentioned is an experiment on the fracture of brittle substrates such as silicon wafers which sometimes happened during the measurement.

KEY WORDS: adhesion, vacuum evaporation, thin films, epoxy

The adhesion of thin evaporated films to their substrates is one of the important problems in physics as well as in surface coating techniques. Many methods [*1*][2] have been proposed to determine the adhesion qualitatively or quantitatively; for instance, peeling [*2*], shearing off [*3*], pulling off [*4,5*], centrifugal [*6*], and scratching [*7,8*]. Each method has its own advantages accompanied by inherent limits of application.

One of the authors and others [*4*] devised a method to measure the adhesion: to pull down a rod which is bonded to the film surface with epoxy cement. In this paper, some experiemental results obtained by using this

[1]Professor, assistant, and graduate student, respectively, Faculty of Engineering, Hiroshima University, Hiroshima, Japan.

[2]The italic numbers in brackets refer to the list of references appended to this paper.

method are described along with the results obtained by use of a new method to twist a rod bonded to the film surface. The subject is restricted to the examination of the nature of the adhesion and the characteristics of the methods.

Methods

Two types of adhesion measurement are used. One is the pulling-down method reported previously [4]. Figure 1 illustrates the concept of the method. A cylindrical rod or column is bonded at its bottom with epoxy cement to the surface of the film evaporated onto the substrate. A gradually increasing force, *F*, is applied to the top of the rod at right angles to the rod axis. The critical force when the rod is pulled down is taken as a measure of the adhesion.

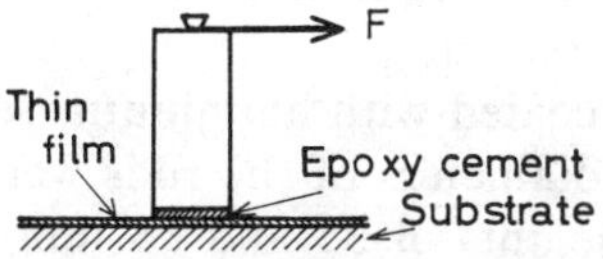

FIG. 1—*Pulling-down method for adhesion measurement.*

Another type of measurement devised is similar to the former type but the working force to strip off the film from the substrate is definitely different. As illustrated in Fig. 2, a gradually increasing torque, *L* is applied to the rod cemented to the film surface. The critical torque when the rod is twisted off from the substrate is taken as a measure of the adhesion. The torque is supplied through a box wrench fixed to the rod before the measurement.

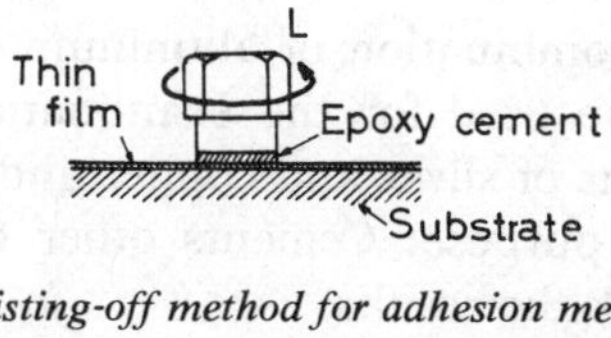

FIG. 2—*Twisting-off method for adhesion measurement.*

Both methods have an inherent restriction such that the adhesion to be measured should be lower than the adhesion of epoxy cement to the film and the rod, or to the cohesion of epoxy cement itself. According to our experimental survey, one of the necessary conditions for the improvement of the adhesion characteristics of the epoxy cement is to keep the atmosphere at an adequate temperature and humidity during the process of cementing and measurement. The results following were those obtained under the conditions of 16 °C (61 °F) temperature and humidity below 60 percent. Such control of the atmosphere also effectively contributes to the reproducibility of epoxy cement adhesion as well as to the suppression of the dispersion of measured values.

Experimental Procedures

To study the nature of the aforementioned methods of pulling down and twisting off, the following procedures were employed. Substrates of mild steel, 10 by 50 by 50 mm^3, were finished to a mirrorlike surface by polishing with emery paper followed by buffing. The substrates were cleaned by an ultrasonic bath of acetone and pure water. Aluminum films about 1000 Å thick were evaporated onto the room temperature substrates at pressures around 2×10^{-5} torr. Four substrates were coated simultaneously by a single run of the evaporation.

On the substrate thus coated with aluminum, several rods were bonded with epoxy cement. The diameters of the rods were 6, 8, 10, and 12 mm. Two sorts of rods with heights of 35 and 15 mm were prepared for each diameter. Each taller rod had a pin on its top surface for pulling; each shorter rod had a box head to fix a box wrench for twisting. All the rods, one kind for each diameter, were arranged on the substrate. Another substrate without coating was prepared for the purpose of measuring the adhesion of epoxy cement itself, on which the rods were arranged in the same way as on the coated substrates.

The rods were pulled or twisted one by one, in predetermined order, by a motor-driven force or torque through a chain of transmission members. The force or torque was detected and recorded by strain gages mounted on one of the members. When the adhesion happened to break (pulled down or twisted off), the stylus of the recorder fell suddenly. The peak value was taken as a measure of the adhesion.

The aforementioned combination of aluminum films on mild steel substrates is a standard one used for the examination of the nature of the measurement itself. Films of silver and copper and substrates of glass were also used for the same purpose. Cements other than epoxy and sticking pastes were used to check the results.

Results and Discussion

Relationship Between Adhesion and Rod Diameter

The force, F, to pull down the rod bonded with epoxy cement to the nude surface of the mild steel substrate is shown in Fig. 3 as a function of the rod diameter, D. From the slope of the curve it is deduced that

$$F \propto D^n, \qquad n \sim 2.8 \tag{1}$$

The torque, L, to twist off the rod cemented to the nude surface of the mild steel substrate is shown in Fig. 4 as a function of the diameter of the rod. It is deduced in this case also that

$$L \propto D^n, \qquad n \sim 2.9 \tag{2}$$

The force to pull off the aluminum film from the mild steel substrate is plotted in Fig. 5 as a function of the rod diameter. In this case the measured force is appreciably low and scatters very much compared with the result shown in Fig. 3. The overall trend, however, is the same

$$F \propto D^n, \qquad n \sim 2.6 \tag{3}$$

It is noted that these data are obtained from the total of several runs of evaporation. The sample standard deviation of a single run is about one half of these results.

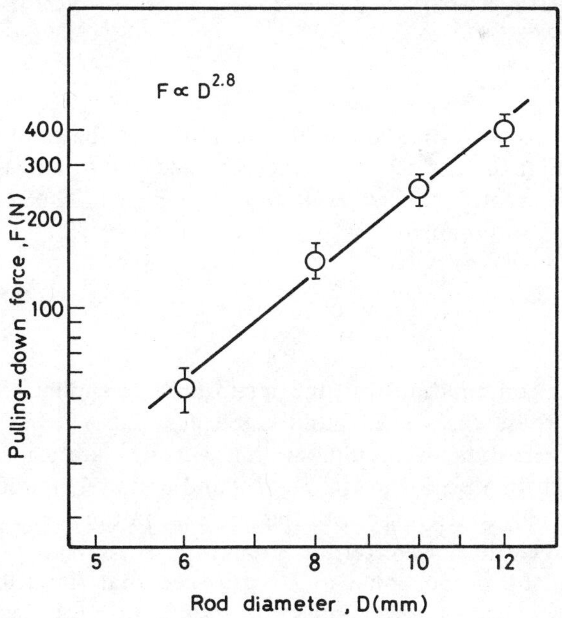

FIG. 3—*Force to pull down a rod epoxied to nude surface of mild steel substrates as a function of rod diameter.*

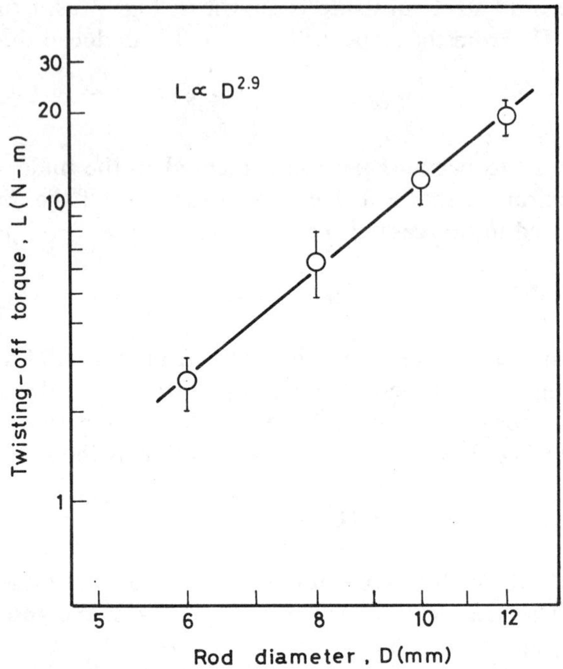

FIG. 4—*Torque to twist off a rod epoxied to nude surface of mild steel substrates as a function of rod diameter.*

The torque to twist off the aluminum film from the mild steel substrate is shown in Fig. 6. It is seen that the obtained values and their scattering are not so different from the result shown in Fig. 4. The relation between the torque and rod diameter is

$$L \propto D^n, \qquad n \sim 2.8 \tag{4}$$

From other combinations of film to substrate, a similar relationship was deduced. In these cases n is found to be near 2.5 ~ 2.7. Considering the scattering of the data, we could say that F and L are approximately proportional to D^n in all cases; n is 2.5 ~ 2.9 and definitely lower than 3. There is a tendency of n to decrease with the decrease in the adhesion.

A brief calculation based on the strength of materials shows that F and also L should be proportional to D^3 provided that the adhesive strength (tensile or shear) is uniform under the rod. It does not seem possible to determine the adhesion in the dimension of force per unit area from this measurement, because the calculation inevitably includes the relation of D^3.

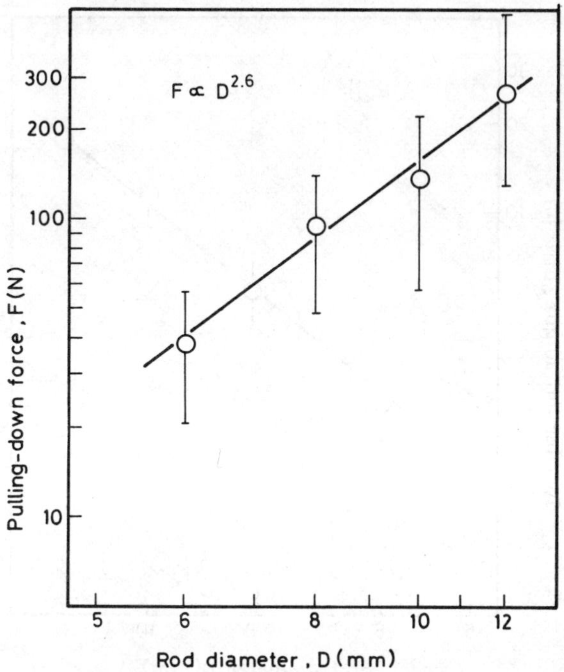

FIG. 5—*Force to pull down a rod epoxied to aluminum film (1000 Å) evaporated onto mild steel substrates as a function of rod diameter. The aluminum film is stripped off by the rod from the substrate.*

The deviation of n from 3 might be caused by local fluctuations of adhesion or the nature of the adhesion itself, or both.

Relationship Between Dispersion and Adhesion

Looking at the scattering or dispersion of the values obtained so far in various measurements, we can recognize that there is a rather definite tendency of the dispersion to increase with the decrease in the adhesion value measured. Clear evidence of this is shown in Fig. 7, where the relative dispersion (sample standard deviation divided by the mean value) is plotted against the mean value for the force to pull down the rod of 10 mm diameter. In the same figure, results obtained from other cements and paste are also shown.

It is clearly seen that, in the case of thin film adhesion, the dispersion rapidly increases with the decrease in the adhesion. In the case of cement or paste, the dispersion is far lower than that of thin film and has little tendency to increase with the decrease in the adhesion. The same relation

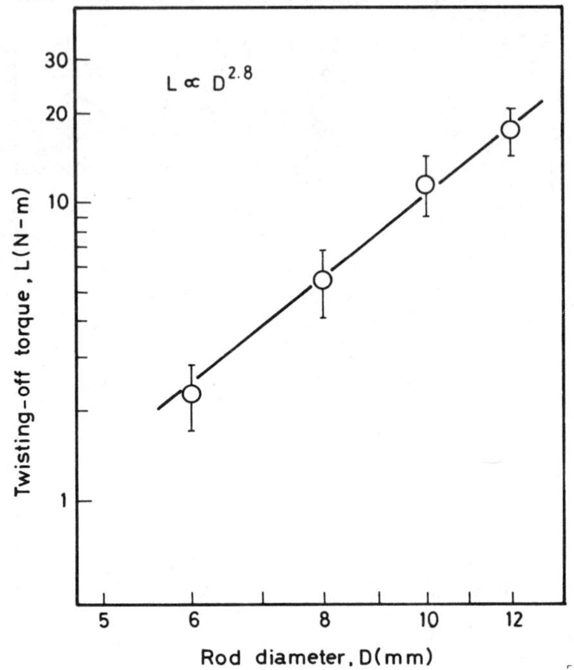

FIG. 6—*Torque to twist off a rod epoxied to aluminum film (1000 Å) evaporated onto mild steel substrates as a function of rod diameter. The aluminum film is stripped off by the rod from the substrate.*

was also obtained for the twisting-off torque. In the case of thin evaporated films, local contaminations of the substrate surface would cause local fluctuations of adhesion and this would be one origin of the dispersion. The reason why the dispersion sharply increases with the decrease of the adhesion is not easily understood. The weak adhesion is presumably very sensitive to the state of the substrate surface, especially of contamination and its local fluctuation.

Relationship Between Ratio of Tensile to Shear Strength and the Adhesion

The ratio of twisting-off torque (L) to pulling-down force (F) for 10-mm-diameter rod is plotted in arbitrary units as a function of L in Fig. 8. This ratio is proportional to the ratio of shear to tensile strength of the adhesion layer. As seen in the figure, the ratio for thin films increases sharply with the decrease in the adhesion, while the ratio for the usual cement and paste scarcely changes. This suggests again that the mechanism of adhesion is different between the film and the cement.

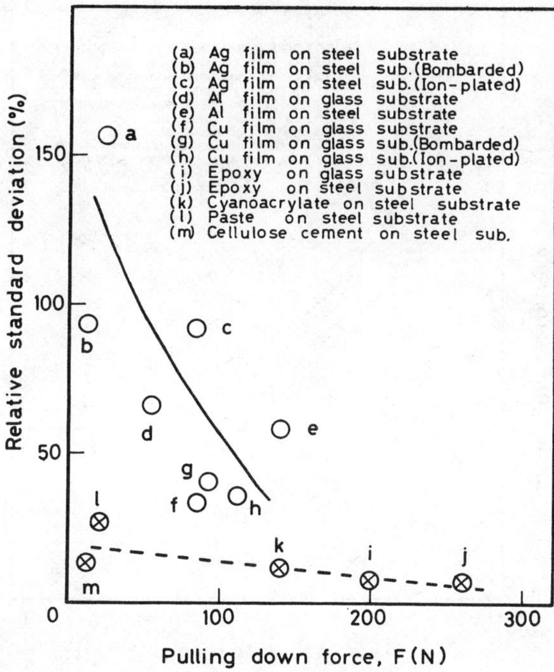

FIG. 7—*Relative dispersion (standard deviation divided by mean value) of force to pull down a rod of 10 mm diameter as a function of mean value.*

Fracture of Brittle Substrates During Measurement

When the measurement is made on brittle substrates such as glass or silicon wafers, the surface under the rod is sometimes broken and a part of the substrate is brought away by the rod at the instant of pulling down. It has been confirmed by supplemental experiments that this fracture is caused by minute cracks already present under the surface.

This phenomenon, in turn, offers a method of inspecting the imperfection of silicon wafers. Imperfect silicon wafers having minute cracks under the surface cannot serve as a good base for the formation of integrated circuits. To inspect the layer of cracks, a silicon wafter extracted from a lot is cemented to a steel substrate, and a rod is cemented on the silicon wafer. The rod is pulled down by applying a force in the same manner as used in the adhesion measurement. If the surface of the silicon wafer is broken, it definitely has some imperfections under the surface. It has been proved that this inspection method is a handy, effective, and satisfactory way of deciding whether enough chemical polishings have been performed to remove the layer of cracks.

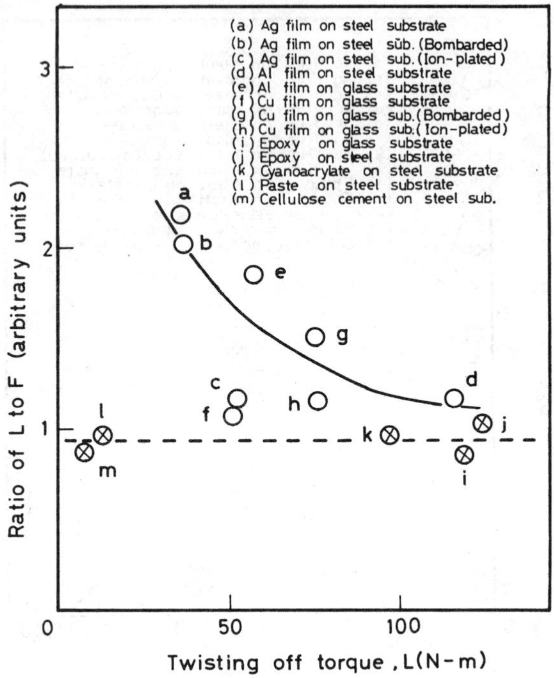

FIG. 8—*Ratio of twisting-off torque to pulling-down force as a function of mean value of twisting-off torque, for rod of 10 mm diameter.*

References

[*1*] Strong, J., *Review of Scientific Instruments*, Vol. 6, 1945, p. 97.
[*2*] Lin, D. S., *Journal of Physics D: Applied Physics*, Vol. 4, 1971, p. 1977.
[*3*] Mittal, K. L., *Electrocomponent Science and Technology*, Vol. 3, 1976, p. 21.
[*4*] Kuwahara, K., Nakahara, S., and Nakagawa, T., Supplement to *Transactions*, Japan Institute of Metals, Vol. 9, 1968, p. 1034.
[*5*] Butler, D. W., *Journal of Physics E*, Vol. 3, 1970, p. 979.
[*6*] Beams, T. W., *Technical Proceedings of the American Electroplaters Society*, Vol. 43, 1956, p. 211.
[*7*] Benjamin, P. and Weaver, C., *Proceedings of the Royal Society*, London, Vol. A254, 1960, p. 177.
[*8*] Greene, J. E., Woodhouse, J., and Pestes, M., *Review of Scientific Instruments*, Vol. 45, 1974, p. 747.

DISCUSSION

W. D. Bascom[1] (*written discussion*)—Since you have recognized that the failure of your metal film is controlled by flaw growth and propagation, may I recommend that you modify your test to one that measures fracture directly as we described in our earlier papers?

Kaizo Kuwahara et al (*author's closure*)—We agree with Dr. Bascom's suggestion that the failure of a metal film could be related directly to fracture stress. In order to substantiate the failure mode of a metal film, we are currently analyzing the morphological aspects of such fracture surfaces using a scanning electron microscope. Based on these results, we intend to modify our test apparatus to obtain more essential data that control the failure mode of a metal film.

[1]Adhesion Section, Naval Research Laboratory, Washington, D.C. 20375.

Adhesion Measurement of Thick Films

T. T. Hitch[1]

Adhesion Measurements on Thick-Film Conductors*

REFERENCE: Hitch, T. T., **"Adhesion Measurements on Thick-Film Conductors,"** ***Adhesion Measurement of Thin Films, Thick Films, and Bulk Coatings, ASTM STP 640,*** **K. L. Mittal, Ed., American Society for Testing and Materials, 1978, pp. 211–232.**

ABSTRACT: The paper briefly reviews the measurement of adhesion of conductor film materials and describes in particular the requirements of thick-film technology. The test methods used to measure thick-film adhesion strength, which have been previously described by other workers, are treated from a fundamental basis and are discussed with reference to the usefulness and practicability of the test both for research and for more routine purposes.

The second part of the paper reviews two thick-film adhesion tests and their use at RCA Laboratories over the past several years. The first test is the thermocompression bonded peel test, which has been used for the adhesion strength measurement of gold- and silver-based conductor films. This test was developed at RCA and has proved useful for films of a wide range of adhesion strengths.

The second test description reviews RCA Laboratories' use of the soldered-wire, peel test. It also treats our progress in eliminating subjectivity from this test by limiting the use of hand operations and by controlling time-temperature cycles required in the assembly of test specimens. For both these tests, more than one failure mode has been observed. The meanings ascribed both to the failure modes and to test data, which are taken when the modes occur, are treated.

Data illustrating the variation in adhesion strength with firing temperature and film thickness are shown for several ink types. These data are correlated with the composition of the inks by grouping the gold and silver inks into three bonding classifications—frit-bonded, reactively bonded, and mixed-bonded.

KEY WORDS: adhesion, adhesion tests, reviews, bond classification, adhesion test classification, thick films, gold conductors, silver conductors, peel tests

*The research reported in this paper was supported in part by the Department of the Navy, Naval Air Systems Command, under Contract Nos. N00019-74-C-0270 and N000A-75-C-0145.

[1]Member of the technical staff, David Sarnoff Research Center, RCA Laboratories, Princeton, New Jersey 08540.

The major film material classes in thick-film[2] technology are conductors, resistors, and dielectrics. Of these, conductors have been the greatest source of concern regarding adhesion strength. By surviving several process refirings, dielectric and resistor films demonstrate sufficient adhesion to perform most hybrid circuit functions, but the conductors must resist added stresses. They form the electrical attachment points for silicon and titanate chip device additions to the hybrid circuit and must endure any mechanical stresses imposed by the differential thermal expansions of the chips and the hybrid substrate. The conductor films are similarly stressed by packaging designs in which the electrical connections from the hybrid substrate to other electronic circuits and to the outside world are also the mechanical supports for the substrate.

The principal object of adhesion tests on thick-film conductors has been to predict whether the films will stand up in use. Therefore, many of the adhesion test configurations and methods of making attachment to the films for testing have been chosen to simulate the attachments of electrical/mechanical connections of a hybrid substrate with the outside world. Only recently have attempts been made to choose test methods to study more fundamental aspects of film adhesion strength.

An object in some adhesion testing has been to make individual conductor film materials seem to be more adherent than they are. Examples of test methods used to accomplish this are noted.

The amount of scatter of data about the mean strength that is expected for a test method is important because a high scatter severely limits the ability of a test to detect changes in adhesion strength resulting from variations in materials or processing. The amount of data scatter to be expected with several adhesion tests is discussed.

In this paper several aspects of adhesion testing are presented, followed by a review of some of the tests being used in research and industry to measure thick-film adhesion strength. A few other workers have reviewed thick-film adhesion testing [*1–3*].[3] In related fields, adhesion strength testing has received much more attention (see, for example, Refs *4–10*).

Two tests currently being used to measure thick-film adhesion strength at RCA are described in the last part of the paper. RCA has developed a classification scheme for gold and silver thick-film conductors based on the correlation of adhesion strength behavior and the chemical constituency of the film materials. To illustrate these classes and the two adhesion tests in use for thick films at RCA Laboratories, the variation of adhesion strength with film thickness and firing temperature is shown by reduced three-di-

[2] In this paper the term "thick film" will denote material deposited on a ceramic substrate by screening or stenciling, dried, and fired at a temperature between 600 and 1100 °C (1112 and 2012 °F).

[3] The italic numbers in brackets refer to the list of references appended to this paper.

mensional plots for selected ink materials, and chemical analyses are presented for these inks.

Types of Adhesion Tests

If an ideal adhesion test existed, it would (*a*) be useful over a wide range of substrate and film material types; (*b*) give quantitative, reproducible data over a wide range of adhesion strengths; (*c*) not be influenced by the variability of any materials other than the specimen film under test; (*d*) be implemented easily and not be sensitive to operator experience, dexterity, or individual technique; and (*e*) allow the measurement of adhesion strength distinct from other strength parameters such as cohesion strength [*11*]. Mittal also has stipulated criteria for adhesion test ideality [*10*,*12*].

In the real world, however, adhesion test results fall into two main groups, both of which are far from ideal. Qualitative or threshold tests are the first group and the least ideal. Such measurements are not designed to quantify adhesion strength except to assert that the adhesion is greater than some arbitrary (and often ill-defined) limiting strength. The Scotch tape test and the razor blade scratch test, as normally used, are threshold tests.

The second group of tests is designed for quantitative measurement of adhesion strength. Although such tests may not fulfill even one of the above criteria of ideality, their use continues because there is a mandatory requirement to quantify adhesion strength in order to detect harmful changes in the substrate, the film material, or in their processing. The value of such tests is that they indicate the robustness or safety factor of adhesion strength in a film-substrate laminate.

Often tests that are designed to be quantitative cause failure in some location other than the film-substrate interface. The failure load recorded in such a test may be taken to indicate an adhesion strength threshold. As Chapman has noted [*5*], it is the opinion of some workers that adhesion test methods are so poor that only "use tests" have value; for example, if the film-substrate assembly can survive fabrication into a useful article, this is taken to indicate that the article will bear the loads to be imposed in service. A use test is clearly most applicable when the maximum stresses anticipated on the film will be those incurred during the manufacture of articles, but this is not usually the case. Since use tests seldom provide a scale of quality, they indicate principally when a manufacturing process is grossly out of control.

Ideally, adhesion would be a fundamental material property, like surface free energy. Using tests similar to those used to measure surface energy by cleavage propagation [*13–15*], Mostovoy [*16*] and Freiman et al [*17*] have developed adhesion tests to measure the fracture energy of composite structures. The configurations of their test specimens have been adapted to thick-film adhesion measurement at the Naval Research Laboratory (NRL)

[*18*], and the use of these adaptations is described elsewhere in this publication [*19*]. An advantage of the techniques is that they measure fracture energy and thus allow correlation of the locus of crack propagation with the characteristic fracture energies of material classes. A disadvantage is that test specimen preparation is time-consuming.

Other Considerations in the Choice and Interpretation of an Adhesion Test

Stress Raisers—Discontinuities in the material or shape of a loaded structure can strongly raise the stress in their vicinity. Discontinuities of certain geometries and their effects have been treated in detail (for example, see 20–22). In adhesion testing, flaws in the substrate-film interface will lower the adhesion strength and cause scatter in the data about the mean. Flaws in the solder or other materials used to grip a film to test its adhesion to a substrate may also cause data scatter, but there it is an artifact of the test specimen assembly. Stress raisers are used to advantage in several adhesion test specimen designs. In those tests, stress concentration is used to force the locus of failure into the film-substrate interface.

Test Machine Selection—There are several concerns in choosing the method of loading adhesion test specimens and recording the failure data. One is the "beam hardness" of the test machine. This is simply an expression of the spring constant of the system comprising the test machine and load cell. A principal source of soft-beam response is the use of spring balance gages to measure the applied force in cheaper test equipment. A principal concern is the energy stored in the machine with a particular load applied to the specimen. In hard-beam machines with hydraulic or electrical resistance strain-gage load sensors, the load can be greatly relieved by a small strain in the specimen. With soft machines, the greater stored energy in the machines will contribute to the continued extension of cracks, which have begun to propagate. For a fracture propagation test like peeling, this may change the apparent load to maintain peeling.

Finally, machines with electronic load measurement offer the following advantages: A permanent record of the test is easily recorded; the full-scale load range can be changed during the test, if required; and more precise (reproducible) data are usually obtained.

Effect of Surface Roughness—Up to this point it has been implied that a smooth surface was under the film being subjected to adhesion testing. In fact, technical surfaces have significant roughness, and this has several ramifications. First, the increased surface area due to surface roughness, in comparison with the (smaller) projected area of a surface, will give increased adhesion because of (*a*) the additional area and (*b*) the fact that the adhesion test loading will not always be perpendicular to the surface being separated. These effects have been discussed in terms of fracture mechanics [*8*].

Another effect of surface roughness is that interfacial stresses in the film cannot so strongly assist the delamination forces applied by the adhesion test (*a*) because the substrate and film interfaces are keyed together by surface roughness and (*b*) because the nonplanarity of the interfacial stresses may have allowed greater stress relaxation by plastic flow than in the planar interface case.

Specific Tests of Interest

Uniaxial Tension Tests

Workers in many technologies have used tension testing to measure the adhesion strength of their films. Most of the reviewers treat tensile adhesion, but Jacobson and Kruse [*23*] gave detailed attention to analyzing the results of their test method on thin films and to the inherent meaning and reliability of their data.

Several groups have used tension tests to measure the adhesion of thick films [*24–27*]. The innovative techniques for the volume manufacture of tension adhesion test specimens developed by Leven [*24,25*] deserve particular mention. Each of the groups has been successful in showing the effects on the adhesion strength caused by their experiments. However, except for histograms from Leven [*24*] and some mention of failure modes changing, virtually no indication of the data scatter has been reported for these test methods.

Tension adhesion test procedures designed for central loading suffer scattering of their data to a greater or lesser extent, depending on the quality of fabrication of the test part, the compliance of the member attaching the test specimen to the test machine, the axiality of loading, and flaws inherent in the film-substrate laminations themselves. Because the load is distributed over the conductor pad area, until crack initiation causes it to concentrate, tension adhesion tests exhibit high fracture loads per unit area of the thick-film conductor pad.

Wire Bond Test

The wire bond test probably began as a use test. When the films are poorly adherent, thermocompression (TC) bonding of 1-mil (0.025-mm) wire can cause visible lifts in thick-film gold conductors, particularly if the bonds are at the edge of the film. By increasing the wire diameter, it becomes possible to tear out segments of the film (see Fig. 1). Although a large fraction of the peak load is believed to be due to overcoming cohesive forces and tearing the gold thick film, study of the appearance of fracture can suggest methods of eliminating some of this error [*28*]. Surprisingly, it was noted in limited testing at RCA that the peak bond load was not significantly changed by reversing the bond geometry from that shown in Fig. 1 to one in which the part of the wire to be pulled is placed at the edge of the bond pad. Scatter of data about the mean value is high for this test.

One other aspect impedes the estimation of adhesion strength from the test results in the results of this test. The applied forces and thermal shock caused at the substrate-film interface by the bonding operation set up stresses that act to shear the substrate-conductor film interface and form a flaw. Such delaminations will result in test failures at reduced loads. A high intrinsic interfacial adhesion strength (strong chemical bond), interfacial roughness, or the contribution of interpenetrating phase microstructures (such as glass fibers) in withstanding the shear stresses, will reduce the tendency for delamination.

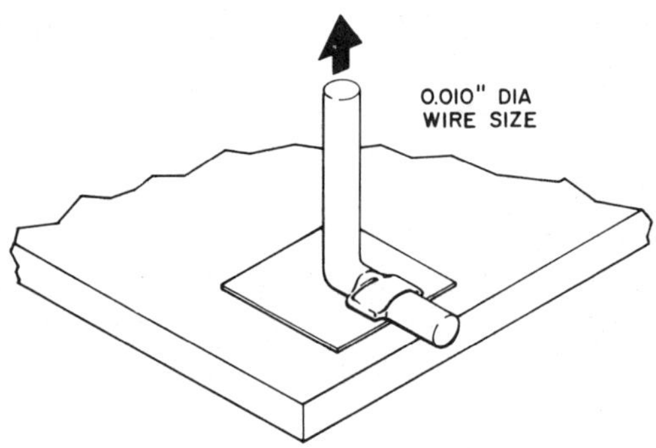

FIG. 1—*Wire bond adhesion test used by RCA in early work (oversize flying wire test).*

The wire bond test has the advantage of usually producing adhesion-type (substrate-metal film separation) failures. In RCA studies, such interfacial failures were produced in all the specimens tested in a number of gold and silver, frit-bonded, thick-film inks on 96 weight percent alumina substrates fired over a range of temperatures [*28*].

Parallel Gap Welding

Johnson and Knutson of Sandia Laboratories have developed an adhesion test based on the parallel gap welding of a 0.002 by 0.15-in. (0.051 by 3.8 mm) cross section gold ribbon to gold and to platinum-gold thick-film conductor films. Using the method, they have measured the effects of welding voltage, film thickness, film type, firing temperature, and furnace belt speed on the adhesion strength of four gold-platinum and two gold conductors [*29*]. It is significant that they found a marked increase in adhesion strength with thicker films, for example, 200-mesh screen double print versus single print films. However, it is not clear whether the differences in strength are due to the increased adhesion of the thicker films or to the ability of the thicker film to dissipate the bonding strains by plastic flow within the thickness of the film and thus reduce the shear stress level at the interface. Cohesive stresses partially obscure adhesion strength information in this test, as they do in the wire bond test. Despite these limitations, the reproducibility and speed of the test make it a likely candidate for continued use.

Peel Tests

One of the most widely used peel tests is the Scotch tape test shown in Fig. 2. Using a strip from a fresh roll of cellophane-backed tape, one presses a strip firmly onto the film to be tested. Briskly lifting the tape applies a force of 6 ± 0.5 lb/in. (107 ± 9 g/mm) of width [*1*]. Any removal of the film by the tape (as shown in Fig. 2) indicates that the combined adhesion and cohesion strengths of the film were too weak to hold the film intact. By varying the angle of pull, the test can be made quantitative [*5*]. The threshold level of the test is too low to allow it to serve as a generally useful proof test; however, its speed and simplicity assure that it will continue to be used in the rejection of poorly adherent films.

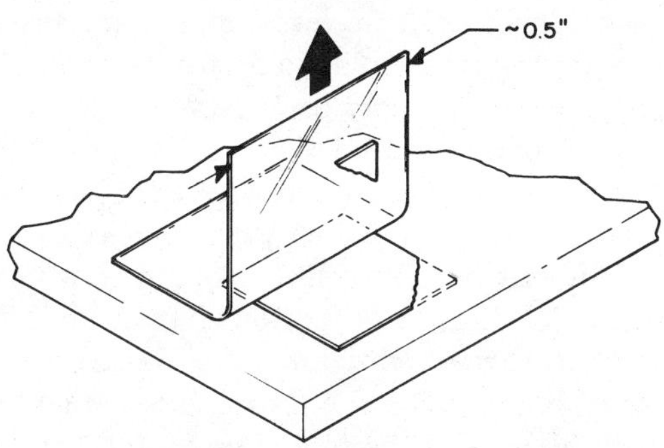

FIG. 2—*Scotch tape test.*

The precutting of a strip of printed-circuit board copper, peeling it from the board at a prescribed rate [usually 0.5 in./min (1.3 cm/min)], and measuring the average load to maintain peeling, constitute a classic peel test. Such a test offers freedom from extraneous materials that may interfere with the adhesion measurement, as solder often does. The forced direction of failure propagation and averaging of the successive failure loads contribute to a small scatter of data about the mean. Thin films are often tested by peeling, if their adhesion strengths are not extremely high [*30,31*].

RCA has developed a peel test for gold and silver thick films in which a strip of gold or silver foil is thermocompression bonded to the end of a fired-film gold or silver conductor land. The configuration of the test is shown in Fig. 3; the method is described later in the paper.

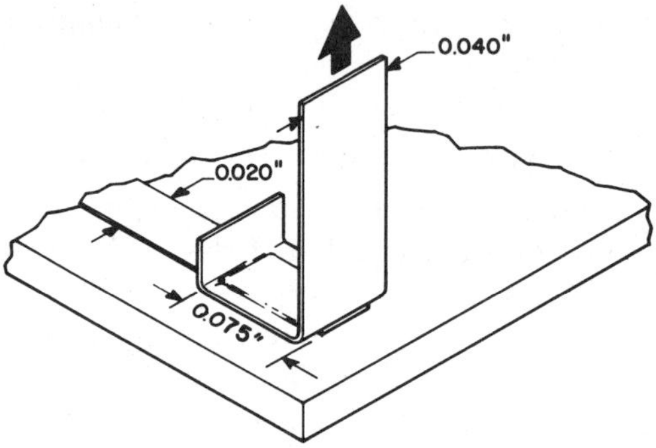

FIG. 3—*RCA peel test for gold and silver thick films (oversize beam-lead test).*

A number of theoretical analyses and reviews [*32*] of the peel test have been made; however, most are derived for the case of elastic or viscoelastic elements. Chen and Flavin [*33*], however, claim to have analyzed the important case of peel testing where plastic flow does occur.

When attempts are made to apply peel tests to other thick-film conductor materials such as gold-platinum and silver-palladium, a problem arises in that their adhesions are often too great and their cohesions too small to allow removal by pulling the film alone. Furthermore, TC bonding to most of those alloys is difficult or impossible. Soldering a copper strip to the films often will allow the film to be peeled, but in limited testing at RCA the data obtained by this method exhibited wide scatter. One reason may be that the copper ribbon used in the testing has a much higher expansion coefficient than the alumina substrate. On cooling after soldering, such specimens must incur large stresses that act to lower the apparent adhesion strength. Another factor which may have influenced the test results was an irregular moment arm in the pull strip. The thickness of the solder layer between the thick film and copper ribbon is likely to be critical for such test specimens. Anjard found a similar test inferior to the soldered-wire peel test [*2*]. However, soldered strip peel tests are still being used in the thick-film industry.

Soldered-Wire Tension-Peel Test

The soldered-wire peel test, illustrated in Fig. 4, is the test most widely used for measuring the adhesion of thick films [*2,34*], but it is not yet well standardized despite efforts by several groups [*2,35,36*[4]]. Variations in wire size, wire hardness, pad dimensions, the type and amount of solder, aging of the soldered assembly before testing, and the geometry and stresses produced by bending the pull wire up can all affect the test results. Several modifications of the specimen design shown in Fig. 4, which offer somewhat higher failure loads, have been privately reported to the author by ink vendors. Flattening the wire where it contacts the test pad reduces the tendency for the wire to pull out of the solder. Another of these modifications requires twisting the wire to form a loop, soldering the two free ends of the wire across a (0.1-in.)^2 [(0.25-cm)^2] or (0.080-in.)^2 [(0.20-cm)^2] conductor pad, and testing by pulling the loop at 90 deg to the substrate surface.

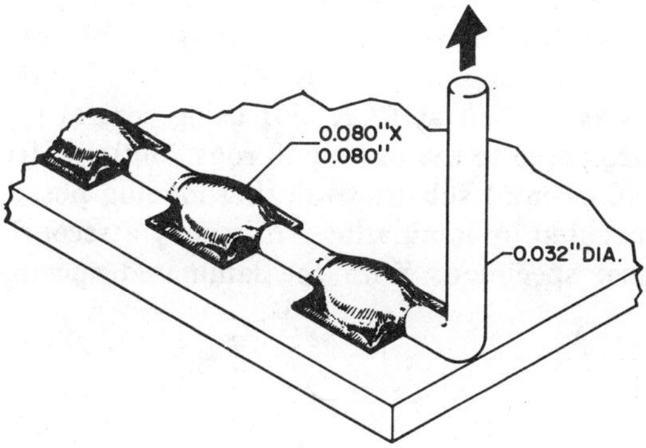

FIG. 4—*Soldered-wire, tension-peel adhesion test.*

Dissolution of the thick film by the solder during test specimen fabrication will alter the average failure load. Less obviously perhaps, the spacing of the wire within the solder from the thick-film pad and substrate can change the failure load. This is particularly true when the specimens must undergo thermal shock, since the stresses induced by differential thermal expansion can be more easily relaxed by plastic flow if solder joints are thicker. The plastic flow in the solder prevents flaw formation (or complete joint failure) in the specimens before they are tested.

Harder solders than 63Sn-37Pb[5] or 62Sn-36Pb-2Ag can be substituted to reduce wire pull-out, but the mechanics of the test are so complex that only empirical correlations are possible between data taken with hard and with

[4] RCA is further studying the method under a Naval Avionics Facility, Indianapolis, contract (No. N00163-76-C-0287).

[5] All solder constituencies are expressed in percent by weight.

soft solders. Metallurgical effects of increased pad dissolution or brittle phase formation with harder solders may also complicate the problem.

However, by replacing the eutectic tin-lead (or 62Sn-36Pb-2Ag) solder by indium-containing and other alloys, workers have successfully used the soldered wire peel test for measurements on gold-based conductor films [*37,38*].

IBM Dot Test

Shown in Fig. 5, the test is similar to the topple test of Butler et al [*39*]. The dot test has the advantages of simply fabricated test parts and straightforward testing. It partially bypasses the problem of flaws by forcing fracture to initiate in a pre-selected area; that is, only flaws in that area are likely to influence the test. The test part itself contains a notch, and this should further reduce the action of flaws. A full description of the test has not been published, but the test has been used at IBM to measure the adhesion of electroless platings to ceramic [*40*].

The dot test was studied at RCA [*36*] using several types of thick-film conductors soft-soldered to the brass pull rods. Initially, fracture across the 0.025-in.-thick (0.64-mm) substrates during loading posed a problem, but that was circumvented by laminating with epoxy a second substrate to the back of each test specimen. With the laminated specimens, conchoidal

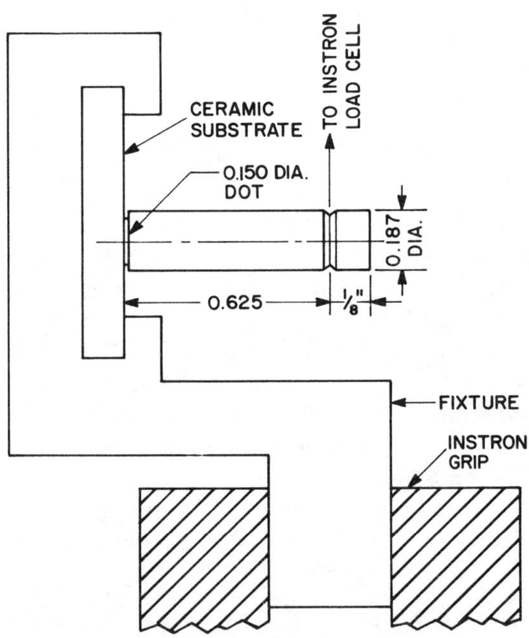

FIG. 5—*IBM dot test (nominally after Ameen and Ellis* [40]).

(shell out) fracture in the ceramic underneath the adhesion pad was found with increasing frequency for inks fired on 96 weight percent alumina substrates, when the failure load exceeded about 5 lb (2.3 kg), as it does for most useful conductors. Because a more quantitative measure of the adhesion of high-adhesion-strength thick films was required in our material/process evaluations, the dot test was eliminated from further study.

Experience with Two Tests at RCA Laboratories

TC Peel Test

The TC peel (thermocompression bonded peel) test has been used at RCA to measure adhesion strength values of gold- and silver-based thick-film conductors with a wide range of adhesion strengths [*11*,*28*,*32*,*41*]. Several failure modes can occur with the test and some of the more interesting ones are shown as stages of failure in Fig. 6.

For well-sintered, weakly bonded gold films, the failure mode is by peeling as shown in Fig. 6, Part (*c*)3. The mean stress (grams per millimetre of width) to maintain peeling at the machine strain rate of 1.3 cm/min (0.5 in./min) is quite reproducible, that is

$$V = \sigma/\overline{X} \leq \sim 10 \text{ percent} \qquad (1)$$

where V is the variation, σ the standard deviation, and $\overline{X}$ the arithmetic mean of the data.

An initial peaking of the stress (Part 1 of the stress-strain curve as shown in Fig. 6*b*) usually occurs before peeling begins. When delamination from the TC-bonding process is severe, the initial "peak" load can be lower than the peeling load.

Gold films of medium-high adhesion strength will not peel but instead, fail by tearing a section of the conductor land off the substrate and exhibit only a peak stress. This is principally an adhesion failure, although some influence of cohesion strength appears likely. Silver thick films usually fail by this mode and seldom peel. If bonding conditions and geometries are carefully maintained, the V's of peak data often approach 10 percent.

For most gold thick-film conductors the peak and peel loads are directly related. Figure 7 shows data taken with a variety of gold films bonded with silver strips and from which the following empirical relation was derived:

$$0.8 \times (\text{peak load, g}) \cong \text{peel stress, g/mm} \qquad (2)$$

By use of Eq 2, plots of adhesion strength have been constructed which cover adhesion strength variations of two orders of magnitude.

Other failure modes besides peak and peel occur. Bond delamination

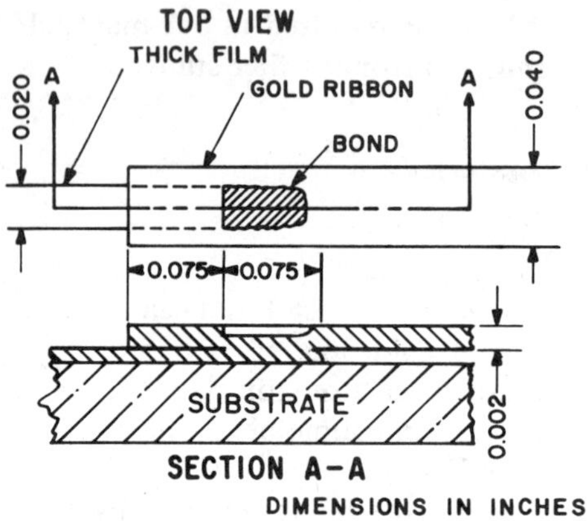

(a) Layout of the test part

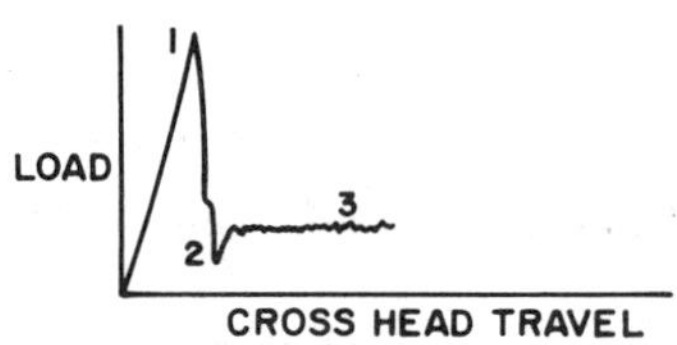

(b) Typical load-elongation test curve

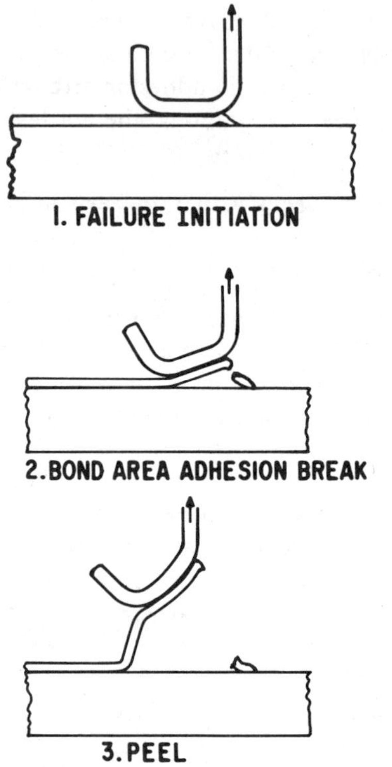

(c) The failure stages which correspond to the features of (b).

FIG 6—*RCA peel test.*

failures are found, usually at very high loads. Cohesive failures also are seen in which sections of the thick-film conductor are torn out without revealing the substrate underneath the film. Distinguishing which of these two failure modes has occurred must be done metallographically. Particularly with a soft gold pull tab strip, tearing of the strip around the bond is seen. Only threshold values of adhesion strength can be inferred from the peak loads exhibited by test specimens failing by these last three modes.

Using the TC peel test, we have studied the bonding mechanisms in silver- and gold-based inks. Examples of adhesion data for example ink materials, chemical analyses of the materials, and the classification system resulting from the correlation of composition and adhesion are presented in the following.

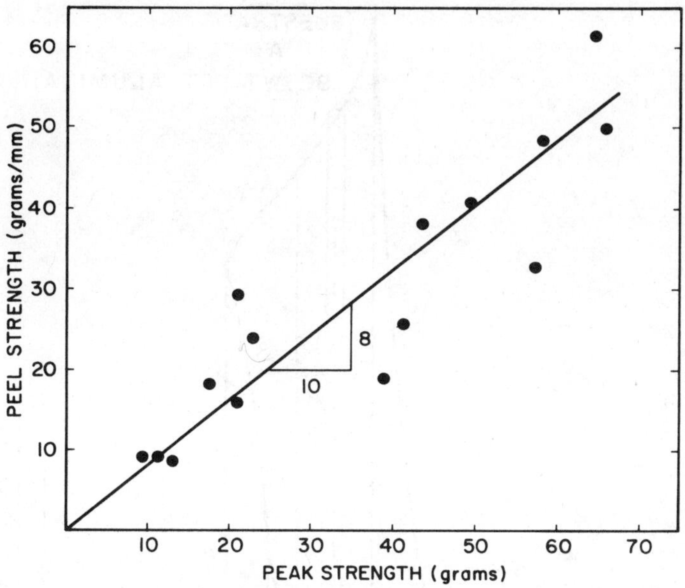

FIG 7—*Peak and peel adhesion strength values for gold thick-film specimens.*

Reactively Bonded Gold and Silver Films—Figure 8 is a graphical representation of the variation of the adhesion strength (indicated by vertical line segment lengths) of TC-bonded peel specimens as a function of dried print thickness and firing temperature. This ink developed the highest adhesion strengths we measured for gold inks, and did so after firing at temperatures approaching the melting temperature for gold (1060°C [1940°F]). The nominal composition of the inorganic solids of this ink, determined at RCA, is given in Table 1. The ink is further described in a U.S. patent [*42*]. In an early variation of this material, omission of the cadmium resulted in much lower adhesion strengths for the films, unless they were fired at temperatures even closer to the melting point of gold. We have designated inks containing copper oxide with or without additions of cadmium oxide or zinc oxide and similar materials as "reactively bonded" because we believe that the formation of chemical compound(s) (which probably includes copper aluminum oxide [$CuAl_2O_4$], a spinel) is important to strong bond formation. Reactively bonded, silver-based inks have been studied at RCA [*41,43*] and elsewhere [*44*]. Similar behavior was found in the adhesion strengths of reactively bonded silver- and gold-based thick-film conductors when the firing temperatures had approached the melting temperatures of the silver or gold, respectively.

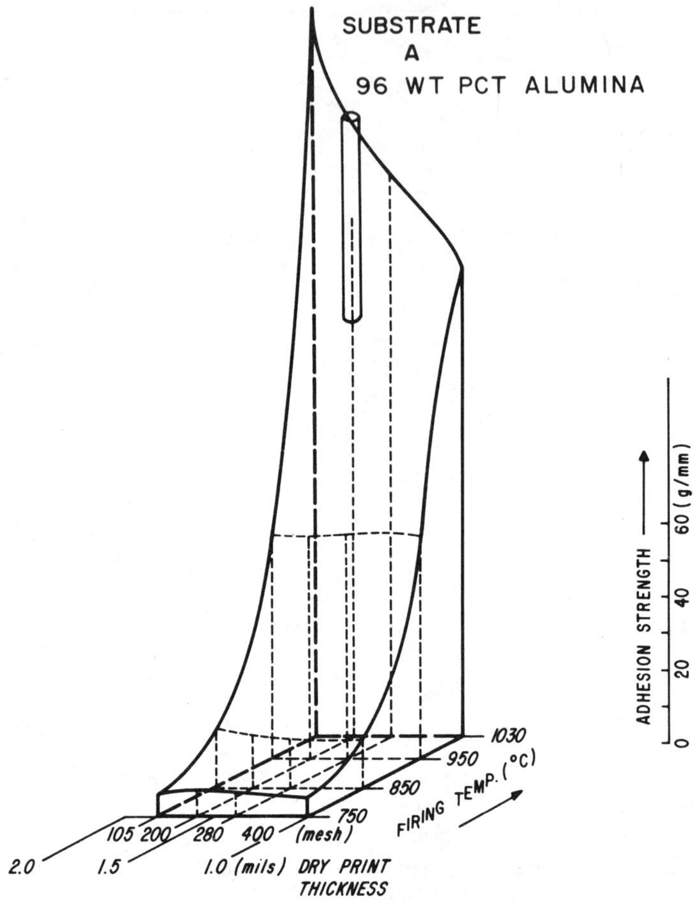

FIG. 8—*Adhesion strength of gold ink H-1 as a function of firing temperature and print thickness.*

Frit-Bonded Inks—Figures 9 and 10 show the adhesion strength curves for gold- and silver-based inks of another bonding class. The compositions of their inorganic solids are given in Table 1. It is important to note that they contain no copper. We classed them as "frit-bonded," by which we indicate that their adhesion is obtained through the wetting action of glass and bismuth oxide additions to the thick-film inks. These form an interlayer between the noble metal film and the alumina substrate. For highest adhesion strength, the frit phase structures interpenetrate the metal films [*28*]. The maxima of adhesion strength for these materials are much lower than for reactively bonded inks. However, many frit-bonded inks are still

TABLE 1—*Inorganic chemical constituencies of example thick-film conductor inks.*

Code	Metal, weight percent	Total Binder, weight percent	Binder Constituents
			reactively bonded
H-1	Au—98.3	1	CuO, CdO—4:1
			frit-bonded
E-2	Au—91.8	7	4/5 Bi_2O_3; also SiO_2, B_2O_3, Al_2O_3, P_2O_5
H-2	Ag—86	10	Bi_2O_3, PbO, SiO_2-5:2:1; also B_2O_3, CdO
			mixed-bonded
D-2	Au—95.8	3	2/3 Bi_2O_3; remainder CuO and CdO 1:4
C-7	Ag—91	11	Bi_2O_3; CuO, PbO, – 4:3:2; also ZnO, SiO_2, B_2O_3

NOTE—Binder constituents are listed in order of decreasing weight fraction. More detailed analyses of these and related inks are available [*11, 36*].

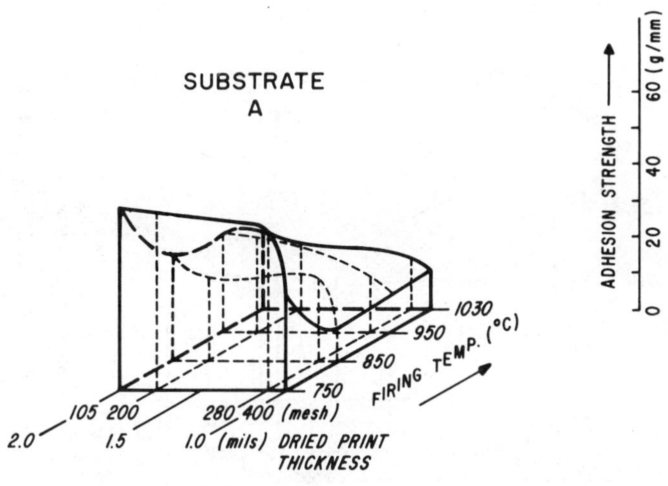

FIG. 9—*Adhesion strength of gold ink E-2 as a function of film thickness and firing temperature.*

in use. Useful adhesion strengths can be obtained with frit-bonded films over a wide range of firing temperatures (~500 to 1100°C [932 to 2012°F]) by the proper choice of frit.

Mixed-Bonded Inks—The last class includes, as our designation for it implies, thick-film conductor inks containing additions of materials for both frit bonding and reactively bonding. Figures 11 and 12 show the adhesion strength versus firing temperature and film thickness data for gold- and silver-based, mixed-bonded inks. The compositions of the inks are given in Table 1. The principal advantage of these materials over frit-bonded inks is their higher adhesion strength. Compared with reactively

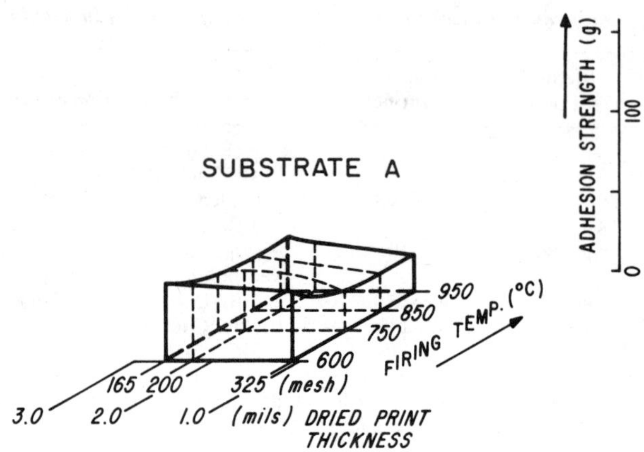

FIG. 10—*Adhesion strength of silver ink H-2 as a function of film thickness and firing temperature.*

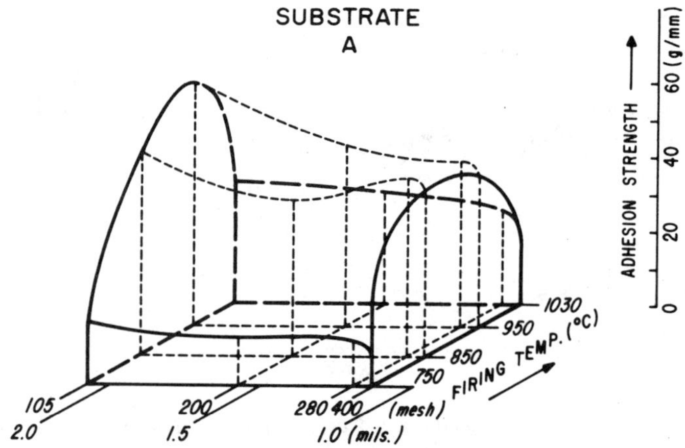

FIG. 11—*Adhesion strength of gold ink D-2 as a function of film thickness and firing temperature.*

bonded inks, mixed-bonded inks display comparable adhesion strengths, but achieve their adhesion maxima at lower firing temperatures. Thus their preferred firing temperature lies more within the range of firing temperatures commonly used for the manufacture of hybrid microcircuits. Also, they usually exhibit a smaller change of adhesion strength for a change in the process firing temperature than is expected with a reactively bonded ink.

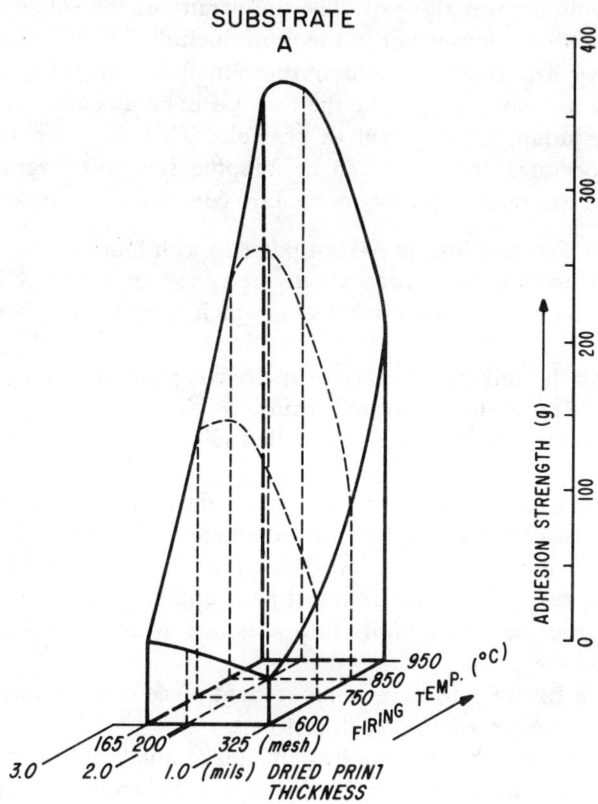

FIG. 12—*Adhesion strength of silver ink C-7 as a function of film thickness and firing temperature.*

Soldered-Wire, Tension-Peel Test at RCA Laboratories

Like the TC peel test, testing soldered-wire, tension-peel specimens has produced a variety of failure modes. For low to medium-low adhesion strength films, the minor stresses caused by loading specimens into the test machine cause many failures; there, invariably the film, solder, and wire break from the substrate in one piece. With medium-strength films (2 to 4 lb [1 to 2 kg] for a (0.1-in.)^2 [(0.25-cm)^2] pad), the failed part has about the same appearance, but most specimens can be tested and yield quantitative data. When the adhesion strength is high, the wire often pulls out of the ductile solder, giving, at best, threshold values of adhesion strength. This failure mode can be suppressed to a limited extent by increasing the amount of solder on each pad. Any benefit from increasing the amount of solder is lost when the solder begins to run from the pad along the wire, changing the solder volume over the pad. The uniformity of the solder lump joining the pad and wire is important to the reproducibility of test results.

At RCA we are striving to achieve reproducibility in this test. A principal rationale we are using is to make the test specimen assembly as independent of operator judgment and skill as possible. This will help to assure that any test procedure developed can be adopted straightforwardly by others. Our current specimen assembly procedure consists of the following steps:

1. Screen, dry, and fire special test patterns with four (0.1-in.)^2 [(0.25-cm)^2] test pads symmetrically spaced about the center of a 1 by 1 by 0.025 in.^{3} [2.5 by 2.5 by 0.063 (cm)^3] substrate, which usually is 96 weight percent alumina.
2. Burnish the substrate surface to assure complete wetting by the solder.
3. Dip in flux (mildly activated rosin).
4. Immerse in a temperature-controlled solder pot for 5 s (62Sn-36Pb-2Ag solder).
5. While the solder is still molten, wipe all the excess from the substrate.
6. Place two annealled 0.032-in.-diameter (0.81-mm) OFHC copper wires and the substrate into a low-thermal-mass jig that positions the wires, each against two of the thick-film test pads and centered over them.
7. Place the jigged assembly into a holder which positions it over the heat source.
8. Place a fixed-volume, vee-shaped-bent piece of 62Sn-36Pb-2Ag solder wire over the copper wire on each test pad.
9. Flux each pad with a controlled amount of mildly activated rosin flux.
10. Heat the jigged assembly through a controlled temperature-time cycle, using hot air as the heat exchange medium.
11. Remove flux residue by soaking in solvent.
12. Reject before testing any soldered-wire test pads which do not appear uniform.
13. Bend up the wires to conform to the test geometry.
14. Cut the copper wires from between adjacent pads.
15. Wait at least 24 h to allow stress relaxation in the solder.
16. Test in an Instron at 0.5 in./min (1.3 cm/min) crosshead speed.

We are continuing to work with the soldered-wire peel test. Methods to reduce the time required for Step 10 of the foregoing and to eliminate operator influence from Steps 2 and 13 are under study. Correlation of data from this test and the TC peel test will also be attempted.

We have successfully measured the adhesion strengths of several copper- and gold-platinum-based commercial thick-film inks, using procedures similar to that just described, in the performance of Naval Air Systems Command (NASC) contract studies [*41,45*]. An example of the adhesion data for a copper ink is shown in Fig. 13. An analysis of the scatter of data for the gold-platinum work indicated that if the average failure load is greater

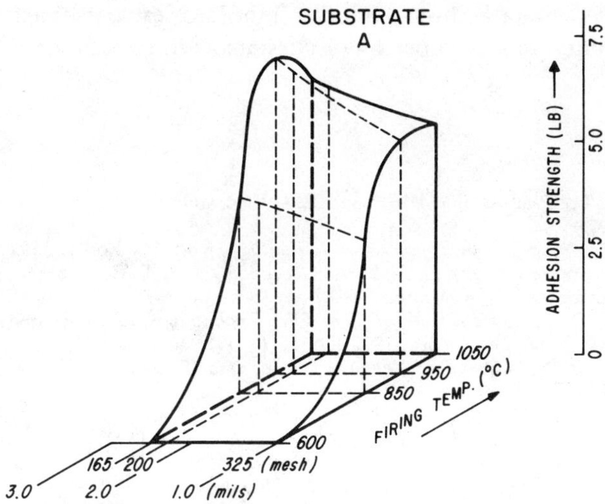

FIG. 13—*Adhesion strength of copper ink H-3 as a function of film thickness and firing temperature.*

than 4 lb (1.8 kg) the data scatter, *V*, for a specimen size of 12 tests will be below 25 percent, and for average failure loads greater than 6 lb (2.7 kg) *V* would fall to approximately 10 percent.

Conclusions

No adhesion test for thick-film conductors has been found to approach ideality. Each of the tests described has good and bad points. The search for better test methods and for methods of standardizing existing tests will continue, but new or modified methods will be widely adopted only if they demonstrate significantly improved value.

Thick-film technologists must interpret adhesion strength data in terms of the failure modes observed and the microstructural aspects of the materials. If they do, significant improvements in the understanding of the materials and of the test methods can result.

Acknowledgments

The author gratefully acknowledges the support of the Navy and of RCA for this work. The chemical analyses were carried out by the Materials Characterization Group at RCA Laboratories. Much of the data acquisition was performed by W. I. Rogers and E. J. Conlon. My supervisor Dr. G. L. Schnable; colleagues, John Vossen and Kenneth Bube; members of the

NASC and Naval Research Laboratory staffs; representatives of several thick-film ink manufacturers; Sandia Laboratories' personnel; and others have contributed to this paper by discussions with the author.

References

[1] Jacobson, L., *Proceedings,* IEEE and EIA Electronic Components Conference, 1971, pp. 474-479.

[2] Anjard, R. P., *Microelectronics and Reliability,* Vol. 10, No. 4, 1971, pp. 269-275.

[3] Savage, J. in *Handbook of Thick Film Technology,* P. J. Holmes and R. G. Loasby, Eds., Electrochemical Publications Ltd., Ayr, Scotland, 1976, pp. 108-112.

[4] Campbell, D. S. in *Handbook of Thin Film Technology*, L. I. Maissel and R. Glang, Eds., McGraw-Hill, New York, 1970, pp. 12-(3-49).

[5] Chapman, B. N., *Journal of Vacuum Science and Technology,* Vol. 11, No. 1, 1974, pp. 106-113.

[6] Bullet, T. R. and Prosser, J. L., *Progress in Organic Coatings,* Vol. 1, 1972, pp. 45-71.

[7] "How to Test Adhesive Properties," *Materials Engineering,* Vol. 77, No. 3, 1972, pp. 60-64.

[8] Anderson, G. P. et al, *Journal of Colloid and Interface Science,* Vol. 47, 1974, pp. 600-609.

[9] Mittal, K. L. in *Properties of Electrodeposits: Their Measurement and Significance,* R. Sard, et al, Eds., The Electrochemical Society, Princeton, N.J., 1975, pp. 273-306.

[10] Mittal, K. L. in *Electrocomponent Science and Technology,* Vol. 3, 1976, pp. 21-42.

[11] Hitch, T. T. and Bube, K. R., *Basic Adhesion Mechanisms in Thick and Thin Films,* Naval Air Systems Command Contract Final Report by RCA, Princeton, N.J., No. AD-A011906, 31 Jan. 1975, available from the National Technical Information Service (NTIS), Springfield, Va. or from the Defense Documentation Center (DDC), Alexandria, Va.

[12] Mittal, K. L., this publication, pp. 5-16.

[13] Gilman, J. J., *Journal of Applied Physics,* Vol. 31, 1960, pp. 2208-2218.

[14] Westwood, A. R. C. and Hitch, T. T., *Journal of Applied Physics*, Vol. 34, 1964, pp. 3085-3089.

[15] Westwood, A. R. C. and Goldheim, D. L., *Journal of Applied Physics*, Vol. 34, 1964, pp. 3336-3339.

[16] Mostovoy, S. et al, *Journal of Adhesion*, Vol. 3, 1971, pp. 125-163.

[17] Freiman, S. W., Mulville, D. R., and Mast, P. W., *Journal of Material Science*, Vol. 8, 1973, pp. 1527-1533.

[18] Becher, P. F. et al in *Proceedings*, ISHM International Microelectronics Symposium, 1975, pp. 279-286.

[19] Bascom, W. D., this publication, pp. 63-79.

[20] Drucker, D. C. in *Fracture of Solids,* D. C. Drucker and J. J. Gilman, Eds., Interscience, Wiley, New York, 1963, pp. 3-50.

[21] Marsh, D. M. in *Fracture of Solids,* D. C. Drucker and J. J. Gilman, Eds., Interscience, Wiley, New York, 1963, pp. 119-142.

[22] Timoshenko, S., *Theory of Elasticity*, McGraw-Hill, New York, 1934, pp. 75-82.

[23] Jacobson, R. and Kruse, B., *Thin Solid Films*, Vol. 15, No. 1, 1973, pp. 71-77.

[24] Leven, S. S., "Qualification Requirements for Thick-Film Networks," ECOM Contract Final Report No. ECOM-73-0326-F by Westinghouse, Baltimore, Md., Oct. 1975.

[25] Leven, S. S., in this publication, pp. 269-283.

[26] Buckthorpe, A. C. in *Proceedings,* IERE Conference on Hybrid Microelectronics, Sept. 1973, pp. 57-70.

[27] Lemon, T. H. in *Proceedings*, IERE Conference on Hybrid Microelectronics, Sept. 1975, pp. 23-32.

[28] Hitch, T. T. in *Proceedings*, ISHM International Microelectronic Symposium, 1971, pp. 7-7-(1-11).

[29] Johnson, D. R. and Knutson, R. E., *IEEE Transactions on Parts, Hybrids, and Packaging PHP-12*, No. 3, 1976, pp. 187-194.
[30] Von Harrach, H. G. and Chapman, B. N., *Thin Solid Films*, Vol. 13, No. 1, 1972, pp. 157-161.
[31] Poley, M. M. and Whitaker, H. L., *Journal of Vacuum Science and Technology*, Vol. 11, No. 1, 1974, pp. 114-118.
[32] Hardy, A. in *Aspects of Adhesion 1*, D. J. Alner, Ed., University of London Press, London, England, 1965, pp. 47-65.
[33] Chen, W. T. and Flavin, T. F., *IBM Journal of Research and Development*, Vol. 16, No. 1, 1972, pp. 203-213.
[34] Hoffman, L. C., Bacchetta, V. L., and Frederick, K. W., *IEEE Transactions on Parts, Materials and Packaging PMP-1*, No. 1, 1965, pp. s-(381-386).
[35] "Method of Test for Wire Peel Adhesion of Soldered Thick Film Conductors to Ceramic Substrates," *The Thick Film Handbook*, E. I. duPont de Nemours & Co. Inc., Photo Products Department, Wilmington, Del., March 1971.
[36] Hitch, T. T. and Bube, K. R., *Basic Adhesion in Thick and Thin Films*, Naval Air Systems Command contract final report by RCA, Princeton, N.J., No. A0 23091, Jan. 30, 1976, available from the National Technical Information Service or from the Defense Documentation Center.
[37] Smith, B. R. and Dietz, R. L. in *Proceedings*, ISHM International Microelectronics Symposium, 1972, pp. 2-A-5(1-8).
[38] Zeien, R. H. in *Proceedings*, ISHM International Microelectronics Symposium, 1974, pp. 7-15.
[39] Butler, D. W. et al in *Aspects of Adhesion 6*, D. J. Alner, Ed., University of London Press, London, England, 1971, pp. 55-63.
[40] Ameen, J. C. and Ellis, T. L., *Electronic Packaging and Production*, Vol. 14, No. 1, 1974, pp. 124-136.
[41] Hitch, T. T., *Journal of Electronic Materials*, Vol. 3, No. 2, 1974, pp. 553-577.
[42] Smith, B. R., U.S. Patent No. 3,799,891, March 1974.
[43] Hitch, T. T. and McCurdy, T. E., U.S. Patent No. 3,962,143, June 1976.
[44] Loasby, R. G. et al, *Solid State Technology*, Vol. 15, No. 5, 1972, pp. 46-50, 72.
[45] Hitch, T. T. and Bube, K. R., "Basic Adhesion Mechanisms in Thick and Thin Films," Naval Air Systems Command Contract No. N00019-76-C-0256, First Quarterly Report by RCA, April 1976.

DISCUSSION

J. J. Bikerman[1] *(written discussion)*—The peel test is widely used in the science of adhesive joints and several theories have been published concerning its meaning. Did you compare your data with any of these theories? It seems to me that the results obtained in the study of adhesive joints are not sufficiently utilized by those studying coatings or attached films.

T. T. Hitch (author's closure)—As indicated in the paper, I had hoped to make such comparisons with the theory of Chen and Flavin, but it has

[1] Department of Chemical Engineering, Case Western Reserve University, Cleveland, Ohio 44106.

not yet been published in a usable form. I know of no other workers, besides them, who have seriously attempted to treat plastic work which occurs in the measurement of adhesion. In the two tests described, the TC peel test and the soldered-wire, tension peel test, plastic work in the thick film itself and in the metal attachments to the film for testing can require large amounts of strain energy.

L. H. Sharpe[2] (*written discussion*)—Why don't you call your test a peel test and report data as peel strength—which is what it is—rather than report data as adhesion strength, which implies that you're learning something fundamental about the interface—which you're not?

T. T. Hitch (*author's closure*)—I believe that the TC peel test, despite its limitations, is as good a measure of the adhesion in thick films as there is today. If the questioner has meaningful data from an improved test which are applicable to these materials, I would be pleased to see them and his test method.

[2] Engineering and Development, Bell Laboratories, Murray Hill, N.J. 07974.

Adhesion of Thick Films to Ceramic and Its Measurement by Both Destructive and Nondestructive Means

REFERENCE: Morey, R. L., "**Adhesion of Thick Films to Ceramic and Its Measurement by Both Destructive and Nondestructive Means,**" *Adhesion Measurement of Thin Films, Thick Films, and Bulk Coatings, ASTM STP 640*, K. L. Mittal, Ed., American Society for Testing and Materials, 1978, pp. 233-250.

ABSTRACT: Techniques for the measurement of adhesion of thick films to alumina have been evaluated. One method is to solder a connection into a thick film pad and measure strength using a force gage. By this method it is not possible to differentiate between inherent adhesion problems and soldering effects. Another method is to use a wire peel test, but it requires optimized procedures or the effects of soldering can mask the adhesion characteristics. Metallurgical sectioning with use of a scanning electron microscope (SEM) and energy dispersive X-ray (EDAX) attachment is an invaluable aid for examining the interface and determining the characteristics of glass and metal components. Bondability tests are also a means of evaluating adhesion but are affected by bonding technique and wire size. Standardized methods are needed for adhesion measurements which, coupled with a comprehensive quality-assurance procedure, can give full process control.

KEY WORDS: adhesion, thick films, tests, evaluation, bondability, quality control, substrates, dielectrics, measuring instruments

In a thick-film multilayer board (MLB), reliability problems can occur due to poor adhesion of the metallization to the ceramic substrate or to the multilayer dielectric film. In instances where the thick film is used for interconnects or for bonding discrete components, adhesion can be of paramount importance. Unfortunately, the achievement and maintenance of good adherent bonds of thick film to a ceramic surface is not a simple matter [*1,2*].[2] There have been repeated discussions concerning the mechanisms of the attachment of the thick film to the ceramic surface. Hailes

[1]Staff engineer, C. S. Draper Laboratory, Inc., Cambridge, Mass. 02139.
[2]The italic numbers in brackets refer to the list of references appended to this paper.

and Crossland [3] have stated that for good adhesion a thin layer of glass is required at the interface between ceramic and thick film for mechanical interlocking. Hitch [4] also states that the glass bonds chemically to the ceramic and wets the metal conductor to form some degree of mechanical interlocking or keying with the sintered metal layer. There is intermixing of metal and glass at the interface of thick film and ceramic, and the amount of penetration of the glass into the ceramic is a function of the rate and degree of sintering of the metal particles and of the viscosity and wetting ability of the glass during firing. Schneider et al [5] claim that this glass layer is not necessary and is present only where overfiring or excessive thickness of gold existed. In these cases glass was extruded to the periphery of the conductive film. The bonding of the gold layer to the alumina appeared to be basically mechanical with interlocking of glass and gold phases.

In an earlier report [6], we concluded that the presence of a glassy layer between the thick-film pad and the ceramic is undesirable and is detrimental for adhesion. The glassy elements are essential for a cohesive thick-film pad, but the diffusion or extrusion of the glass out of the pad to form a continuous layer between ceramic and pad can lead to disaster.

To effect greater reliability of finished product and to determine the effects of processing variables, a study of the adhesion of gold-based thick film conductors was undertaken. This paper is concerned with evaluating various procedures for the measurement of adhesion.

Experimental

In the course of this study it was not possible to investigate fully all processing variables due to scheduling limitations, and, in general, tests were run on production specimens. Substrates of 96 percent by weight purity were used. The finished multilayer board consisted of five layers of thick film printed circuitry separated by double-screened dielectric layers. There were 46 pads on one edge for bonding bifurcated terminals for vertical interconnects and 42 pads around holes on two edges to allow for swaging and bonding of vertical pins for lateral connections. Substrates were of two types. One was laser-cut to size from American Lava AlSiMag 614 stock by Lasermation Inc., and the other was stamped from green tape and fired at Comco Inc. Laser-cut substrates were ground on one side to meet a camber tolerance of 10 μm/cm (0.001/in.). Certain substrates from each type were routinely processed whereas other substrates were printed with thick-film pads and then multifired to simulate the addition of dielectric and conductor layers and gold vias.

Initial screening was with ESL 5835 platium/gold with an overlay of ESL 8835 gold for the external pads. Two layers of each were screened and fired at 950°C (1742°F). The gold on internal layers was ESL 8831, and the dielectric was ESL 4608 FB. To control camber, firing of these layers was at 850 °C (1562 °F).

In the initial production cycle, bifurcated terminal pads were screened front and back. Due to alignment problems and the reduced adhesion to the ground back side as opposed to the as-fired top surface, the back pads were subsequently eliminated and replaced by epoxy. Testing of the solder joint, however, was made prior to the addition of epoxy. Bifurcated terminals were installed by screening 80/20 gold-tin paste onto burnished pads and reflowing the solder by use of either a B.T.U. Engineering continuous-reflow furnace Model GPM/1 or a Browne hydrogen flame soldering system. The vertical pins were installed by first hot-swaging pins into the hole through the ceramic and then by using an 80/20 gold-tin preform for an electrical bond to the thick film pad. A typical finished multilayer board is as shown in Fig. 1.

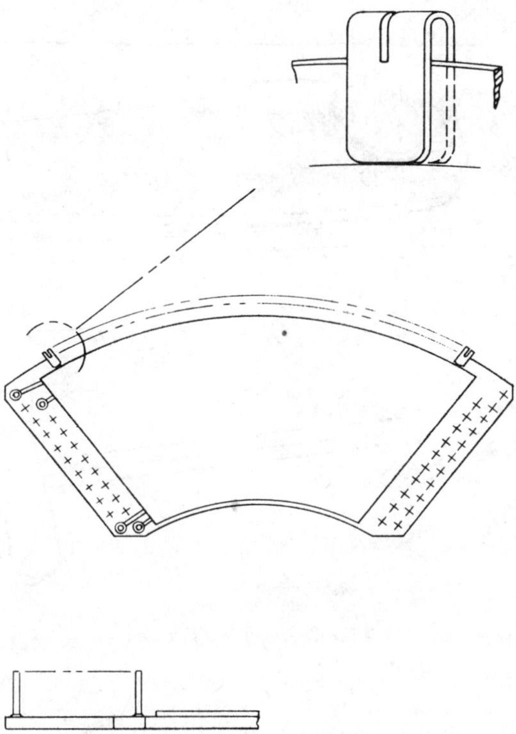

FIG. 1—*Thick-film multilayer board.*

Adhesion Measurement Using a Force Gage

The adhesion of the bifurcated terminals and pad to the ceramic was measured initially by use of a force gage as shown in Fig. 2. A calibrated spring feeler gage with an extension arm was fabricated so that the end was formed to fit over the U-terminals. The gage and arm is allowed to slide in a slotted base. The multilayer board is clamped at the other end of the fixture and must be rotated so as to allow a perpendicular push on each terminal. The gage can either be preset to allow a known force to be applied or else the force to cause destruction can be established. It was initially thought that this test would measure both the attachment of clip to pad and the pad to the ceramic. However, the attachment of the pad to the ceramic was always the limiting factor, and failures could not be induced in the solder joint. Soldering does have an effect on adhesion of the pad due to leaching of the platinum-gold and gold pad, but this can be determined by monitoring penetration of the tin component of gold-tin into the thick-film pad.

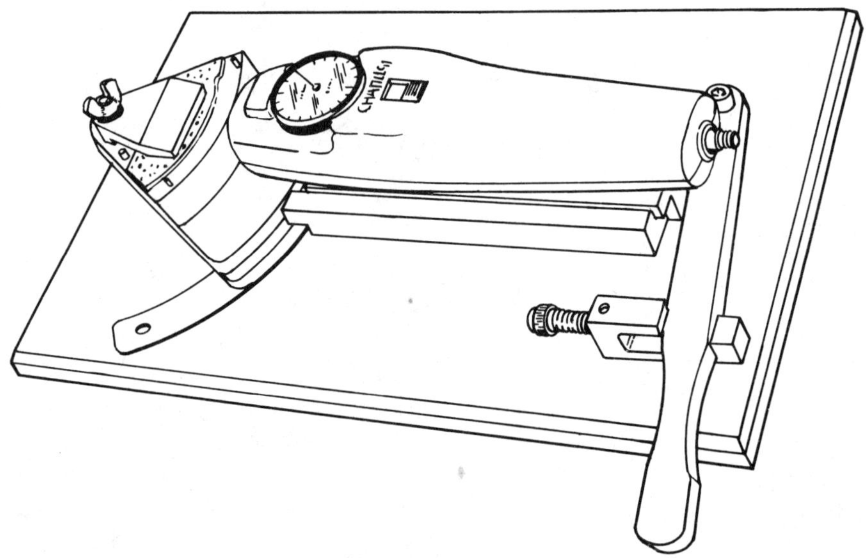

FIG. 2—*Push-test fixture.*

The modification of this method is shown in Fig. 3, in which the ceramic board is clamped and the vertical load is lowered onto the clip.

An Instron universal test machine can be used for comparison of the results with the force gage data. Again the multilayer board is clamped, the rate of loading can be varied, and the effects of time at load (creep) can also be studied. Generally the load before failure causes deformation of the terminal, producing a bending moment in addition to the shearing force.

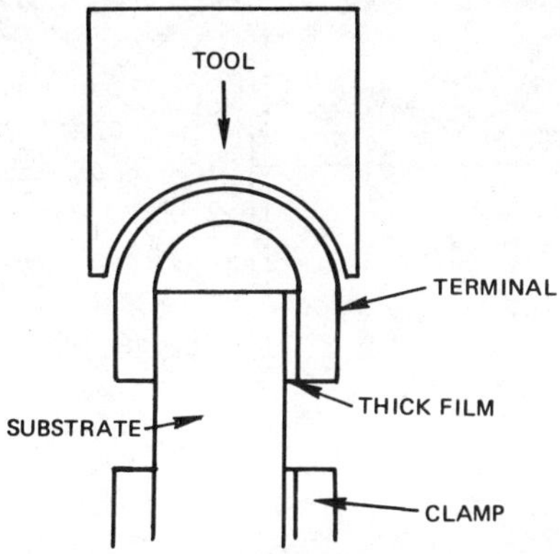

FIG. 3—*Push-test fixture modification.*

It was found that the rate of loading was not a significant variable. A push test model was built with a motor-driven mechanism, but this was not used for routine production because hand pushing was found adequate. Great care had to be taken that no tangentially applied forces occurred as this would lead to premature failure. The machine which applied the load vertically was more prone to this type of problem and was discarded.

The horizontal push test can be modified for use when an assessment of the electrical characteristics of the joint is required. In this case electrical connections could be soldered onto clips and monitored during the push test operation. A representation of this method is shown in Fig. 4. Here the pushing is made away from the dielectric. The dielectric can in some instances act as a barrier to the push operation and could give artifically higher results. A correlation was made using a setup as shown in Fig. 5, but results were essentially similar. It was easier in production to use a push test than a pull test, and a standardized test method was inaugurated.

For a nondestructive test, the gage can be preset to a given force value and all pads can be proof-tested. A visual examination at ×20 minimum is then required to verify that lifting of the pads has not occurred. It is desirable to test all pads to destruction, on a specimen board, to determine the range of values obtained. If an unacceptable distribution, pattern, or range of values is obtained, then the solder screening operation, the variation in temperature laterally across the furnace, and the soldering reflow operation need to be evaluated to determine the controlling variables. A

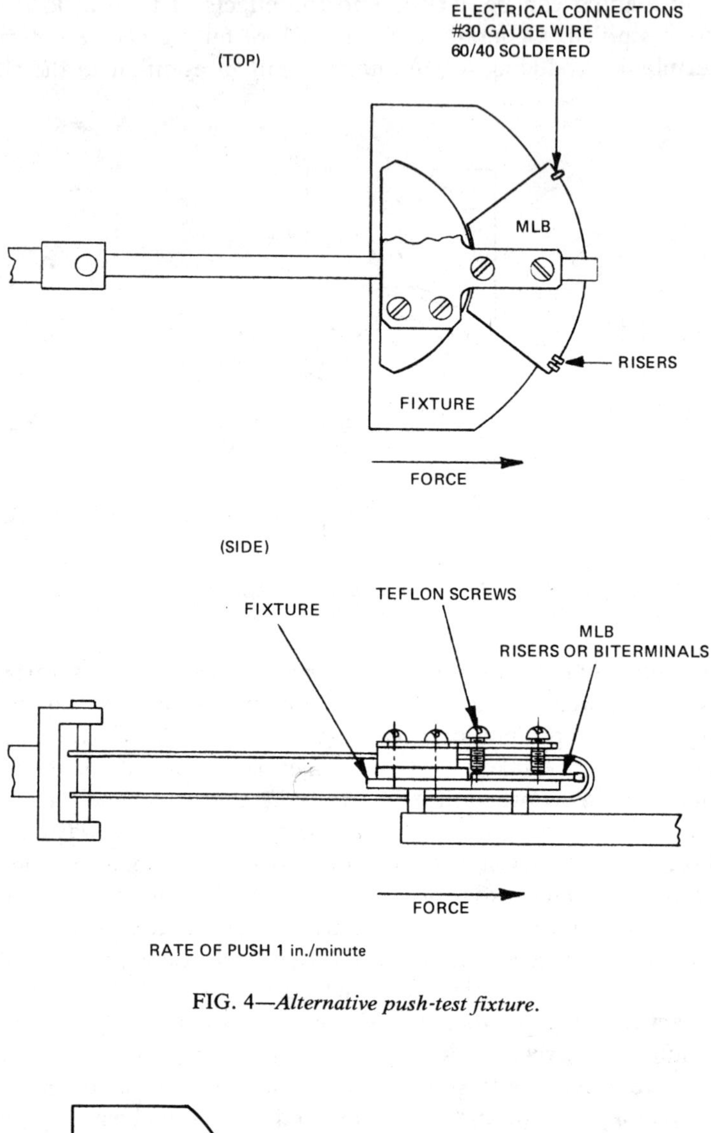

FIG. 4—*Alternative push-test fixture.*

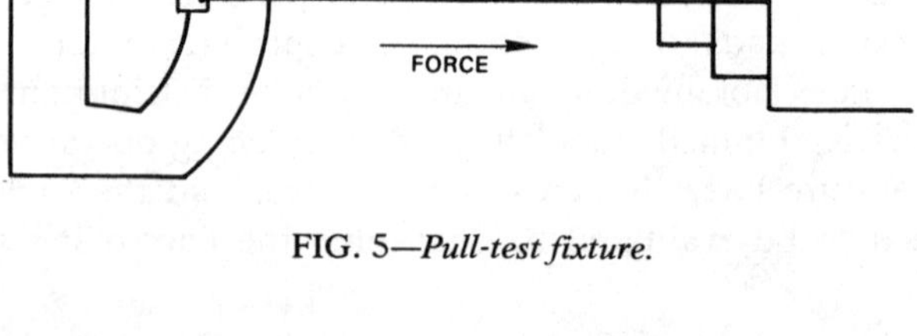

FIG. 5—*Pull-test fixture.*

regular schedule was established to monitor parameters, and verification of product was made using this test procedure.

Adhesion Method Using a Wire Peel Test

In order to produce a multilayer thick-film board, multiple firings are required. For a six-conductor layer circuit, approximately 20 firings are required, and, in cases where pads are to be used for interconnects as shown in Fig. 1, it was essential that the adhesion did not degrade.

A modification of the duPont method of testing [7] was used to determine the bond strength characteristics of the thick film. Due to persistent adhesion problems, verification specimens were processed with each batch of material at the time of initial screening of interconnect pads. These specimens were then given a simulated total screening operation by being passed through the furnace 20 times. Wire pull tests were then run on the pads to ascertain the range of pull strength values. If the value was less than 1.36-kg (3-lb) pull, then the batch was rejected. For surviving batches a correlation test was also run at the end of the normal production cycle to verify the initial screen results and to determine if changes in the furnace profile over the period required to process the batch had caused problems.

In the duPont test the breaking strength is defined as the maximum load for a specified bond area and bending distance required to separate one member from the other over the adhered surfaces. The load is applied in tension and perpendicular to the substrate plane, and the rate of separation, as stated in the duPont method, should be 0.02 cm/s (0.5 in./min).

The duPont method calls for the use of a specific test pattern and substrate, 2.6 cm by 2.6 cm by 0.66 mm (1 × 1 × 0.025 in). In our case there was a possibility of variations in the ceramic material, and each batch had been identified, so specimens of the same configuration batch and materials were utilized. The thickness of the specimen was a nominal 0.76 mm (0.030 in.) and the test pads were actual interconnect pads. The test bond configuration was as shown in Fig. 6; 20 AWG tinned copper wire was used. The flux used was MILROS 611 Activated Rosin meeting MIL-Spec F14356. The wire was cut 7.8 cm (3 in.) long and pretinned with 60/40 solder using an Ungar soldering iron with a 37.5 W tip. The element was heated through a Variac set at 100 V. The pads were swabbed with isopropyl alcohol and the wire held to a prefluxed pad and heated with the Ungar iron. The bonds were aged 16 h and pulled at room temperature using a Hunter Spring Gage with a rate of speed of 0.04 cm/s (1 in./min).

A schematic of the test fixture used is shown in Fig. 7, and results for three specimen boards from a particular lot are shown in Fig. 8.

This modified wire peel test method is highly dependent upon the soldering operation and the preferable solution is to use a solder bath as outlined in the duPont method. However, acceptable results can be obtained using

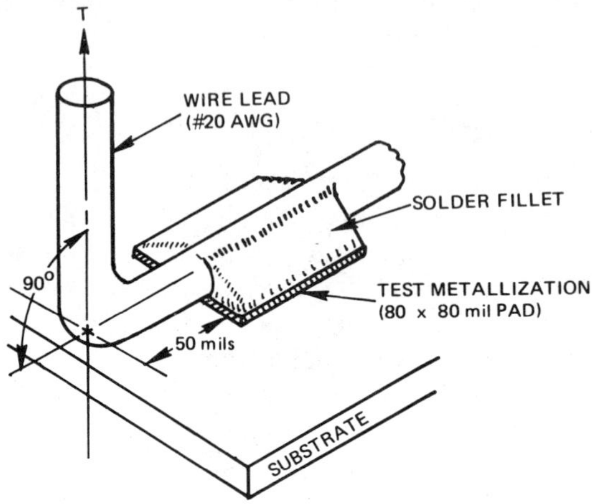

FIG. 6—*Adhesion test bond configuration.*

skilled operators and a controlled hand-soldering operation. The results in Fig. 8 show a typical Gaussian distribution. A correlation study was carried out at several facilities to determine the reproducibility of such a test, and it was found acceptable.

Pull-on-Loop Method

In general, a necessary requirement is that certain pads be suitable for wire bonding, that is, thermocompression or ultrasonic. In our case the bond was ultrasonic aluminum to gold. Again, adhesion problems can be encountered, and for components bonded onto the top layer the gold-platinum pad is screened onto a dielectric layer rather than onto ceramic. The test called for 22 g minimum, using 38-μm (0.0015-in.) aluminum wire and the bond pull-on-loop method per MIL-STD-883, Method 2011, Test Condition D. Care must be taken with the geometry of such a bond. As shown in Fig. 9 and by Ref *8*, changes in loop configuration can lead to marked variations in results. Certain power components require heavier wire connections, for example, 0.2 mm (0.008 in.), and, if this is the case, the adhesion of the pad requiring the use of heavy wire needs to be evaluated. If the thickness of the pad is insufficient or the adhesion is marginal, the bond will fail.

Metallurgical Sectioning

When adhesion problems are encountered, an examination of the interface between pad and ceramic can be an invaluable aid. At low magnifi-

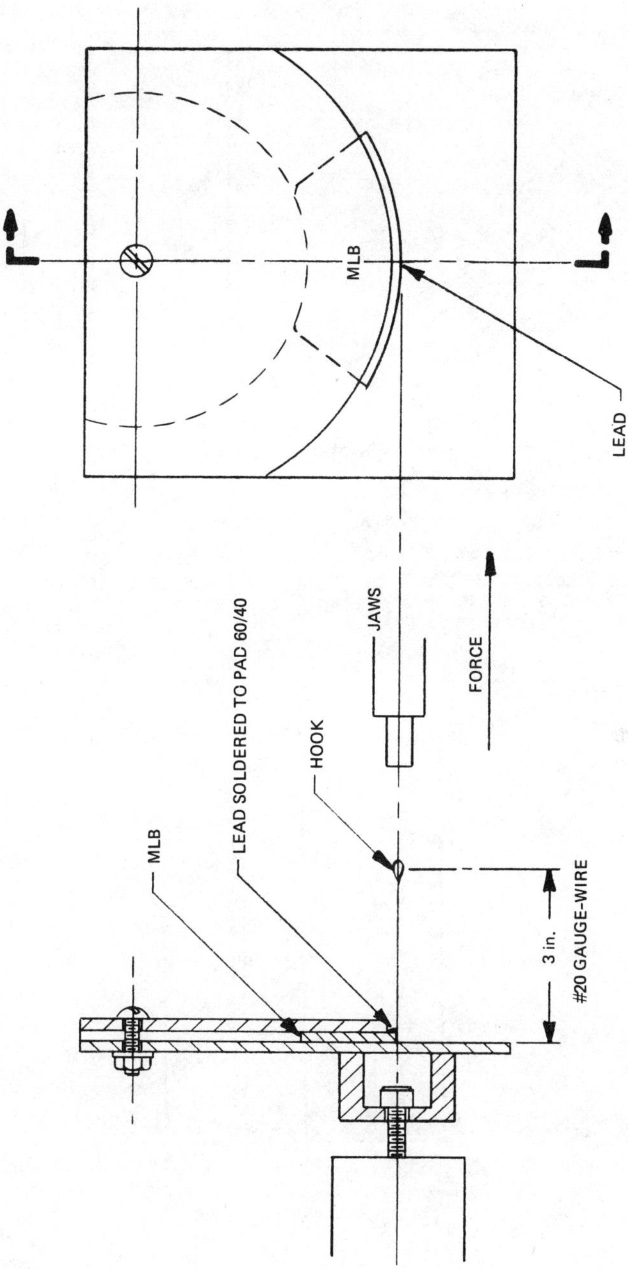

FIG. 7—*Peel-test fixture.*

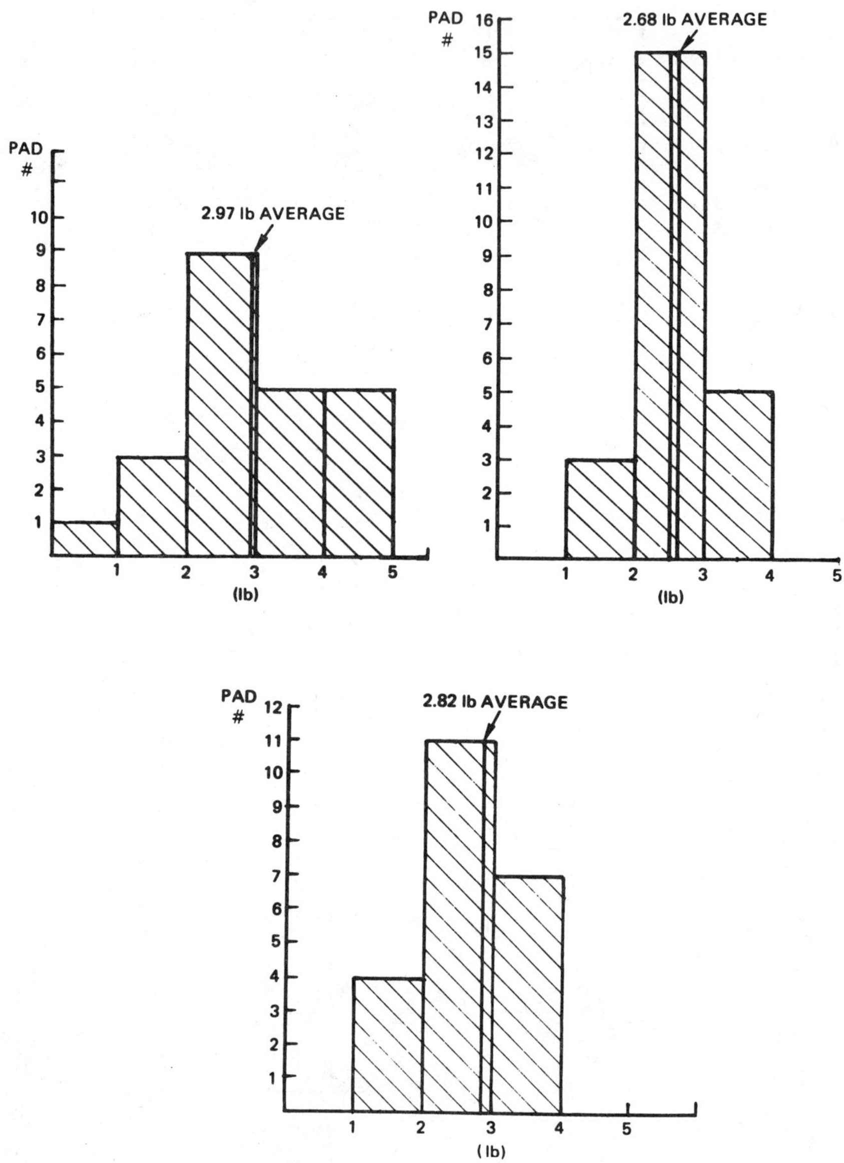

FIG. 8—*Peel-test results.*

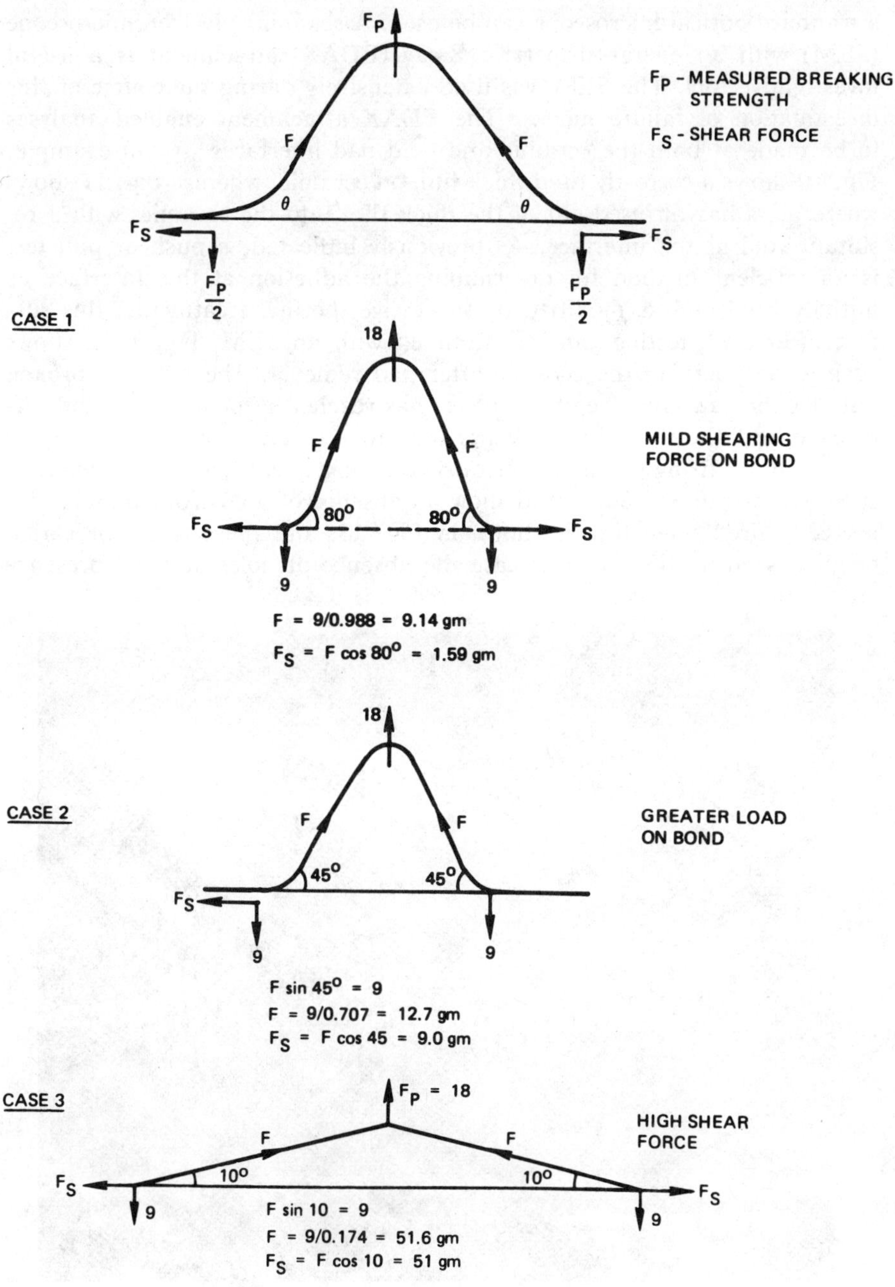

FIG. 9—*Geometry of pull test.*

cation a stereo microscope is useful and for magnifications up to ×2000 a standard optical microscope can be used. A scanning electron microscope (SEM) with an energy dispersive X-ray (EDAX) attachment is a useful investigative tool. The SEM was used extensively during the course of this investigation of failure modes. The EDAX attachment enabled analyses to be made at both the ceramic and gold pad interfaces. As an example, Fig. 10 shows a correctly fired thick film on ceramic, whereas Fig. 11 shows where glass has diffused out of the thick film into the ceramic, with a resultant void at the interface. As previously indicated, a push or pull test is an efficient method for determining the adhesion at this interface as initially fired and as modified by successive thermal treatments; this surface after such testing can be examined with an SEM. Figure 12 shows such a layer left on the ceramic after pad removal. The holes or broken bubbles indicate that the glassy phase has reached a molten state and diffused out of the thick film. This is indicative of an overfired state either during initial firing conditions or during a subsequent firing. An EDAX of such a substrate surface would show an absence of gold. An example of a correctly fired thick film is shown in Fig. 13, and there is no indication of glass stratification. In this case the absence of holes and the presence

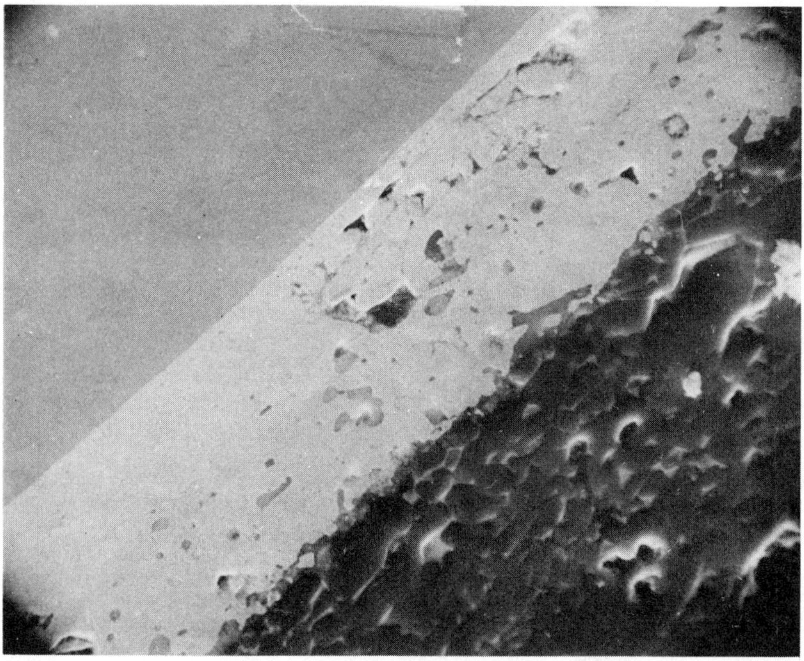

FIG. 10—*Correctly fired thick film* (×*1000*).

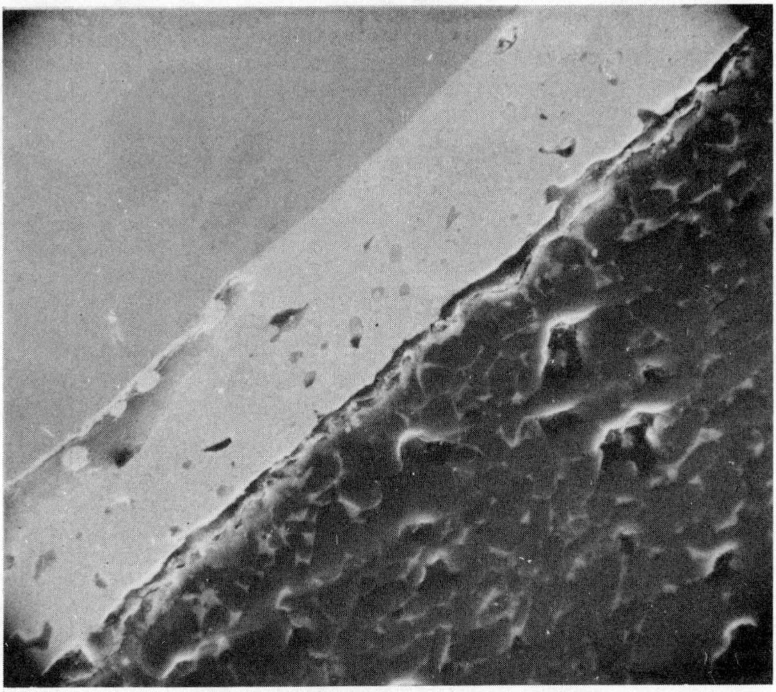

FIG. 11—*Overfired thick film* (× *1000*).

of fractures with well-defined peaks and deep valleys show that a good locking action is present between the metal and glass phases in the thick film and the ceramic. An EDAX of such a structure would show the presence of gold on the ceramic surface. Often, in cases of good adhesion the fracture will occur in the ceramic, for example, Fig. 14.

Metallographic sectioning is recommended when furnace profiles are being established to insure that overfiring or even underfiring is not present.

Solderability

Thick-film pads can be leached by the solders used to bond components or interconnects. The problem can be kept under control by carefully monitoring process variables. Solderability is also affected by the choice of solders and the particular metalization system chosen. If sequential soldering operations are necessary in particular areas, for instance, soldering wires into slotted biterminal clips previously soldered to a thick-film pad, then relatively high melting point solders have to be used for the initial soldering operation. Obviously an automated system poses less problem than a manual-

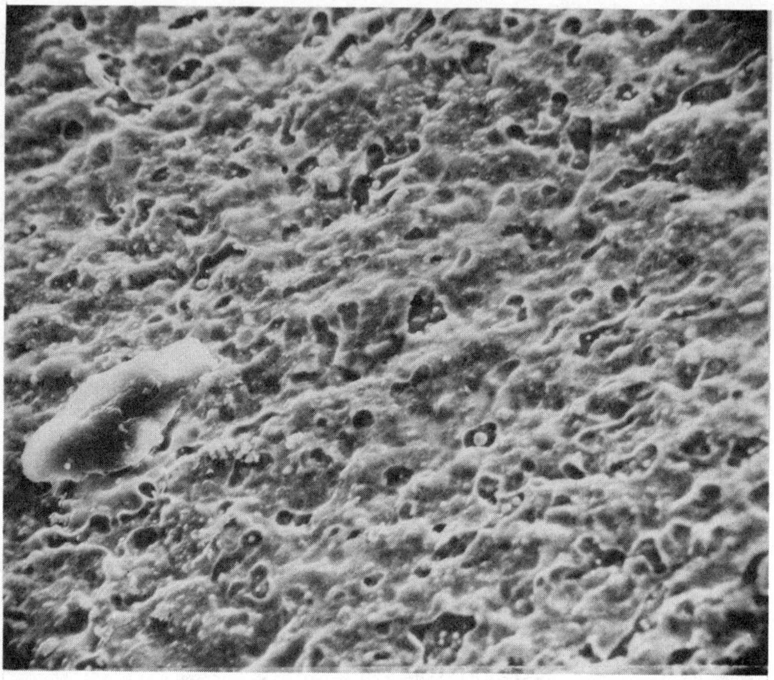

FIG. 12—*Overfired glassy layer* (×*800*).

type operation. Metallurgical sectioning and SEM analysis can help in establishing process limits. The use of element maps for tin can show the penetration of this component of gold-tin and lead-tin alloys into the thick-film pad. This procedure can show how much dissolution of the platinum-gold pad has occurred. By modification of the time/temperature process and use of the SEM, an optimized procedure can be achieved.

Quality Control

To ensure that the end product is acceptable, it is important that any reliability and quality assurance (R&QA) program achieve the following:

(*a*) Monitoring of quality,
(*b*) Data evaluation,
(*c*) Assurance that the design intent is met, and
(*d*) Recommendations for corrective action.

Such a program would serve two functions in that controls on the following would be accomplished:

1. Materials, process, and equipment:
   (*a*) Material test evaluation,
   (*b*) Process equipment evaluation,
   (*c*) Materials and equipment specifications, and
   (*d*) Process specification.
2. Statistical function:
   (*a*) Data analysis,
   (*b*) Failure mode and failure effects analysis,
   (*c*) Reliability analysis, and
   (*d*) Summary data reports.

The key to this successful program was the early and continuous involvement of an R&QA group to ensure reliability and to see that the imposed quality requirements were met. The procedure used for this thick-film MLB operation was as follows:

(*a*) Select thick-film material and test for screenability. Check screened coupons for electrical properties, for example, maximum allowable capacitances, minimum allowable breakdown resistance, and see that conductivity requirements are met.

(*b*) Test processes to be utilized; for example, establish correct furnace profiles.

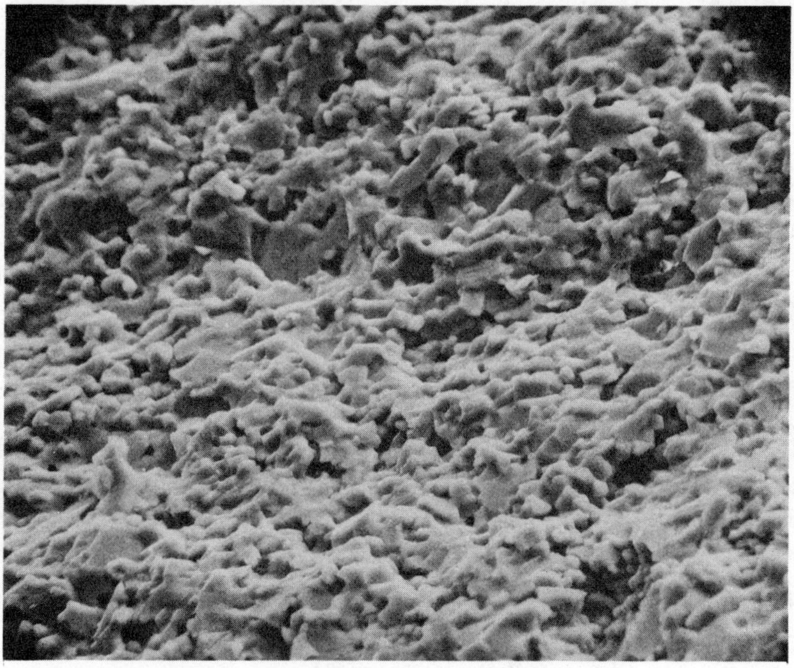

FIG. 13—*Correctly fired glassy layer* (×*800*).

FIG. 14—*Ceramic breakout* (×60).

(*c*) Establish traceability for substrates, inks, and equipment.

(*d*) Monitor procedures, noting all deviations from the process, using a traveling record card for each self-contained batch.

(*e*) Establish acceptance criteria and procedures for all processes and materials.

(*f*) Integrate an adhesion test into the acceptance criteria.

(*g*) Publish failure analysis summaries.

(*h*) Ensure that all rework performed is according to written procedures and is documented.

(*i*) Ensure that a complete data package is provided with each batch of screened and fired thick-film boards.

A goal that is needed is the establishment of procedures (preferably American Society for Testing and Materials) to cover the adhesion testing of thick films.

Summary

The five experimental methods of adhesion testing described herein were utilized in setting up process controls for the manufacture of a complete thick-film MLB on ceramic. When connections are soldered onto thick-film pads the adhesion of such connections to pads and of the pad to ceramic can be measured by use of a force gage. It was shown that the attachment of the pad to the ceramic was the limiting factor. It was shown also that soldering has an effect on the adhesion of the pad to the ceramic due to leaching, and that a force gage is useful for routine monitoring of product. It was further shown that no appreciable differences in results were found in utilizing a push- or a pull-type gage.

It was found that wire peel tests are an invaluable tool, particularly on substrates that have been run through firing profiles to simulate the addition of all the multilayers. Once a firing profile had been established, then push or pull tests can be utilized to test the strength of thick-film pads.

This combination of a wire pull test and push/pull test serves both as an adhesion and solderability test and allows one to determine if solderability affects adhesion.

Pull-on-loop wire adhesion tests are useful for determining the bondability characteristics of pads and of pad to dielectric. It was found that careful attention has to be paid to the geometry of the loop, as marked changes in results can occur.

The use of destructive metallurgical sectioning and analysis of failures using an optical microscope or SEM was found to be essential in identifying causes of poor strength. A rigid quality-control procedure was established to ensure that a reliable product is produced.

References

[1] Becher, P. F., Bascom, W. D., Bitner, J. L., and Murday, J. S., *Proceedings,* International Society for Hybrid Microelectronics Symposium, 1975, p. 279.

[2] Hitch, T. T., Conlon, E. J., and Rogers, W. I., *Proceedings*, International Society for Hybrid Microelectronics Symposium, 1975, p. 287.

[3] Hailes, L. and Crossland, W. A., *Electronic Packaging and Production,* 1972, Vol. 12, No. 6.

[4] Hitch, T. T., *Journal of Electronics Materials*, 1974, Vol. 3, No. 2, p. 553.

[5] Schneider, B., Vinikman, V., Samuel, A., and Kaplan, E., *Proceedings*, International Society for Hybrid Microelectronics Symposium, 1974, p. 16.

[6] Morey, R. L. and Pijoan, P. J., *Proceedings,* International Society for Hybrid Microelectronics Symposium, 1975, p. 252.

[7] "Method of Test for Wire Peel Adhesion to Soldered Thick Film Conductors to Ceramic Substrates," Bulletin-A-74672, E. I. DuPont de Nemours and Co., Inc., Electrochemical Department, Wilmington, Del., 1971.

[8] NBS Technical Note 743, p. 29 and 767, p. 16 (1972), 1973 U.S. Department of Commerce, National Bureau of Standards.

DISCUSSION

K. L. Mittal[1] (*written discussion*)—In your push test, is there a pad thickness limitation? If so, what are its effects or results?

R. L. Morey (*author's closure*)—*The pad thickness of platinum-gold is 18* μm (0.0007 in.) with a gold overlay of 13 μm (0.0005 in.) together with a gold-tin solder layer of approximately 13 μm to 0.4 mm (0.0005 to 0.0015 in.). The thickness of the pad was not a limitation, the only criterion being that the head of the push tool to be aligned accurately with the terminal so that the load was applied axially.

[1]IBM Corporation, Hopewell Junction, N.Y. 12533.

G. J. Ewell[1]

Evaluation of Methods for Performing Adhesion Measurements of Thick-Film Terminations on Chip Components

REFERENCE: Ewell, G. J., "**Evaluation of Methods for Performing Adhesion Measurements of Thick-Film Terminations on Chip Components,**" *Adhesion Measurement of Thin Films, Thick Films, and Bulk Coatings, ASTM STP 640,* K. L. Mittal, Ed., American Society for Testing and Materials, 1978, pp. 251–268.

ABSTRACT: The nailhead lead tension test is shown to be a very useful method of measuring the adhesion strength of thick film, solderable end terminations on chip resistors, and multilayer monolithic ceramic capacitors. Alternative test concepts are considered and detailed comparisons are presented. Specimen preparation and test parameters are defined, and the effects of variation in termination configuration are illustrated. It was found that attachment parameters must be well defined and followed so as not to introduce other variables in measured strengths. The test criteria defined in this paper can be applied to other specific devices, but limits for their use must be individually established. The data provide application limits for some commonly used chip components.

KEY WORDS: adhesion, adhesion tests, resistors, ceramic capacitors, terminations, passive components, tension tests, shear tests, thick-film inks

Recent studies of the adhesion of thick-film inks and coatings [*1–5*][2] indicate, first, that adhesion can only be defined practically by means of the type of test used for adhesion measurement and, second, that there is no completely satisfactory method now available to measure adhesion. The latest review of techniques for adhesion measurement [*3*] concentrates upon preparing carefully selected specimens on substrates for testing and does not mention the testing of adhesion of thick-film terminations on completed components. Determining adhesion on completed components rather than on special specimens is necessary, however, to test for changes

[1] Head, Inorganic Materials Section, Technology Support Division, Hughes Aircraft Company, Culver City, Calif. 90230.

[2] The italic numbers in brackets refer to the list of references appended to this paper.

in adhesion resulting from the specific processing given ceramic substrates or to monitor the quality of individual production lots. Some component vendor documents and military specifications require nailhead lead tension specimens (Fig. 1) to evaluate termination adhesion. Hughes Aircraft Company recently began a vendor surveillance program to survey silver-terminated and palladium/silver-terminated capacitors and platinum/palladium/gold-terminated resistors. As part of that program it was necessary (*a*) to compare alternative termination adhesion test methods with the nailhead lead test and (*b*) to define specimen preparation and test parameters necessary for use with the optimum test method.

All parts studied in this program were terminated with thick-film, glass-fritted inks, fired in ambient atmospheres. The terminations were all of a wraparound configuration, selected for solderability, leach resistance, and reliable use experience. Parts from six vendors were studied, as listed in Table 1.

This study first considers a comparison of nailhead tension and ribbon lead shear tests, and second the required nailhead specimen preparation and test parameters.

Comparison of Nailhead and Ribbon Test Methods

Selection of Ribbon Lead Shear Test as Alternative

Seven criteria were used to select an alternative test method to the nailhead tension test. Listed in Table 2, the first criteria relate to the sensitivity of the test method to adhesion degradation caused by thermal exposure or improper processing and to compatibility of resultant adhesion data to the large amount of information contained in the part vendor's past data. Two criteria relate to the attachment of these parts to substrates by reflow soldering. Direct attachment of devices to substrates results in high shear stresses being applied to the termination-to-part interface closest to that substrate as are tensile stresses to the part's end. The test method should approximate both of these stress loading conditions.

The chip component industry has only one alternative test method that meets the majority of the test criteria. The ribbon lead shear test, shown in Fig. 2, offers ease both of specimen preparation and of testing by the same equipment used to test nailhead lead specimens. Furthermore, this test method approximates the stress loading conditions that the parts would experience in service by duplicating both force directions and magnitudes.

Specimen Preparation Techniques

Ribbon and nailhead lead specimens of both chip resistors and capacitors were prepared. Table 3 lists the materials and techniques chosen

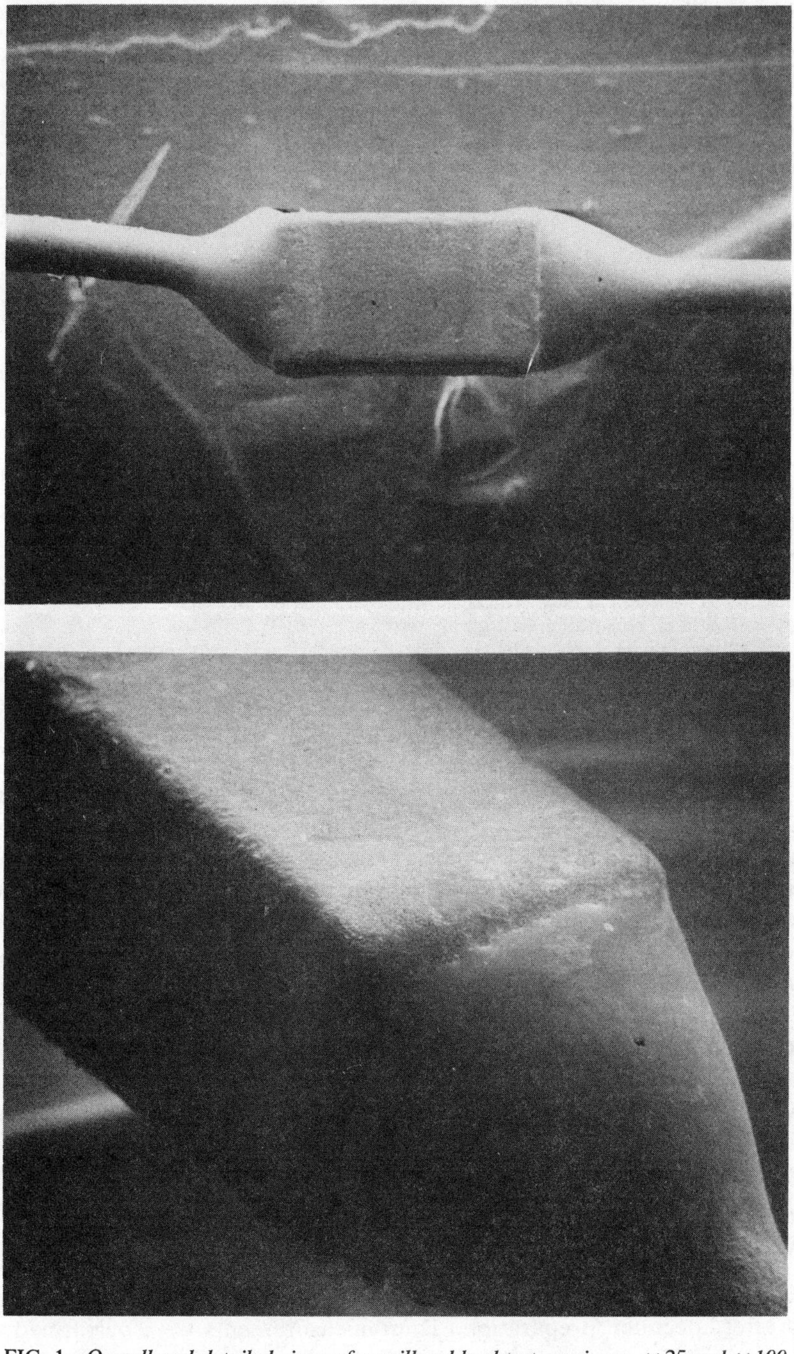

FIG. 1—*Overall and detailed views of a nailhead lead test specimen, ×25 and ×100.*

TABLE 1—*Chip components evaluated.*

Component	Vendor	Termination	
		Material	Coverage
Thick-film resistor	A	Pt-Pd-Au	ends, top, bottom
	B	Pt-Pd-Au	ends, four sides
	C	Pt-Au/Ni	ends, top, bottom
Multilayer chip capacitor	D	Ag + Barrier Layer No. 1	ends, four sides
	E	Ag + Barrier Layer No. 2	ends, four sides
	F	Pd/Ag	ends, four sides

TABLE 2—*Criteria for test method evaluation.*

Sensitivity to adhesion degradation
Compatibility of test information with vendor data
Ease of specimen preparation and testing
Reproducibility of test results
Ability to test adhesion at most critical substrate-termination interface
Effect of specimen preparation on part properties
Ability to approximate stress loading conditions termination will experience in service

for specimen preparation. These methods of preparation were chosen to approximate those commonly used in the industry. All specimens were prepared without fixtures, but visual alignment guides were utilized for lead attachment. Nailhead lead specimens were prepared by holding the leads, one at a time, in contact with the termination, and then touching the soldering iron to the lead at a point about 1 cm from the nailhead until the solder wetted the lead. Ribbon lead specimens were prepared in a similar manner, except that the leads were held flat against a table top during soldering, making alignment much easier. Strengths were determined with an Instron, using crosshead speeds of 0.13 cm/min for resistors and 0.05 cm/min for capacitors unless otherwise noted. The different speeds were selected to allow comparison of the test data to data of the chip manufacturers.

Effects of Specimen Preparation on Electrical Characteristics

Electrical parameters of a number of specimens were measured before and after specimen preparation. These measurements were performed as a result of experience indicating that parameter changes are often associ-

FIG. 2—*Top and side view of a nickel ribbon lead test specimen,* ×*27.*

TABLE 3—*Nailhead lead and ribbon specimen variables.*

Variable	Nailhead Specimen	Ribbon Specimen
Lead material	tinned Cu	ni
Lead dimensions	nailhead 0.09 ±0.005 cm	thickness 0.013 cm
	wire 0.041 cm	width 0.25 cm
	length 4.0 cm	length 4.0 cm
Solder	Kester 63-37 eutectic[a] Sn/Pb	Kester 63-37 eutectic Sn/Pb
Solder flux		
composition	Alpha 511 [b]	Alpha 511
temperature of use	room temperature	room temperature
Flux solvent	1,1,1-trichloroethane	1,1,1-trichloroethane
Soldering iron		
model	Telvac Instruments	Telvac Instruments
	Model 54204-T	Model 54204-T
	18 W maximum	18 W maximum
tip type	conical	conical
tip temperature	316°C (600°F)	316°C (600°F)
Soldering time	3 s	3 s

[a] Per QQ-S-571.
[b] Per Mil-F-14256.

ated with soldering-caused material interactions that also affect termination adhesion. Tables 4 through 6 present results. The results for the resistors (Table 4) indicate that percent change in resistivity is a function of both vendor and type of specimen. Analysis of the Vendor A specimens, however, indicates that soldering of nailhead leads to specimens may be a more severe process than the soldering of ribbon leads. Comparison of parameter changes for NPO[3] capacitors (low temperature coefficient of capacitance) (Tables 5 and 6) indicates that changes in both instances are less than 1 picofarad, with the nailhead test having a slightly greater change. Interestingly, ribbon lead soldering appears to cause a slight *increase* in capacitance, while nailhead lead soldering results in a slight *decrease* in capacitance. The insulation resistances of the capacitors also appear to change in opposite directions for the two tests.

Termination Adhesion

The two test methods were compared for adhesion strength, extent of data scatter, correlation of measured strength with part size, and sensitivity to adhesion degradation. Adhesion data for resistors, both immediately after specimen preparation and after exposure at 150°C (302°F) for 72 h following preparation, are given in Table 7. One half of each set of 20 speci-

[3]Having a temperature coefficient withing 30 ppm/°C within the temperature range of −55 to +125°C.

TABLE 4—*Effect of specimen preparation on chip resistors.*

Vendor	Specimen No.	Resistance at 25°C, kohm		Resistance change, kohm
		Presolder	Postsolder	
A	R1[a]	148.2	152.9	+ 4.7
	R2	146.0	149.3	+ 3.3
	R3	145.8	148.4	+ 2.6
	R4	145.3	149.4	+ 4.1
	R5	147.0	152.7	+ 5.7
	avg	...	...	+ 4.1
A	NH1[b]	145.3	170.0	+ 24.7
	NH2	147.8	150.6	+ 2.8
	NH3	148.2	256.2	+ 108.0
	NH4	148.5	205.5	+ 57.0
	NH5	147.2	147.7	+ 0.5
	avg	...	...	+ 38.6
B	R1[c]	10.06	10.23	+ 0.18
	R2	9.950	9.966	+ 0.016
	R3	9.948	9.948	0
	R4	9.956	10.12	+ 0.164
	R5	9.956	10.02	+ 0.064
	avg	...	...	+ 0.085
C	R1[c]	9.928	9.926	− 0.002
	R2	9.945	9.947	+ 0.002
	R3	9.921	9.916	− 0.005
	R4	9.951	9.951	0
	R5	10.02	10.02	0
	avg	...	...	− 0.001

[a] R: ribbon lead shear test.
[b] NH: nailhead lead tension test.
[c] Specimens for nailhead lead testing were not available.

mens were measured immediately after preparation; the other half were tested following exposure in an oven at 150°C (±3°C). The data indicate that the nailhead tests show, first, a greater decrease in strength following the thermal exposure and, second, have a larger standard deviation, that is, result in more data scatter, than the ribbon lead specimens. Both test methods resulted in comparable pull strengths of as-prepared specimens.

Table 8 contains pull strength data for various sizes and types of capacitors tested both immediately after specimen preparation and after 100 h at 125°C (257°F). The standard voltage conditioning temperature and time for multilayer ceramic capacitors, are, respectively, 125°C (257°F) and 100 h, just as 150°C (302°F) is the standard temperature for conditioning chip resistors. Standard deviations were not calculated for the nailhead specimens because only two specimens of each type were tested. The nailhead lead specimens result in higher pull strengths (except in one instance) than the ribbon lead specimens. The mean strengths of the ribbon lead specimens obviously increase with increase in part size; the same trend is not obvious for the nailhead specimens.

TABLE 5—*Effect of ribbon lead specimen preparation on NPO capacitors.*

		Before Preparation			After Preparation		
Vendor	Specimen No.	Capacitance, pF	Dissipation Factor, %	Insulation Resistance at 25°C, ohm	Capacitance, pF	Dissipation Factor, %	Insulation Resistance at 25°C, ohm
D[a]	1	177.9	0.0015	0.31×10^{12}	177.8	0.0015	10.0×10^{12}
	2	180.4	0.0006	...	180.5	0.0013	...
	3	181.9	0.0006	7.14	182.2	0.0012	...
	4	185.9	0.0002	...	186.2	0.0012	11.1
	5	173.6	0.0005	12.5	173.8	0.0013	16.7
E[b]	6	10.9	0.0009	...	11.0	0.013	...
	7	11.1	0.0011	...	11.2	0.013	...
	8	11.0	0.0010	...	11.1	0.0135	7.69
	9	11.0	0.0010	...	11.0	0.013	...
	10	11.6	0.0010	...	11.7	0.013	0.42
F[b]	11	11.1	0.0008	...	lead broke during measurements		
	12	10.8	0.0010	...	11.0	0.0154	...
	13	broke during measurement			lead broke during measurements		
	14	180.4	0.0002	...	180.4	0.0020	5.0
	15	175.4	0.0002	...	175.7	0.0012	16.7

[a] 180 pF ±5% parts; length 0.254/0.165 cm; width 0.170/0.089 cm; thickness 0.140 cm maximum.
[b] 10 pF ±5% parts; dimensions as above.

TABLE 6—*Effect of nailhead lead specimen preparation on NPO capacitors from two vendors.*[a]

		Before Preparation			After Preparation		
Vendor	Specimen No.	Capacitance, pF	Dissipation Factor, %	Insulation Resistance at 25°C, ohm	Capacitance, pF	Dissipation Factor, %	Insulation Resistance at 25°C, ohm
D[b]	1	11.0	0.001	90.9×10^{12}	10.2	0.0010	...
	2	11.7	0.0009	62.5	10.8	0.0010	...
	3	11.3	0.0009	66.7	10.5	0.0010	...
	4	11.4	0.0009	71.4	10.5	0.0010	12.5×10^{12}
	5	11.5	0.0010	62.5	10.7	0.0009	...
	6	11.6	0.0006	55.6	10.7	0.0010	...
	7	11.8	0.0010	71.4	10.8	0.0010	...
E[b]	1	13.5	0.0130	...	12.0	0.1160	...
	2	10.8	0.0007	100	10.0	0.0007	...
	3	11.2	0.0010	83.3	10.4	0.0010	...
	4	11.0	0.0010	...	10.2	0.0010	...

[a] Data on Vendor F are not available due to a lack of test specimens.
[b] 10 pF ±5% parts; dimensions as noted in Table 5.

TABLE 7—*Resistor end termination adhesion strength.*

Resistor Vendor	Part Condition[a]	Nailhead Strength, *N*		Ribbon Strength, *N*	
		$\overline{X}$	σ	$\overline{X}$	σ
A	1	18.3	7.16	21.5	3.02
	2	12.8	3.56	18.5	2.14
B[b]	1	42.4	6.00	23.7	4.27
	2	27.2	8.45	18.6	4.05
C	1	21.4	5.78	21.9	2.31
	2	11.1	2.40	19.1	3.47

[a] 1. After specimen preparation.
2. After exposure at 150°C (302°F) for 72 h.

[b] The Vendor B specimens were all on 0.0635-cm-thick Al_2O_3 substrates; the other specimens were on 0.038-cm-thick substrates, which may introduce a new variable in this test.

TABLE 8—*NPO capacitor end termination adhesion strength.*

Vendor	Capacitance, pF	Physical Size	Nailhead Strength, *N*[a]	Ribbon Strength, *N*	
				$\overline{X}$	σ
		AFTER SPECIMEN PREPARATION			
D	10	A[c]	>30.7	18.0[b]	2.80
	180	A	30.3	17.3	2.09
	470	C[d]	>28.9	34.6	5.38
E	10	A	26.7	10.9	3.25
	470	C	25.8	20.1	5.29
F	10	A	24.5	17.7	4.36
	180	A	32.9	15.6	5.43
	470	C	38.7	28.2	4.85
		AFTER 100 H EXPOSURE AT 125°C (257°F)			
D	10	A	30.2	...	...
	470	C	>29.8	...	...
E	10	A	26.7	...	...

[a] Average of only two specimens, so standard deviations were not calculated.
[b] Average of 7 to 20 specimens.
[c] Length 0.254/0.165 cm; width 0.170/0.089 cm; thickness 0.140 cm maximum.
[d] Length 0.495/0.420 cm; width 0.241/0.152 cm; thickness 0.190 cm maximum.

Fractographic Analysis

Studies were made in a scanning electron microscope (SEM) of the fracture surfaces of both nailhead and ribbon test specimens of the resistors and capacitors. Figures 3 through 6 illustrate the differences in fracture modes. The nailhead lead resistor specimens shown in Figs. 3 and 4 fractured primarily at and adjacent to the termination/ceramic interface. Capacitor specimens failed in a similar manner. Interface failure indicates

FIG. 3—*Overall view of fracture surface of nailhead lead tested chip resistor,* ×*52.*

the validity of correlating associated pull strengths with termination adhesion. The majority of ribbon lead test specimen fractures exhibited some degree of ceramic substrate fracture. The alumina substrates of the thick-film resistors showed very little shearing but the titanate substrates of capacitors (Fig. 5 and 6) exhibited very complex shearing modes. Pull strengths associated with such failures primarily reflect substrate strengths and can have little correlation with termination adhesion.

Conclusions of Comparison Tests

The nailhead tension test method proved to be optimum in meeting the requirements specified in Table 2. High sensitivity to adhesion degradation is shown in Table 7; the low sensitivity to adhesion degradation when the ribbon lead shear test method is used may indicate that pull strengths measured by this technique are affected by substrate defects. The nailhead method is compatible with vendor data and specimens can be quickly and easily prepared. While standard deviations are higher for the nailhead test compared with the ribbon test, they fall into acceptable ranges. Effects of specimen preparation on electrical parameters were shown to be

FIG. 4—*View of nailhead lead fracture surface of specimen shown in Fig. 3.* ×*50.*

relatively small, and fractographic analyses indicated that strengths were directly related to termination adhesion.

Definition of Nailhead Specimen Preparation Parameters

The effects and interrelationships of both operator and test variables were studied on one lot each of capacitors from Vendors D and E. All strengths given are the mean of at least five test specimens.

Five sets of tests were preformed. The tests were to determine the effects of the following operator and test variables:

1. Lead alignment.
2. Lead wire diameter.
3. Test pull speed.
4. Nailhead diameter/chip thickness ratio.
5. Time to form solder joint.
6. Amount of flux used in solder joint formation.
7. Extent of termination "wraparound."

The first set of tests was to determine the interrelated effects of the first three variables just listed. Lead alignment was tested by preparing two

FIG. 5—*Appearance of fracture surface of ribbon lead tested chip capacitor,* ×*47.*

sets of specimens, one set with nailhead leads aligned as axially as possible, the other set with the first lead axially oriented and the second lead soldered at an angle of 20 to 30 deg off alignment. The effects of lead wire diameter were determined using two sets of nailhead leads with similar head diameters, one set with 0.041 cm lead diameter and 0.094 cm head diameter, the other set with 0.064 cm lead diameter and 0.107 cm head diameter. The effect of test pull or crosshead speed was evaluated by using pull speeds of 0.05 and 0.5 cm/min. Test results in Table 9 indicate that misalignment will reduce measured strengths for specimens with large-diameter leads, but is of minor or negligible importance for specimens with small (0.041 cm) wire diameter leads. Crosshead speed is only of minor importance.

The second set of tests measured the effects of two nailhead diameter/capacitor thickness ratios (1.06 and 0.5) on adhesion strength measured at a 0.05-cm/min pull speed. Wire diameter was held as constant as possible. Results in Table 10 indicate that low ratios drastically reduced measured strengths.

FIG. 6—*Appearance of fracture surface of ribbon lead tested chip capacitor,* ×*61.*

TABLE 9—*Effects of crosshead speed, alignment, and nailhead wire diameter on strengths of NPO capacitors.*

Nailhead Lead Diameter	Crosshead Speed		
	0.05 cm/min		0.5 cm/min
	Vendor E	Vendor D	Vendor E
0.064 cm[a]			
Good alignment	71.6 *N*	52.0 *N*	62.3 *N*
Misalignment	31.1	26.7	26.7
0.041 cm[b]			
Good alignment	30.7	. . .	≥30.7[c]
Misalignment	30.7	. . .	>26.7[d]

[a] Head diameter 0.107 cm.
[b] Head diameter 0.094 cm.
[c] All specimens failed in the leads at 30.7 *N*.
[d] Three specimens failed in the leads at 31.1 *N*; two in the end termination at 26.7 *N*.

TABLE 10—*Effects of head diameter/chip thickness ratios on strength of Vendor E NPO capacitors.*

Head Diameter	Head Diameter / Chip Thickness[c]	Nailhead Strength, N	
		$\overline{X}$	σ
0.127 cm[a]	1.06	64.1	11.6
0.058 cm[b]	0.49	29.4	6.67

[a] 0.064 cm wire diameter.
[b] 0.041 cm wire diameter.
[c] Thickness 0.140 cm maximum; about 0.120 cm typical.

The third and fourth sets of tests measured the effects of several operator variables: the soldering time to form a nailhead lead-capacitor joint, and the amount of flux used in forming the joint. These variables were of concern because examination of several specimens that exhibited abnormally low pull strengths indicated that (*a*) voids resulting from entrapped flux and (*b*) leaching (dissolution of precise metals from the end termination during soldering) both contributed to very low pull strengths. The effects of time were studied by preparing two sets of specimens, one with just sufficient soldering iron contact to the nailhead to cause solder to wet and flow up the lead, the second with the soldering iron held in contact with the lead for two seconds after solder wetting occurred. Effects of flux were studied by preparing one set of specimens in the normal way and a second set with the majority of flux blotted up with a cotton swab. Results in Table 11 indicate no significant effects of these variables.

Because the extent of side coverage or "wraparound" of a termination varies from part to part, a fifth set of tests was used to determine the degree that variation in termination configuration affects pull strengths. Table 12 lists strengths determined on Vendor A resistors, for which the wraparounds include top and bottom surfaces but not sides. Significant adhesion decreases are noted only as both wraparound portions of the termination are eliminated.

Conclusions

The nailhead lead tension test has been shown to be a very useful means for measuring the adhesion strength of thick-film, solderable terminations on chip resistors and multilayer, monolithic capacitors. This test method exhibits good sensitivity to adhesion degradation, adequate repeatability, and fracture at the desired interface. While the ribbon lead shear test allowed for better data repeatability and ease of lead alignment, it was inferior in two very important effects. One effect was the relative inability to detect loss of termination adhesion (the major reason for performing these

TABLE 11—*Effects of soldering time and amount of flux on mean strength of Vendor E NPO capacitors.*

Soldering Time				Amount of Flux			
Normal[a]		Normal + 2 s		Normal[b]		Normal −	
$\overline{X}$	σ	$\overline{X}$	σ	$\overline{X}$	σ	$\overline{X}$	σ
62.7 *N*	14.2 *N*	54.7 *N*	22.2 *N*	58.7 *N*	14.7 *N*	47.1 *N*	13.3 *N*

[a] Sufficient time for solder to wet and flow.
[b] Normal is several drops directly on mating surfaces when in close proximity; Normal − refers to the smaller amount produced by using a cotton swab to blot up available flux.

TABLE 12—*Variation of mean strength with termination configuration of chip resistors.*

Test Method	Termination Configuration	Strength, *N*	
		$\overline{X}$	σ
Nailhead	end, top, bottom[a]	20.82	4.00
	end, bottom	22.02	2.54
	end	8.50	2.36

[a] The terminations did not extend to the sides of these parts.

tests). The other effect, indicated by the damaged substrates noted with the ribbon lead test, was that substrate strength was in some way also being measured, *not* just termination adhesion. Fractographic analyses of both resistor and multilayer capacitor nailhead fracture surfaces, to the contrary, indicate that failure does originate at and stay within the termination-to-ceramic interface area for a majority of the fracture surface.

Investigation of testing and specimen preparation variables indicated that lead misalignment was important with large-diameter (elastically stiff) leads. Such misalignment resulted in significantly smaller adhesion strengths. Misalignment of smaller-diameter leads was much less significant. Strengths also decreased as the lead head diameter/chip thickness ratio decreased, again probably as a result of increased shear stresses. Minor variations in crosshead or pull speed, soldering time, and amount of flux utilized appeared to have no significant effects on measured strength. Finally, modification of termination configuration to eliminate the wraparound portions of the termination did significantly reduce measured strength.

Use of relatively small-diameter (0.041 cm) lead wire is recommended along with nailhead diameters of about the same dimension as the thickness of the part to be tested. The axiality of lead alignment should be assured by either use of alignment fixtures or by careful visual examination. Finally, a detailed soldering schedule for specimen preparation and rasonably close adherence to that schedule will assist in the consistent preparation of test specimens that reflect real adhesion strengths of part end terminations.

References

[1] Editor, "How Do You Measure Thick Film Adhesion?" *Circuits Manufacturing*, Vol. 11, No. 9, Sept. 1971, pp. 60-61.

[2] Anjard, R. P., *Microelectronics and Reliability*, Vol. 10, 1971, pp. 269-275.

[3] Hitch, T. T. and Bube, K. R., "Basic Adhesion Mechanisms in Thick and Thin Films," Technical Report PRRL-75-CR-3, RCA Laboratories, Princeton, N. J., 31 Jan. 1975.

[4] Mittal, K. L. in *Properties of Electrodeposits*, R. Sard et al, Eds., The Electrochemical Society, 1975, Chapter 17, pp. 273-306.

[5] Peckinpaugh, C. J. and Tuggle, R. L., *Proceedings*, International Microelectronics Symposium, International Society of Hybrid Microelectronics, 1968, pp. 417-423.

DISCUSSION

K. L. Mittal[1] (*written discussion*)—Why do they call it a nailhead test?

G. J. Ewell (*author's closure*)—The term derives from the use of a lead in the configuration of a nail (small-diameter shank, large-diameter head) and the attachment of the head of the lead to the specimen. Occasionally as in Fig. 7, small nails are actually used as leads.

W. K. Jones[2] (*written discussion*)—What is the usefulness of this test for chip components?

G. J. Ewell (*author's closure*)— This test has been found useful primarily in detecting changes in the adhesion of terminations to components resulting from changes in termination ink properties or firing schedules. The test may also be used by component vendors as an acceptance criterion for termination inks.

[1] IBM Corporation, Hopewell Junction, N. Y. 12533.

[2] Charles Stark Draper Laboratory, Inc., Cambridge, Mass. 02142.

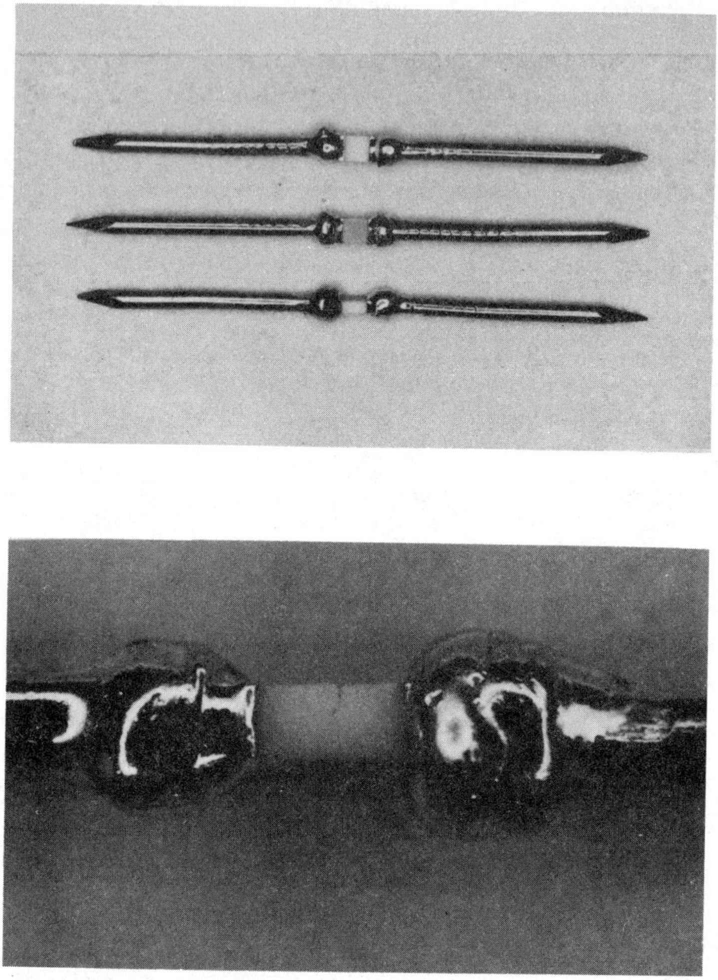

FIG. 7—*Illustrations of "nailhead" test specimens.*

S. S. Leven[1]

Adhesion Measurement Technique for Soldered Thick-Film Conductors

REFERENCE: Leven, S. S., **"Adhesion Measurement Technique for Soldered Thick-Film Conductors,"** *Adhesion Measurement of Thin Films, Thick Films, and Bulk Coatings, ASTM STP 640,* K. L. Mittal, Ed., American Society for Testing and Materials, 1978, pp. 269–284.

ABSTRACT: This paper describes the results of an experiment to determine practical screening and quality-assurance procedures to preclude excessive adhesion degradation, due to solder leaching, in gold alloy/solder systems in thick-film hybrid microcircuits. These results include a comparative rating between the materials tested, along with criteria for rating different materials.

Four environmental stresses were compared to determine which one to use as a quality-assurance test for adhesion decay. The test method consisted of soldering copper pins to conductor pads on the substrate, and then pull-testing the pins to the destruction of the conductor/substrate bonds, both before and after environmental stresses. This method provides a measurement of both tensile adhesion between the soldered conductor and the substrate and adhesion degradation promoted by the environmental stress.

Seven thick-film conductor materials and three solder compositions were used, making a total of 21 conductor/solder systems tested. Pins were soldered to each of the seven conductor materials with each of the three solders. Adhesion measurements were made immediately, again after a minimum three-day shelf life, and also after each environmental stress. The median force-per-unit-area values for each conductor/solder/environmental stress combination were tabulated and plotted. The results are presented to show the relative degradations of the different materials.

Based on the results of the experiment, the temperature cycling stress appears to be the most practical stress to use as a screening procedure for solder leaching. Guidelines are presented for performing the screening test and for determining a rating for any material combination.

KEY WORDS: thick films, adhesion, solder, indium, tension tests, leaching, scavenging, temperature cycling, thermal shock, pull tests

Adhesion of the thick-film conductor to the ceramic substrate is determined, in part, by the substrate and conductor materials. Degradation of this adhesion occurs due to two effects, which, for purposes of this paper, are called scavenging and leaching. Scavenging of the conductor by the

[1] Engineer, Westinghouse Electric Corporation, Defense and Electronic Systems Center, Baltimore, Md. 21203.

solder occurs when the solder is molten, and leaching occurs when the solder has solidified. Previous work [*1,2*][2] has shown that scavenging depends on the solder and conductor compositions, the solder temperature, and the amount of time the substrate is immersed in the molten solder. Scavenging causes a loss of adhesion of the conductor to the substrate, since part of the conductor is dissolved, reducing the conductor/substrate contact area. Solder does not adhere directly to glass or ceramic. This loss of adhesion is evident in the initial pull test measurement as a low value of force required to pull the conductor pad, if it remains, from the substrate.

Solder leaching, on the other hand, occurs after the solder has solidified, and is a diffusion process involving the conductor and the solder. Previous work [*3–5*] has shown that this process will in time destroy an originally good bond. Leaching may be controlled by proper combination of materials, namely, solders and conductors. In order to determine whether or not a particular combination of materials has higher leach resistance, a practical screening test must be devised so that any thick-film network manufacturer can prove to himself and to his customers that his choice of materials is acceptable.

Of the techniques available to determine whether leaching occurs, pull testing is the most readily available and offers the information most directly related to leaching: Has the adhesion of the soldered conductor to the substrate changed?

Under normal ambient conditions, leaching occurs in terms of years, rather than hours or days. Therefore, the life of the bond must be accelerated in order to see evidence of leaching in a reasonable time. The purpose of this experiment, in addition to making some initial determination as to acceptable and nonacceptable material combinations, is to identify an acceptable accelerated life condition to use as a practical screening method.

Summary of Experimental Procedure

Previous work [*6*] discussed an experimental procedure to prepare substrates for pull testing, and an experimental flow diagram was proposed. A review of the procedure is presented in this section.

In performing solder leach tests, many parameters must be controlled, including conductor material, conductor processing, conductor thickness, solder material, soldering temperature and time, area of solder joint, wire material, and substrate composition and smoothness. Steps have been taken in this experiment to minimize variations among these parameters.

Substrates of each conductor material being tested were manufactured using the paste manufacturer's recommended processing in order to minimize processing variations. Conductor thickness for each paste was mea-

[2] The italic numbers in brackets refer to the list of references appended to this paper.

sured. Each substrate has a pattern containing 97 conductor pads, each pad being 0.100 by 0.050 in., so that solder joints are the same size.

The material chosen to be soldered to the thick-film pads was oxygen-free high-conductivity (OFHC) copper, in the form of a pin, with two heads at one end. The pin measures 0.031 in. in diameter and is about 1.5 in. long. A 0.070-in.-diameter head is located at one end of the pin, and a 0.054-in.-diameter head is located 0.05 in. away. The purpose of the dual heads is to allow a relatively uniform amount of solder to adhere to the pin between the heads during dipping in the solder pot.

To hold the pins during subsequent operations, a holding fixture was developed. This fixture was milled out from a Micarta block to the shape of the substrate, and 97 holes of 0.035 in. diameter were drilled to align with the pads on the substrate. A piece of tape was placed across the bottom of the holding fixture, and the pins were inserted into the holes. The tape held the nonheaded ends of the pins tightly enough for the solder dipping which followed.

To ensure uniformity in the soldering operation, a detailed soldering procedure was established and followed [*7*]. This procedure is summarized in Table 1. The result of following this process was a substrate with 97 pins soldered to it, and with beads of epoxy at the tops of the pins, as shown in Fig. 1.

The experiment that was developed using the soldering procedure of Table 1 included the use of seven different thick-film conductors, three different solders, and four environmental stresses.

The thick-film conductors used were two gold pastes, two glassless (frit-less) gold pastes, two platinum-gold pastes, and one palladium-gold paste. These pastes were the standard, off-the-shelf products of the following suppliers: Cermet Division, Bala Electronics Corp., Conshohocken, Pa.; E. I. duPont de Nemours & Co., Wilmington, Del.; Electro-Materials Corporation of America, Mamaroneck, N. Y.; Electro-Science Labs, Pennsauken, N. J.; Electro-Oxide Corp., Palm Beach Gardens, Fla.

The three solders used are listed in Table 2. This table also lists the solder melting point in the case of the eutectic solder, and the liquidus and solidus points of the noneutectic solders. The solidus point is the temperature at which partial melting begins, and the liquidus point is the temperature at which the solder is completely molten. The range between solidus and liquidus temperatures is commonly referred to as the "plastic range."

The environmental stresses tested in this experiment included, in addition to 3-day minimum shelf storage, the following:

(*a*) Temperature cycling.

(*b*) Thermal shock.

(*c*) Step-stress testing—successive bakes at 100, 125, 150, 175, and 190°C (212, 257, 302, 347, and 374°F).

(*d*) Minimum seven-day room ambient shelf storage.

TABLE 1—*Summary of adhesion testing procedure.*

Step No.	Procedure	Controls
1	Tape back of holding fixture	...
2	Insert pins	...
3	Acetone rinse	30 s
	Alcohol rinse	30 s
	Air dry	...
	Flux dip	10 s
	Solder dip	3 s, melting point + 50°C
4	Solder peak flattening	melting point + 50°C
	Flux remover	30 s
	Alcohol rinse	30 s
	Air dry	...
5	Flux dip	10 s
	Remove tape	keep pins horizontal
6	Flux substrate	...
	Push pins through holding fixture	keep flux away from holding fixture
7	Reflow solder	15 to 30 s, melting point + 50°C
	Cool to room temperature	
8	Oil soak	3 min
	Separate pins from fixture	...
	Acetone rinse 1	1 min
	Acetone rinse 2	1 min
	Flux remover	1 min
	Alcohol rinse	1 min
	Air dry	...
	Substrate identification coding	...
9	HCl dip	1 min
	Water rinse	1 min
	Epoxy Dip	...
	Epoxy cure	24 h max
10	Pull testing	

The environmental stresses were being analyzed to determine which stress to use in the eventual screening procedure.

The seven conductor materials and three solders provide 21 different material combinations to be tested. It was determined that five substrates of each conductor/solder combination should be fabricated; thus 15 substrates of each conductor were needed. Substrates were processed as close to the paste manufacturer's specifications as practical, with all the processing parameters recorded.

Pins were soldered to each of the fifteen substrates of each conductor type by the process described in Table 1, five substrates being soldered with each of the three solders of Table 2.

After completion of the soldering operation, the five substrates of each conductor/solder combination followed the flow chart of Fig. 2. Pull

FIG. 1—*Pins soldered to substrate and epoxied.*

TABLE 2—*Solders used in the experiment.*

Solder Composition	Melting Point, °C	Solidus, °C	Liquidus, °C
63% Sn, 37% Pb	183	...	...
90% In, 10% Ag	...	141	237
50% In, 50% Pb	...	180	209

testing was performed using a Chatillon model LTCM Pull Tester, with a Chatillon DPP-5 gage, as shown in Fig. 3. The gage had a 7.3-lb scale limit, and for some measurements the scale was increased to 14.6 lb by the use of a pulley. The pull rate was ½ in./min.

The first pull tests were done 24 h after the completion of the soldering operation. This 24-h waiting period allowed the epoxy to cure and allowed stress relief of the internal forces within the solder. The substrates were then set aside for a minimum of three days, when more pull tests were made. During each of these two pull tests, five pins were pulled from each of the five substrates of each conductor/solder combination, making a total of 25 data points from each combination for each test.

Each conductor pad was given a row-column designation; each substrate had a unique code indicating manufacturer, conductor material, solder, and serial number; and each pull test was given a unique two-digit code. During the pull tests, the substrate code, test code, pad location

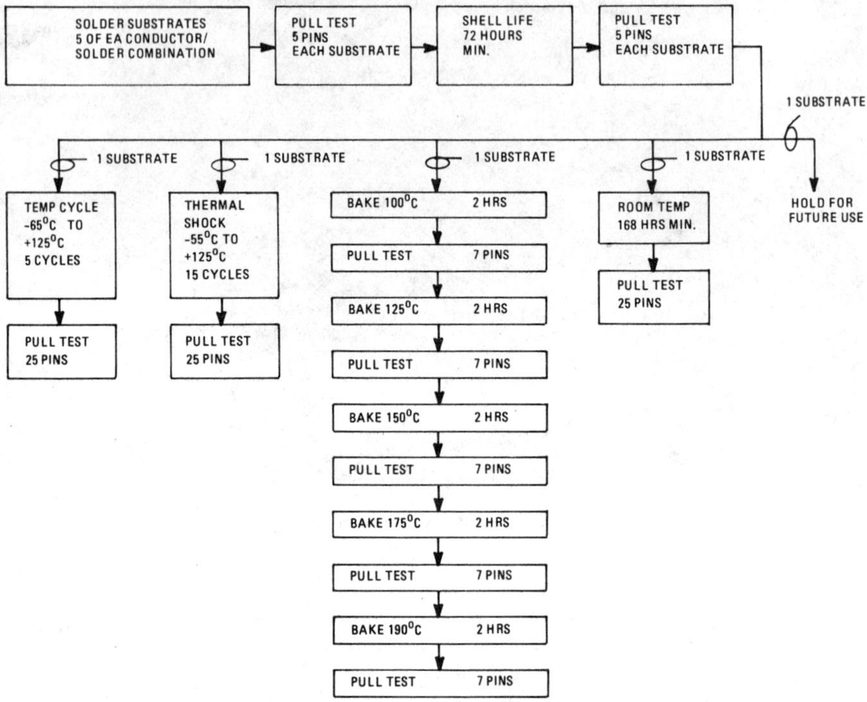

FIG. 2—*Experiment flow diagram.*

code, force value measured, and percentage of conductor pad removed from the substrate were all recorded on computer keypunch transcript forms. The data were later keypunched, and a program was written to sort and average the data.

After the second pull test of five pins per substrate, the five substrates of each conductor/solder combination were separated, one going through each of the branches in the lower portion of Fig. 2.

One substrate of each material combination underwent the temperature cycling test of MIL-STD-202D, Method 102A, Test Condition C. This test provides for five cycles, each consisting of:

$$-65^{+0,-5}\ °C,\ 30\ \text{min}$$

$$25^{+10,-5}\ °C,\ 10\ \text{to}\ 15\ \text{min}$$

$$125^{+3,-0}\ °C,\ 30\ \text{min}$$

$$25^{+10,-5}\ °C,\ 10\ \text{to}\ 15\ \text{min}$$

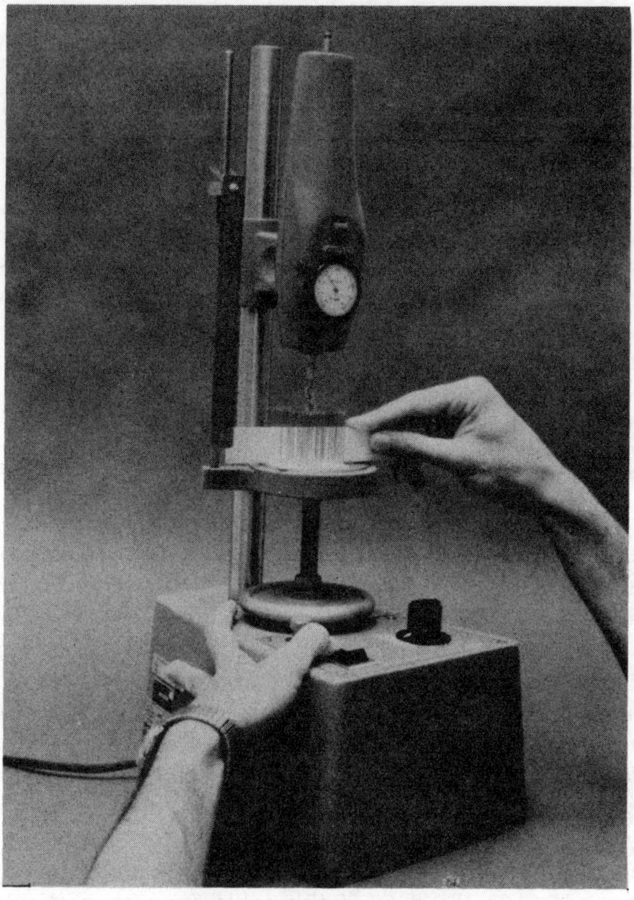

FIG. 3—*Pull-test equipment.*

After completion of the environmental stress, 25 pins were pulled from each substrate, and the data were recorded on the transcript forms.

A second substrate of each material combination underwent the thermal shock test of MIL-STD-883, Method 1011, Condition B. This test consisted of 15 cycles, each cycle consisting of 5 min of immersion in liquids at each of two temperatures: $-55^{+0,-5}$ °C, and $125^{+5,-0}$ °C. The 125 °C (257 °F) bath consisted of glycerin on a hotplate, the temperature being monitored by both a thermometer and an iron-constantan thermocouple. The −55 °C (−67 °F) bath consisted of methanol and dry ice, monitored by an alchohol thermometer and a thermocouple. A 5-s rinse of distilled water was used between baths to avoid mixing of the glycerin and methanol.

After the thermal shock environmental stress, 25 pins were pulled and the appropriate data recorded.

A third substrate of each material combination underwent the step-stress testing, which consisted of 2 h bakes at each of five temperatures: 100, 125, 150, 175, and 190 °C (212, 257, 302, 347, and 374 °F). Seven pins were pulled from each substrate after each of the 2 h bakes. All substrates in this group experienced all five temperatures. The purpose of this step-stress testing was to determine if adhesion decay accelerates rapidly at one specific temperature.

The fourth substrate of each material combination sat at room ambient conditions for a minimum of 168 h (one week) and served as a control group. At the conclusion of this test, 25 leads per substrate were pulled and the data were recorded.

The fifth substrate of each material combination has been held in reserve at room ambient conditions for future use.

Experimental Results

A computer program was written to compute the force-per-unit-area for each measurement made, sort the data, plot histograms, compute the median of each group of data, and prepare tables of the median values. The force-per-unit-area was calculated by dividing the absolute force value in pounds by the percent of pad pulled. This resulted in a force-per-unit-area value in pounds per 0.005 $in.^2$ (the pad area).

After the force-per-unit-area values were calculated for all the raw data, the values were sorted into 210 groups. Each group represented a unique combination of paste (seven used), solder (three used), and specific environmental test (ten used). A histogram of the force-per-unit-area values for each group was plotted.

The median force-per-unit-area values were then summarized in a table. Plots were made from the data of this table for each particular material combination. Four examples of the 21 plots are shown in Figs. 4 and 5. The median of the initial adhesion values was plotted, as well as its change due to the 72-h minimum shelf life. In the figures, this latter value, which is the baseline for all other environmental tests, is shown accordingly as a dotted line. The first arrow represents the adhesion change due to the temperature cycling stress discussed in the foregoing. The second arrow represents the adhesion change due to the thermal shock stress. To the right of these arrows is the curve showing the adhesion change due to the successive 2 h bakes at the temperatures indicated. At the very right is the plot of the change in adhesion between the time of the 72-h shelf life pull test and the 168-h shelf life pull test.

Analysis

One reason for performing the step-stress tests was to determine if the leaching phenomenon begins to occur at or near a specific temperature.

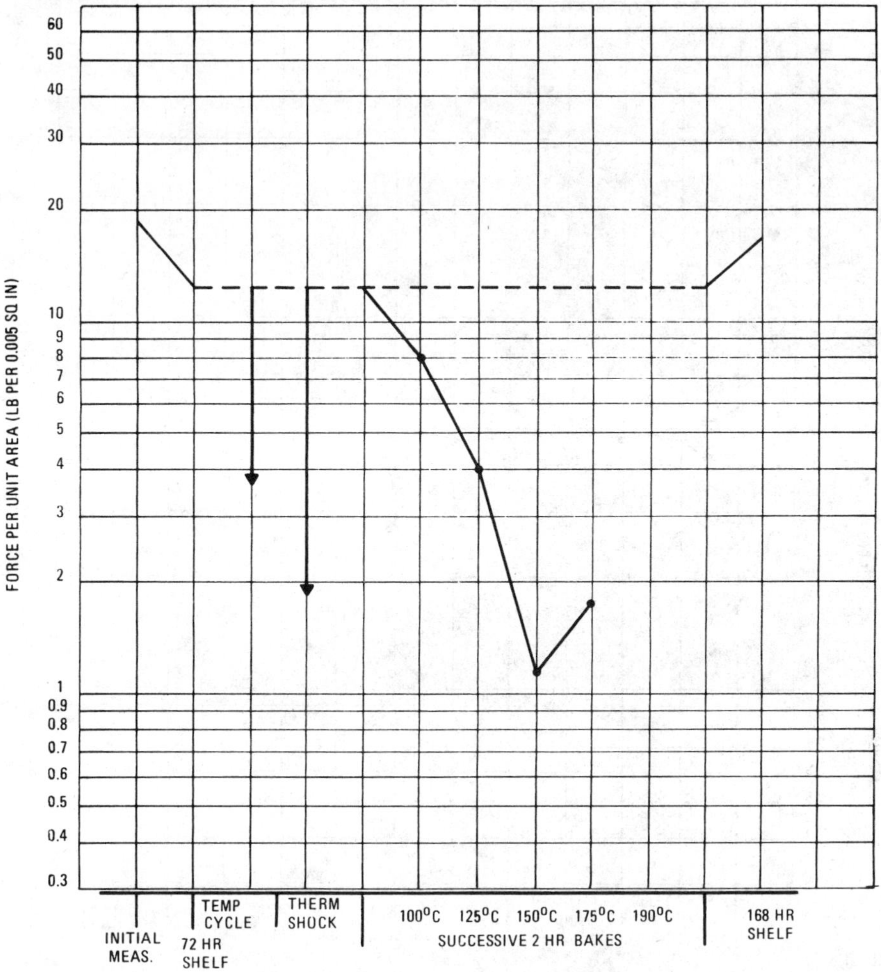

FIG. 4*a*—*Adhesion degradation due to environmental stress. Paste Manufacturer A, gold conductor, 63/37 tin/lead solder.*

Careful study of the data showed that, for most material combinations, there is a temperature where leaching begins to be more prevalent; however, the temperature is different for each material combination. Therefore, the one-temperature bake could not be used as a universal screening procedure for all materials.

Study of the median plots reveals that in some cases, during the successive 2-h bakes, the adhesion actually increases before it degrades. A clear example of this effect is shown in Fig. 5*b* (Paste Manufacturer E, glassless

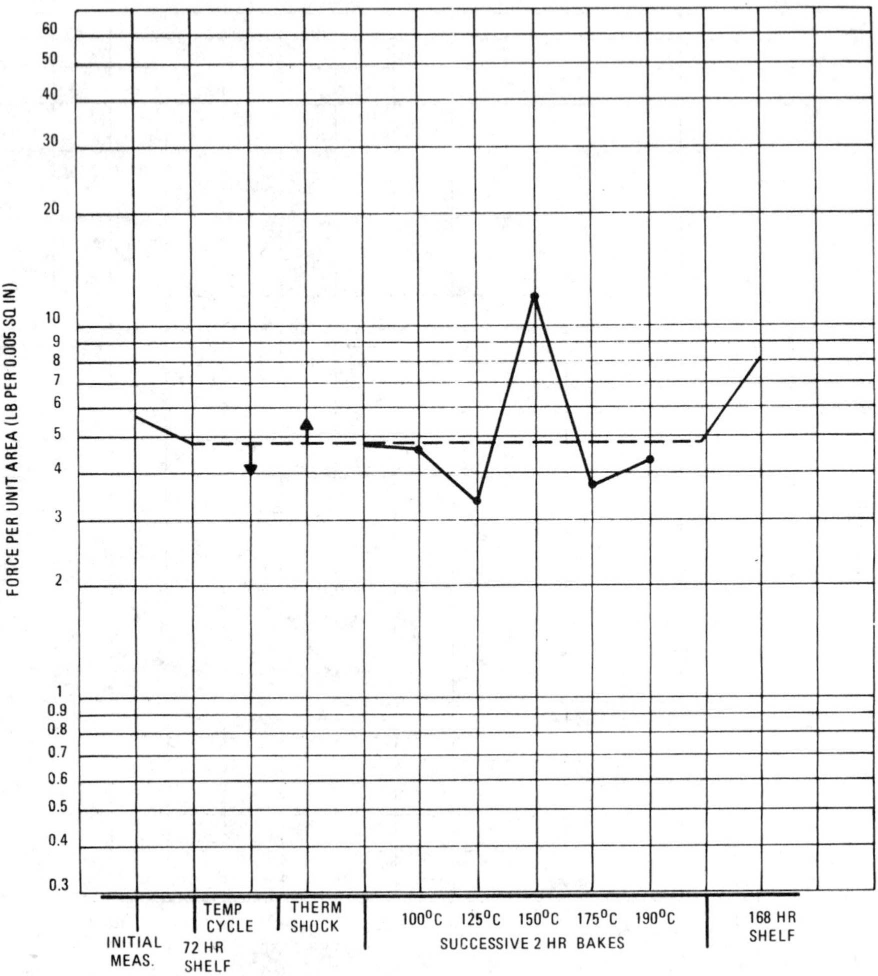

FIG. 4*b*—*Adhesion degradation due to environmental stress. Paste Manufacturer A, gold conductor, 90/10 indium/silver solder.*

gold, 50/50 indium/lead solder). This curve fits the theory concerning solder leaching; the time-temperature effects cause the formation of an intermetallic compound which in some cases is stronger at first than the original solder bond. Eventually, as more of the conductor is leached, the conductor/substrate bond deteriorates.

From the study of the median plots, it becames apparent that the most practical, yet accurate, indicator of the relative adhesion degradation for material systems is the temperature cycling stress. The degradation that

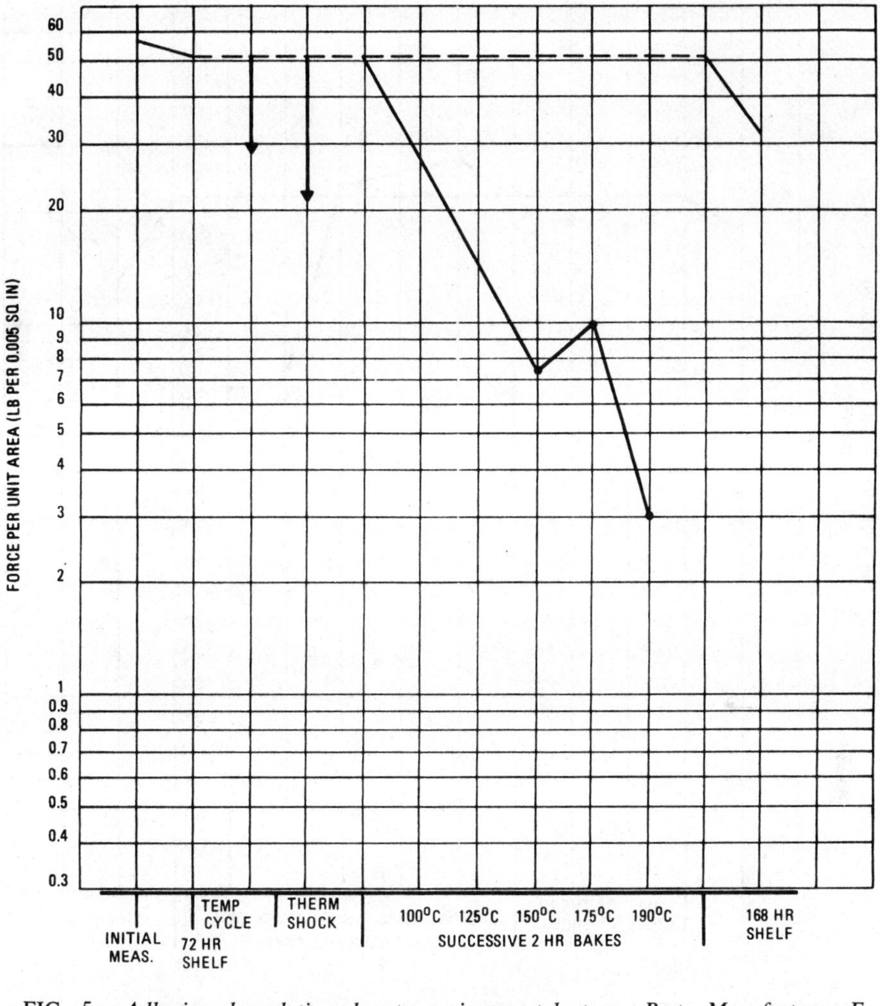

FIG. 5*a*—*Adhesion degradation due to environmental stress. Paste Manufacturer E, glassless gold conductor, 63/37 tin/lead solder.*

occurs during this stress, in most cases, closely follows the degradation which occurs due to the other stresses.

It is apparent that some material combinations have high initial (or after 3-day shelf life) adhesion values that tend to degrade during environmental stress (for example, Fig. 4*a*, A-gold-tin/lead), while other combinations have lower initial adhesion that degrades only slightly (for example, Fig. 4*b*, A-gold-indium/silver). Both the initial value and the amount of degradation are important to consider when selecting a material combination.

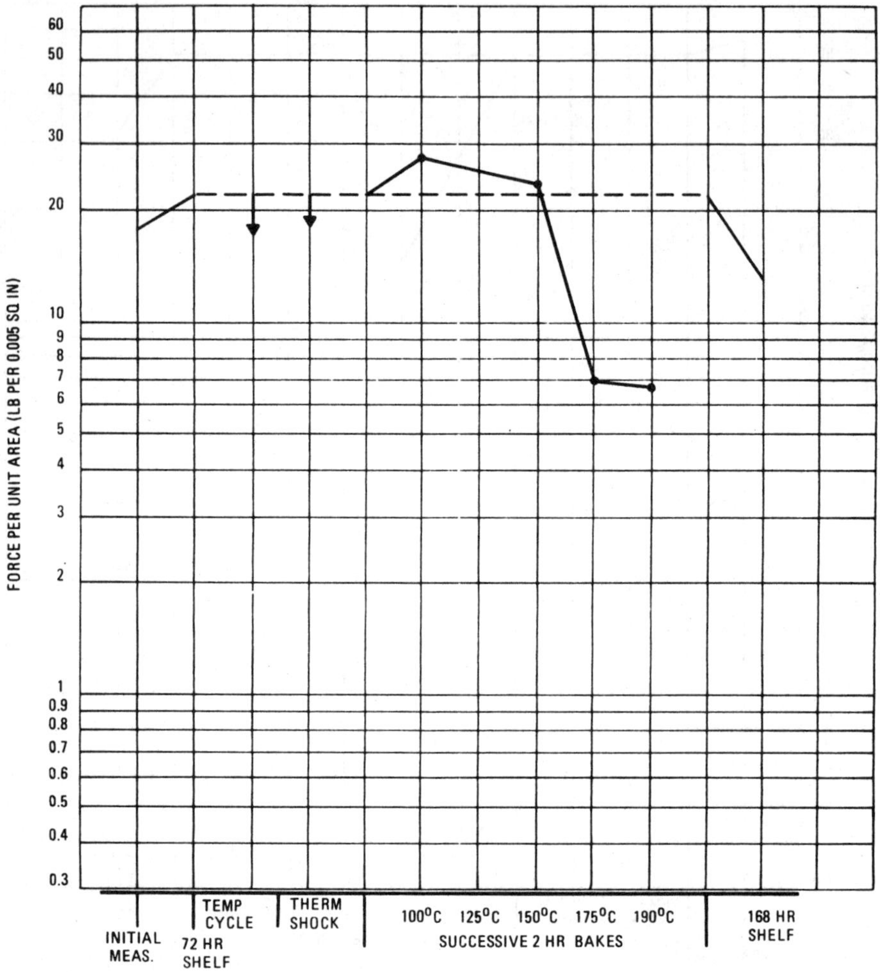

FIG. 5b—*Adhesion degradation due to environmental stress. Paste Manufacturer E, glassless gold conductor, 50/50 indium/lead solder.*

Table 3 lists the after-72-h-shelf-life adhesion value in force-per-unit-area for each of 21 material combinations, the percentage degradation due to temperature cycling, the relative rating (A, B, or C, with A being highest) of the combination, and any applicable remarks. Note that the adhesion of eight of the combinations actually increased as a result of the temperature cycling stress. The ratings listed in Table 3 followed the general criteria listed in Table 4. These criteria were evolved from general observation and grouping of all 21 of the data plots. The remarks column

TABLE 3—*Material combination ratings.*

Material Combination		After 3-Day Shelf Life, lb/0.005 in.^{2}	% Degradation Due to Temperature Cycling	Rating (A is Highest)	Remarks
Conductor	Solder				
A-Gold	Sn/Pb	12.10	70.25	C	excessive degradation
A-Gold	In/Ag	4.79	16.49	C	initial low
A-Gold	In/Pb	11.33	47.66	C	excessive degradation
A-Pt/Au	Sn/Pb	12.50	42.16	B	high degradation
A-Pt/Au	In/Ag	6.00	45.50	C	initial low
A-Pt/Au	In/Pb	8.95	−20.67[b]	A	...
B-Gold	Sn/Pb	34.00	31.76	B	high degradation
B-Gold	In/Ag	4.55	−40.66[b]	C	initial low
B-Gold	In/Pb	10.55	− 9.48[b]	A	...
C-Pd/Au	Sn/Pb	...[a]	...[a]	B	overall test results
C-Pd/Au	In/Ag	8.00	8 88	A	...
C-Pd/Au	In/Pb	6.07	27.51	B	initial low
D-glassless	Sn/Pb	24.00	23.79	A	...
D-glassless	In/Ag	5.93	24.11	C	initial low
D-glassless	In/Pb	8.31	−166.54[b]	A	...
D-Pt/Au	Sn/Pb	8.75	−31.43[b]	A	...
D-Pt/Au	In/Ag	7.14	−22.97[b]	B	initial low
D-Pt/Au	In/Pb	10.17	− 6.19[b]	A	...
E-glassless	Sn/Pb	52.00	45.38	B	high degradation
E-glassless	In/Ag	6.44	−52.33[b]	B	initial low
E-glassless	In/Pb	22.00	19.91	A	

[a] No 3-day shelf life data available.
[b] Adhesion increase (negative degradation).

TABLE 4—*Rating criteria.*

Rating	3-Day Shelf Life Median		% Degradation Temperature Cycling
A	> 8 lb/0.005 in.^{2}	and	$< 35\%$
B	> 6 lb/0.005 in.^{2}	and	$< 46\%$
C	< 6 lb/0.005 in.^{2}	or	$> 46\%$

of Table 3 contains information as to why a particular material combination received other than an A rating. It must be emphasized that these ratings are only relative to each other and are not intended to disqualify a material combination from use in thick-film circuitry.

One other point of interest is apparent from this study. The initial conductor-to-substrate adhesion is greatest using the 63/37 tin/lead solder, but it degrades rapidly. The 90/10 indium/silver-soldered substrates had a low initial adhesion that increased during 3-day shelf life, but did not degrade significantly after that. The adhesion of the 50/50 indium/lead-soldered substrates had relatively high initial adhesion which degraded significantly during 3-day shelf life, but then did not vary much during the succeeding environmental stresses. According to the rating criteria of Table 4, then, these solders would be rated as follows:

63/37 tin/lead C	C
90/10 indium/silver	B
50/50 indium/lead	A

The median force-per-unit-area values from which the foregoing ratings were determined were calculated from the complete set of adhesion values for the particular solder and environmental stress, regardless of conductor material. These ratings do not, however, indicate that one solder is unconditionally better than another, as may be seen from the ratings of Table 3.

For the strict case of solder leaching only, the numbers in the "% Degradation" column of Table 3 indicate the degree of leaching; the higher the number in that column, the worse the leaching.

Conclusions

In summary, the temperature cycling stress appears to be the stress that best differentiates between conductor/solder combinations. Based on the results of this experiment, a material qualification test has been verified, which consists of the following procedures:

1. Soldering of pins per Table 1.
2. Minimum 3-day shelf life.
3. Initial pull testing.
4. Temperature cycling per MIL-STD-202D, Method 102A, Test Condition C.
5. Final pull tests.

Acknowledgment

This work was supported by Contract DAAB07-73-C-0326 entitled "Qualification Requirements for Thick-Film Networks" for the U.S. Army Electronics Command, Fort Monmouth, New Jersey.

References

[*1*] Novick, D. T. and Kroehs, A. R., "Noble Metal Scavenging in Hybrid Circuit Bonding Applications," International Microelectronic Symposium, Beverly Hills, Calif., 16–18 Nov. 1970.

[*2*] Novick, D. T. and Kroehs, A. R., "Gold Scavenging Characteristics of Bonding Alloys," International Microelectronic Symposium, San Francisco, Calif., 22–24 Oct. 1973.

[*3*] Lasch, K. B. and Lynch, K. S., "Integrity of Solder Joints Connecting Nickel Leads to Thick Film Metallization on Ceramic Substrates," International Microelectronic Symposium, Boston, Mass., 21–23 Oct. 1974, pp. 427–440.

[4] Sulouff, R. E., "Time Dependent Relationships in Solder Interconnections," International Microelectronic Symposium, Boston, Mass., 21-23 Oct. 1974, pp. 418-426.
[5] Gillis, T. B. and Schroter, S., "Practical Consideration Affecting Thick-film Conductor Adherence of Solder Systems," 21st Electronic Components Conference, 1971, pp. 487-495.
[6] Leven, S. S. and Feinstein, J. H., "Determination of Practical Screening Procedures for Solder Leaching of Thick Film Gold and Gold Alloy Conductors," Government Microcircuit Applications Conference, Boulder, Col., June 1974.
[7] Leven, S. S., "Qualification Requirements for Thick-Film Networks, Final Report for Period 1 July 1973 to 31 Dec. 1974," Report ECOM-73-0326-F, Oct. 1975, pp. 98-125 and 191-229.

DISCUSSION

H. E. Ashton[1] (*written discussion*)—Did you correlate the results of your accelerated tests with changes that occurred over a reasonable service life? (Three days is not a long-term test.) Too many workers seem to think that because one accelerated test produces a greater change than another it is better, even though the same result does not occur in practice; for example, boiling your soldered joints in aqua regia would cause a greater decrease in strength than your cyclic thermal exposure test but would not be related to service life.

S. S. Leven (*author's closure*)—At the time of the experiment the only long-term data was that seen on field failures with gold and 63/37 tin/lead solder. These types of failures are dependent on a time/temperature relationship. The object is to induce the failure by increasing the temperature (but not to the point of melting the solder) in order to decrease the time so that the failure can be observed in a test of reasonable length. The 3-day test in this experiment was a stabilization test, and the 168-h test was a control test. The temperature cycling test was found to produce results that seemed to match the long-term field failures observed.

W. D. Bascom[2] (*written discussion*)—(1) Have you any idea of how these test results relate to the field requirement of microelectric equipment? (2) We recognize the need to screen test thick-film/solder systems. However, let me point out that if one has quantitatively determined the fracture energy for a connection, including the environmentally induced fracture

[1] National Research Council of Canada, Division of Building Research, Ottawa, Ont., Canada.
[2] Adhesion Section, Naval Research Laboratory, Washington, D.C. 20375.

rate, then there is at least the possibility of using these data to estimate how strong a bond must be in a particular microelectronic device or how to design connections that will not exceed the failure strength of the bond. Moreover, environmental cracking rate data provide some estimate of the lifetime of a device. There is a need in microelectronics to develop a rational approach to design analysis based on appropriate failure (usually fracture) criteria.

S. S. Leven (author's closure)—The degree of solder leaching found in this experiment cannot be directly related to the degree of field failure from leaching. However, the differences between material combinations as measured in this experiment seem to indicate the relative degree of leaching which will occur in the field. No long-term tests were conducted. The purpose of the experiment was to develop a screening procedure for leaching which would offer a comparison of material combinations and which could be performed in a reasonable time and for a reasonable cost.

H. S. Ingham, Jr. [1]

Adhesion of Flame-Sprayed Coatings

REFERENCE: Ingham, H. S., Jr., **"Adhesion of Flame-Sprayed Coatings,"** *Adhesion Measurement of Thin Films, Thick Films, and Bulk Coatings, ASTM STP 640,* K. L. Mittal, Ed., American Society for Testing and Materials, 1978, pp. 285–292.

ABSTRACT: The ASTM Test for Adhesion or Cohesive Strength of Flame-Sprayed Coatings (C 633-69) is a bond test consisting of coating a flat end face of a 2.5-cm-diameter rod of substrate material, attaching this coating with an epoxy adhesive to the face of another rod, and subjecting the assembly to tensile load normal to the plane of the coating. Flame-sprayed coatings include wire, rod, or powder sprayed by combustion flame, plasma spray, arc gun, and detonation processes. Bond strengths of such coatings range from about 0.05 N/m^2 to greater than 0.7 N/m^2. A problem with the test method is a tendency for some adhesives to penetrate the porosity of these coatings and thereby influence strength results. An attempt to determine adhesive penetration by testing different thicknesses of coating is described. Tests of bond strength versus thickness of the thinner coatings gave erratic results. To avoid penetration the procedures require a coating thickness minimum of 0.4 mm, and there is a list of recommended adhesives in the standard method. Other related test procedures, as well as round-robin testing which led to the prescribed procedures of ASTM Method C 633-69, also will be described.

KEY WORDS: adhesion, coatings, cohesion, bonding strength, adhesive strength, epoxy, flame spray, plasma spray

The ASTM Test for Adhesion or Cohesive Strength of Flame-Sprayed Coatings (C 633-69) was first introduced in 1959 [*1*].[2] Shown in Fig. 1, this method, known colloquially as a "bond test," consists of coating a flat end face of a fixture of 2.5-cm-diameter rod of substrate material, attaching this coating with an epoxy adhesive to the face of another similar fixture, and subjecting the assembly to tensile load normal to the plane of the coating. The test became practical when developments with higher-strength adhesive bonding agents made the technique possible. It is particularly adaptable for use with flame-sprayed coatings, which include wire, rod, or powder sprayed by combustion flame, plasma spray, arc gun, and detonation processes.

[1]Manager, R&D Department, METCO Inc., Westbury, N.Y. 11590.

[2]The italic numbers in brackets refer to the list of references appended to this paper.

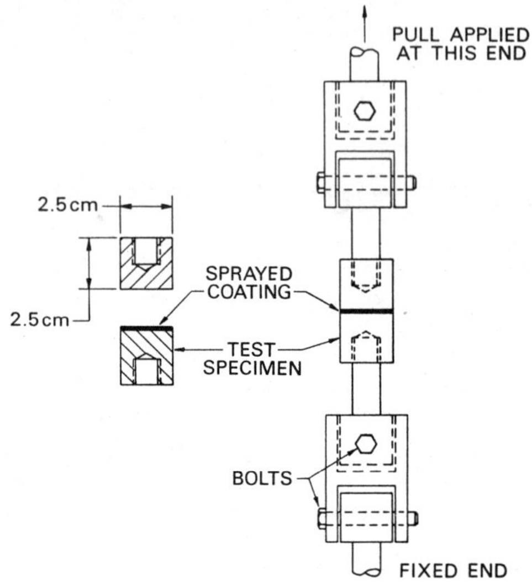

FIG. 1—*ASTM Test for Adhesion or Cohesive Strength of Flame Sprayed Coatings (C 633-69).*

All of these coating processes, also collectively called thermal spraying, produce coatings by melting and propelling particles of metal or ceramic to a substrate. Since the substrate is kept cool, the spray particles deposit and solidify individually to build up the coating, which may range in thickness from 0.02 to 2 mm or greater. The deposition mechanism inherently results in a finite porosity generally from less than 1 to 20 percent, depending on material and process. Adhesion or cohesive strengths range from about 0.05 N/m^2 to greater than 0.7 N/m^2. The latter now represents the limit of the strength of the epoxy adhesive available for the bond test procedure.

During the growing use of this test for flame-sprayed coatings, in addition to the strength limit, one other significant problem was observed by those working in the field. Erratically high strengths with thinner coatings seemed to be related to choice of adhesive bonding agent. The problem was a tendency for some adhesives to penetrate the porosity of the coatings and thereby influence strength results. For example, one particular commercial adhesive came into use for this test because of its high strength and ease of handling. However, bond strength results reported from that adhesive were as high as ten times normally accepted values for some of the weaker coatings. Subsequent investigations showed this particular adhesive to penetrate deeply. Other adhesives have also shown similar effects with very thin flame-sprayed coatings. The solution to the problem was to put a minimum of 0.4 mm on the thickness of the coating tested, and to use only certain adhesives.

The purpose of this paper is to present the considerations involved in testing the bond strength of these special kinds of coatings, that is, flame-sprayed coatings. After a survey of various bond tests, some typical results are provided under the ASTM Method C 633-69.

Other Candidate Bond Tests

To provide a background, the following are descriptions of a few other tests proposed or used at various times.

1. ASTM Tension Test of Flat Sandwich Constructions in Flatwise Plane (C 297-61) is somewhat similar to ASTM Method C 633-69. A portion of a sandwich construction is cemented between two loading fixtures and the strength determined in tension. If used for flame-sprayed coatings, a portion should not be cut from a larger sprayed specimen, as the cutting procedure may weaken the coating at the edges.

2. In an important historical approach [*2*] shown in Fig. 2, a specimen is a cylindrical shaft with a through hole perpendicular to the axis. A stud mounted in this hole has one end machined flush with the curved surface of the shaft. This assembly is coated, and the stud is then pulled out of the shaft in a tension testing machine. This method was successful for many years prior to the development of ASTM Method C 633-69. It is tricky to accomplish, and very thick coatings are required to prevent breaking of the coating. A flat panel may also be used with a hole for the stud [*3*].

3. One of several shear bond tests (Fig. 3) consists of coating a cylinder, cutting a slot in the coating to relieve hoop stress, and applying axial pressure to shear the coating [*2*]. Although it was not clear whether the slotting completely eliminated hoop stresses, reasonable consistency of data indi-

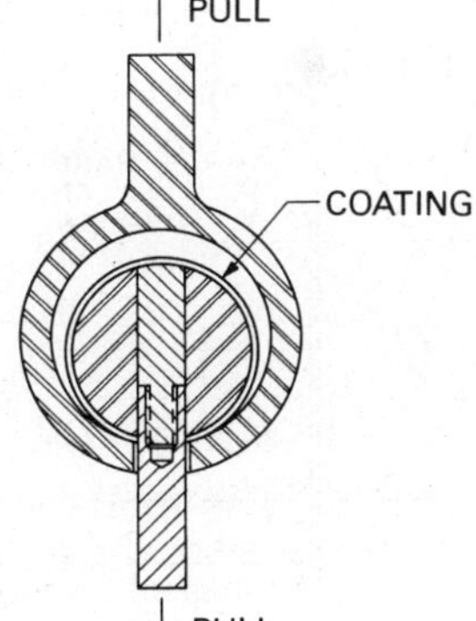

FIG. 2—*An older tensile bond test method for a coating flame sprayed on a shaft with a perpendicular stud* [2].

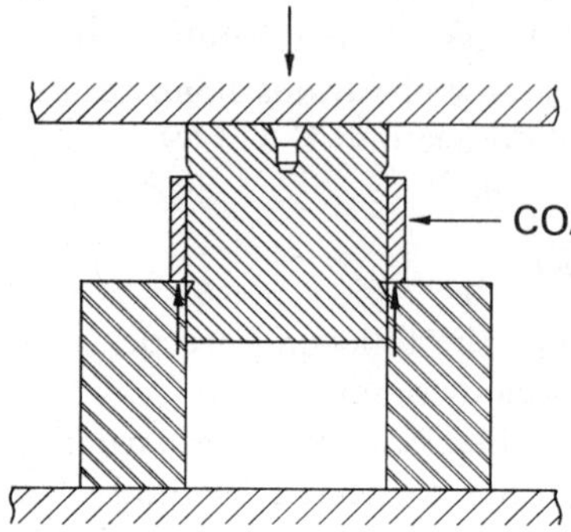

FIG. 3—*Shear bond test of coating on a shaft. Coating is slotted to relieve hoop stress* [2].

cated this problem was not severe. There appeared to be enough of a pattern between shear and tension bond strengths, for different sprayed systems, to indicate that this method was significant. Because of the correlations, it showed also that a simpler tension bond test generally will give sufficient bonding information for most purposes. The shear test is relatively difficult to carry out because it requires particular care in grinding or machining to provide good alignment.

4. In another shear test (Fig. 4) a flat panel is sprayed in two narrow strips on each side near one end [*4*]. The panel is gripped loosely between two edges which just meet the sprayed strips. The panel is pulled between the knife edges so as to shear the coating strips. The results reported appear reasonably consistent, but more experience would be required to establish this procedure.

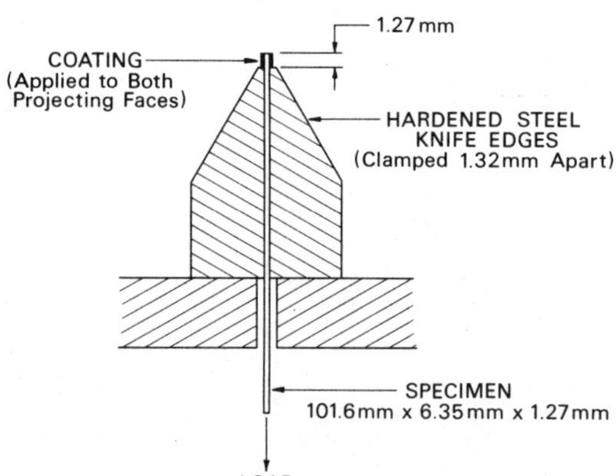

FIG. 4—*Shear bond test of flat panel coated on both sides* [4].

5. A plate with a hole is placed as a mask on a panel and flame sprayed to obtain a coating through the hole (Fig. 5). The panel and plate are pulled apart parallel to their faces to shear the coating in the hole [*5*]. A serious disadvantage is that when a coating is sprayed into a hole, spray particles rebound from the side of the hole, causing weak coating structures.

6. Several tests depend on deforming a coated panel and measuring the degree of lifting of the coating from the panel. In ASTM Test for Adherence of Porcelain Enamel and Ceramic Coatings to Sheet Metal (C 313-59) a steel ball is used to force the test area of the specimen into a die with a recess [*6*]. This version, intended for enamel coatings, is made quantitative by the use of an electric probe to count the number of holes that develop in the coating due to the deformation. It cannot be used with metallic or the already porous flame-sprayed coatings. Qualitative observations of bond failure may be useful if properly interpreted, but these give poor measures of the bond strength of flame-sprayed coatings since the results depend on too many factors. For example, brittle coatings with high internal strength flake off in spite of high bond strength, while a ductile or friable coating with weak bond may stay.

Round-Robin Tests

During the establishment of ASTM Method C 633-69 (Fig. 1) investigations included comparisons between several laboratories and different adhesives. Two coating materials were used. One was aluminum oxide powder (-53 $+15\mu$m) plasma flame sprayed to about 0.6-mm-thick coatings. The other, molybdenum wire (3 mm diameter) was combustion flame sprayed to produce coatings of about the same thickness.

For each series, twelve substrate fixtures were provided by each of three laboratories. All fixtures were SAE 1018 or 1020 steel rods with flat end faces, nominally 2.5 cm diameter, and generally corresponded to requirements of ASTM Method C 633-69. Each substrate (one of the fixtures)

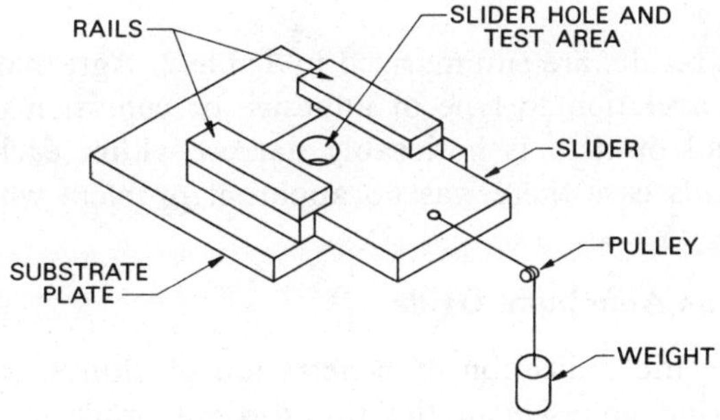

FIG. 5—*Shear bond test of coating sprayed into hole* [5].

was prepared by air blasting at about 60 psi with chilled iron grit (nominally in the 16 to 25 mesh range) just prior to coating. For each series, all substrates were lined up and flame sprayed simultaneously. These were returned back to the corresponding laboratories for testing. Each laboratory separated its 12 coated substrate fixtures into two groups of six. The test was performed on one group with Conap 1222 adhesive, cured at about 149°C (300°F). For the other group a room temperature cure adhesive of the tester's choice was used.

Grinding the surface of the coating before testing had no apparent effect on results. However, lightly grit blasting this ground surface gave widely scattered data.

Aluminum oxide results from the different laboratories are summarized in Table 1. There is close agreement between the three sets using room temperature curing adhesives. The Conap at Laboratory A gave high results, indicating too much penetration of this high-temperature curing adhesive. It is believed that the adhesive becomes more fluid before it cures, and penetration may vary with exact conditions such as rate of temperature size or the age and cure rate of the adhesive.

TABLE 1—*Round-robin results on bond strength of aluminum oxide coatings 600 μm thick. Averages are based on six tests. (Failures were mostly within coatings, with fractional areas of failure at interface or at adhesive.)*

		Room Temperature Curing Adhesive			Conap 1222 Adhesive (cure at 150°C) (302°F)	
Laboratory	Coating Surface	Brand	Avg	Low-High	Avg	Low-High
A	as-sprayed	Armstrong A-12	0.093 N/m^2	0.077–0.110	0.247	0.209–0.270
B	ground smooth	Bondmaster M777	0.106	0.101–0.116	0.091	0.105–0.146
C	as-sprayed	Bondmaster M666	0.100	0.076–0.116	0.119	0.091–0.135

Molybdenum results are summarized in Table 2. Agreement is fair, with no particular correlation to type of adhesive or condition of coating surface. The spread of data is noticeably narrow within each group of six specimens. In this case there was no apparent problem with penetration.

Thickness Test on Aluminum Oxide

Given the specific indication of penetration of aluminum oxide by the temperature-curing adhesive in the round-robin program, an effort was made to measure penetration depth of adhesive into flame-sprayed coatings

TABLE 2—*Round-robin results on bond strength of molybdenum coatings 600 μm thick. Averages are based on six tests. (All failures were within coatings.)*

Laboratory	Coating Surface	Room Temperature Curing Adhesive: Brand	Room Temperature Curing Adhesive: Avg	Room Temperature Curing Adhesive: Low-High	Conap 1222 Adhesive (cure at 150°C) (302°F): Avg	Conap 1222 Adhesive (cure at 150°C) (302°F): Low-High
A	as-sprayed	Armstrong A-12	0.161 N/m²	0.149–0.165	0.175	0.167–0.186
B	ground smooth	Bondmaster M777	0.193	0.189–0.203	0.220	0.207–0.234
C	as-sprayed	Bondmaster M666	0.203 N/m²	0.193–0.210	0.167	0.153–0.183

by applying coatings of different thicknesses onto substrate fixtures and by measuring the results of each thickness. It was hoped that if the strength was significantly greater for those coating thicknesses that were less than some specific critical thickness value, this critical thickness would be con-

TABLE 3—*Bond strengths for different-thickness aluminum oxide coatings with runs different days.*

Coating Thickness, μm	Run[a]	Bond Strength, N/m²: 1	2	3	4	5	6
25	...	...	...	...	...	...	...
50	...	...	...	...	...	...	...
75	...	...	0.18	0.14	...	...	...
100	...	...	...	0.13	...	...	...
125	...	...	...	0.12	0.17	0.12	...
150	...	...	...	0.13	0.17	...	...
175	...	0.10	...	0.09	0.18	0.12	...
200	...	...	...	0.13	0.17	...	...
225	...	...	...	...	0.17	0.11	...
250	...	0.09	0.13	...	0.17	...	...
275	...	...	...	...	...	...	...
300	...	...	0.11	...	...	0.11	...
325	...	...	...	...	...	...	...
350	...	...	...	...	...	...	...
375	...	0.10	...	...	...	0.11	...
400	...	...	...	...	...	...	0.13
425	...	...	...	...	...	...	...
450	...	...	...	...	...	0.10	...
475	...	...	...	...	...	...	...
500	...	...	...	...	...	...	0.13
600	0.13	...	...	...	...	...	...

[a] Round-robin value this laboratory.

sidered a measure of the maximum effective depth of penetration. To test this method, a large number of coatings of aluminum oxide were tested for adherence or tensile strength. The coatings were sprayed and ground finished to various thicknesses. There were three to six specimens comprising a set at each thickness tested at one time. The adhesive used for the test was Conap 1222.

This program was done in a chronology, with the results given in Table 3. Run 1 was a trial that showed no significant variations with coating thickness, so Run 2 included a coating 75-μm thick which barely gives cov-coverage for the size powder used for spraying. This gave a higher value, but the next attempt (Run 3) to find the critical thickness did not even reproduce the high value. The next runs were each consistent within themselves, although different from each other. The latter differences reflect day-to-day differences expected in the coatings.

Conclusions

The two series of round-robin tests are examples of good support of the test. Comparison of results on the molybdenum by the different laboratories and adhesives is comparable to general experience in using this method. The closeness of results on aluminum oxide with room temperature adhesive is even better than normally expected.

The high temperature curing adhesive gave a sporadic value, suggesting penetration to the substrate even with a 600-μm-thick coating. This is believed to be due to a lowering of viscosity before curing takes place. The attempt to establish a practical procedure to determine a minimum thickness was not successful. Thus, the limit in ASTM Method C 633-69 was set conservatively, also based on extensive experience beyond that reported herein. There are specific recommendations on adhesives in the standard.

References

[1] Ingham, H. S., *Summary of Second High-Temperature Inorganic Refractory Coatings Working Group*, WADC TR 59-415, Wright Air Development Center, Dayton, Ohio, June 1959, Discussion 8.

[2] Ingham, H. S. and Shepard, A. P., *Flame Spray Handbook, Wire Process,* Vol. 1, METCO Inc., Westbury, N.Y., 1964, pp. 45–50.

[3] Peterson, J. D., *Materials Engineering,* Vol. 69, No. 5, May 1969, p. 80.

[4] Moore, D. G., Eubanks, A. G., Thornton, H. R., Hayes, W. D., Jr., and Crigler, A. W., "Studies of the Particle-Impact Process for Applying Ceramic and Cement Coatings," (Final Report), National Bureau of Standards, Washington, D.C., Aug. 1961.

[5] Grisaffe, S. J., "Analysis of Shear Bond Strength of Plasma-Sprayed Coatings on Stainless Steel," NASA Technical Note TN D-3113, National Aeronautics and Space Administration, Cleveland, Ohio, Nov. 1965.

[6] "Test for Adherence of Porcelain Enamel to Sheet Metal," Bulletin T-17, Porcelain Enamel Institute, Washington, D.C., July 1953.

S. Schroter[1]

Adherence Measurements and Evaluation of Thick-Film Platinum-Gold

REFERENCE: Schroter, S., **"Adherence Measurements and Evaluation of Thick-Film Platinum-Gold,"** *Adhesion Measurement of Thin Films, Thick Films, and Bulk Coatings, ASTM STP 640,* K. L. Mittal, Ed., American Society for Testing and Materials, 1978, pp. 293–302.

ABSTRACT: The fabrication of thick-film hybrid circuits using the solder system consists of screen printing conductor and resistor networks on alumina substrates and attaching other circuit elements (capacitors, discrete semiconductors etc.) to the conductor using tin-lead solder. The difficulty encountered with this technique has been that, although components are rigidly attached at the time of assembly, deterioration of the solder bonds occurs.

The objective of the investigation was to develop a suitable test method for testing incoming materials, optimizing process variables, and in-process quality control of the product.

After review of methods of conductor adherence testing, a method, pull-peel at 90 deg, was chosen which represented the worst case in actual module application. In order to hold the variables to a minimum, one standard test module was chosen for all experimental work. Part of this test module was incorporated into the design of production modules so that quality control of the product could be directly correlated with the initial testing results and with any process variables. Tests were conducted under controlled conditions which optimized all the assembly steps. Four areas have been found to have marked influence on conductor adhesion: the initial thickness of screened material, the preparation of substrate before soldering, the amount of solder used to make a joint, and the rate of cooling of the soldered joint.

Implementing these controls results in a product with improved conductor adhesion.

KEY WORDS: thick films, adhesion, soldering, manufacturing, variability, temperature cycling, cooling rate, conductivity

[1]Senior chemist, Microwave and Power Tube Division, Industrial Components Operation, Raytheon Company, Quincy, Mass. 02169.

There are two basic systems for attachment of components to thick-film hybrid circuits—the "solder system" and "chip and wire." The reliability of components when using the solder system under various conditions was reported previously.[2] The tests conducted verified that conductor adhesion deteriorated rapidly with life. The following factors were found to have pronounced influences.

1. The amount of solder used to make a joint (10 mg of solder on 0.050 by 0.050-in. pads produced better results than 2 mg of solder).
2. The preparation of substrates before soldering (burnished conductors produced more reliable joints than unburnished conductors).
3. The initial thickness of the screened conductor material (25-μm-thick conductors produced better adhering pads than 12-μm-thick conductors).

This paper describes an additional factor influencing the reliability of soldered conductors, namely, the rate of cooling of the soldered module. It also describes how quality control was built into the final product and the results of product quality over a period of time.

Product Quality Objectives

Since in any testing of conductor adherence to the substrate the readings are spread, it was decided that the governing factor would be the lowest reading of a series of a minimum of 75 and a maximum of 150 readings for each test, depending on the lot size of the manufactured modules. The minimum requirements established are given in Table 1.

TABLE 1—*Conductor adherence test minimum requirements.*

Condition	Minimum Requirements,[a] psi	Method
0 h	1000	peel test 24 h after soldering operation
After 100 temperature cyclings	600	peel test according to Military Standard 202D, Method 107, Condition B III

[a]The minimum requirements were accepted only when the peel test resulted in removal of substrate material together with the conductor. Other types of breakage resulted in 100 percent rejection of the whole lot.

[2]Gillis, T. B. and Schroter, S., "Practical Considerations Affecting Thick-Film Conductor Adherence of Solder Systems," *Proceedings,* 21st Electronic Components Conference, Washington, D.C. 1971.

Processing Steps Prior to Soldering

The processing of typical thick-film substrates was performed according to the following schedule:

1. Substrate preparation.
2. Barrier layer screening and firing.
3. Second conductive layer screening and firing.
4. Resistor screening and firing.
5. In certain cases, crossovers and second conductors screening and firing.
6. Resistor glaze screening and firing.
7. Stabilization of resistors and resistor trimming.
8. Pin attachments.
9. Conductor adherence testing.

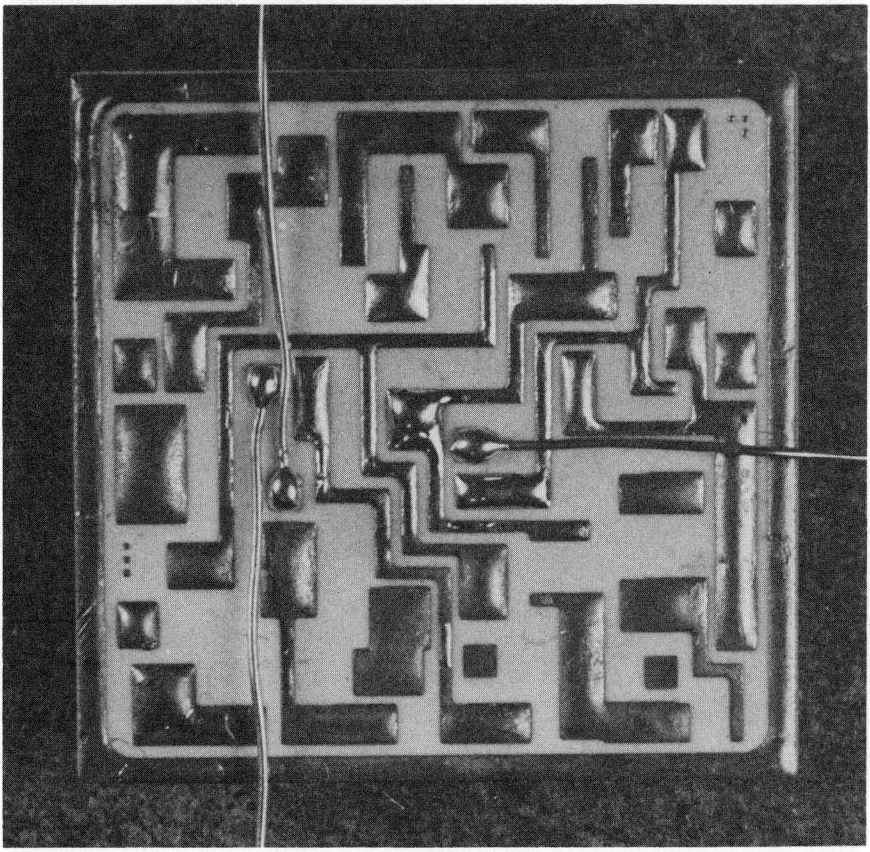

FIG. 1—*Solder pad configuration.*

Method of Testing

Five test pads (0.050 by 0.050 in.) were screened on each production substrate. Following screening, the conductors were presoldered in 60/40 tin-lead solder. The production process used was as follows: Kester 1544 Flux applied, wave soldered, cooled to room temperature, and degreased in trichloroethylene.

Subsequently, tin-plated copper leads (26 Awg) were attached to the soldering pads with 60/40 tin-lead solder using a 12 W iron and with the substrate resting on a hot plate set at 125°C (257°F) to produce the configuration shown in Fig. 1. After 24 h of solder aging time, the copper leads were bent 90 deg. The initial peel test was performed using a Unitek Micropull Tester, Model No. 6-092-01, traveling at the speed of 3 in./min.[3] Figure 2 represents a typical module containing five peel pads after test.

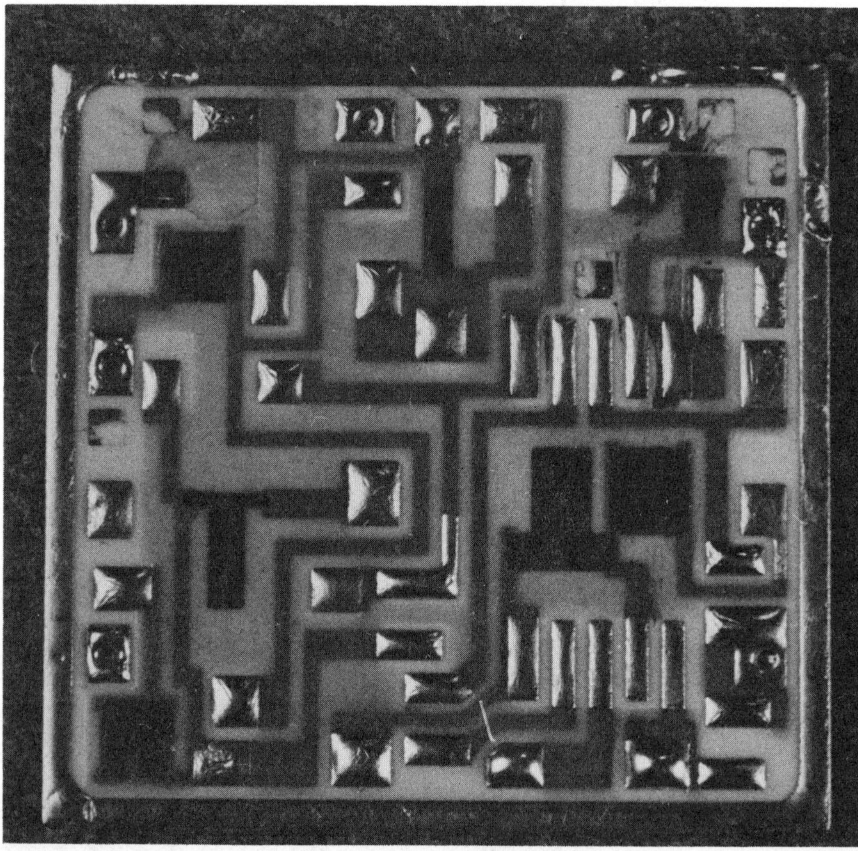

FIG. 2—*Typical 5-pad module.*

[3]Jacobson, L., "Testing For Adhesion of Hybrid Films," *Proceedings,* 21st Electronic Component Conference, Washington, D.C., 1971.

Soldering Revisited

In order to have 100 percent acceptance of processed lots of substrates, the critical parameters of soldering previously reported (see footnote 2) were retested and their validity confirmed. Those were: the amount of solder to make the joint, the initial thickness of the screened conductor material, and the preparation of the substrate before soldering. In addition, several new parameters, not previously tested, were investigated: squeegee screening pressure during paste screening, variation of paste drying conditions after screening, dip soldering versus hand soldering, cooling rate after dip soldering, firing atmosphere dew point, and influence of an operator.

Results

No significant variations of results were detected among the first three new processing steps tested, that is, the squeegee screening pressure, rate of paste drying, and method of soldering. A most significant difference was found between substrates when cooled to room temperature very rapidly, moderately, or very slowly, after wave soldering. Figure 3 represents data of moderate cooling, Fig. 4 of very slow cooling, and Fig. 5 of rapid cooling.

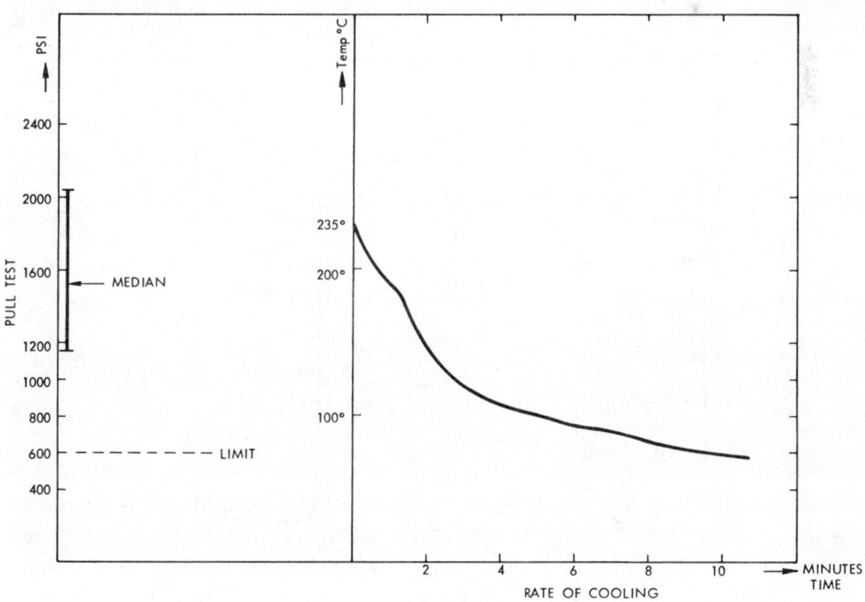

FIG. 3—*Influence of cooling rate cooled under infrared lamps peel data after temperature cycling.*

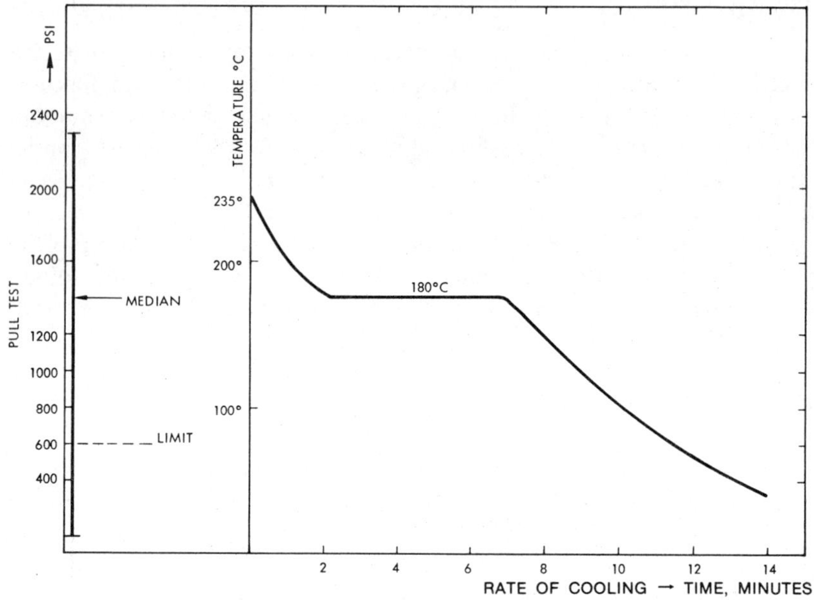

FIG. 4—*Influence of cooling rate very-slow-cooling peel data after temperature cycling.*

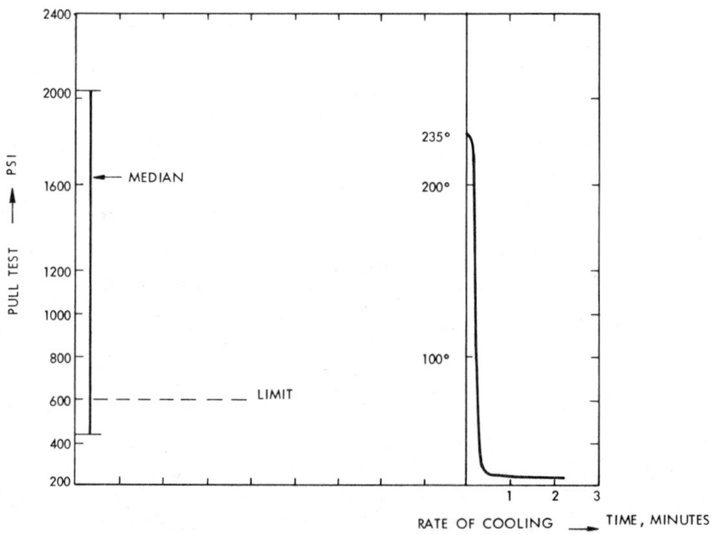

FIG. 5—*Influence of cooling rate rapid-cooling peel data after temperature cycling.*

In cases where the substrates were cooled very rapidly or very slowly, the detrimental influence of extreme variations of cooling rate was indicated: in the case of rapid cooling—most probably due to differences of coefficient of contraction between the substrate and the metallization; in the case of very slow cooling—probably due to accelerated diffusion of metals. Additional variations in adherence quality were detected when the conductors were fired in a very dry atmosphere [−100°C (−148°F) dew point] compared with a moist atmosphere introduced into the firing furnace at approximately 75 percent relative humidity. As shown in Fig. 6, dry air firing resulted in a very high initial peel strength, whereas moist air in the furnace did not have marked influence on adhesion of materials tested. The effect was diminished, but still apparent after thermal cycling.

Influence of Operators

In order to assess the difference in soldering ability between operators, several modules from five lots had wires soldered by two experienced operators. The wires were peeled before and after temperature cycling. The results are shown in Fig. 7.

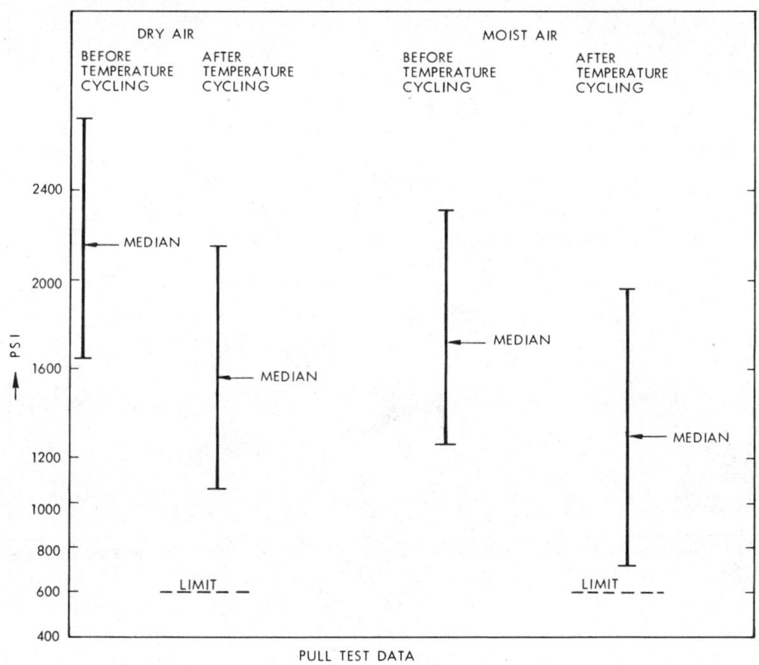

FIG. 6—*Pull test data.*

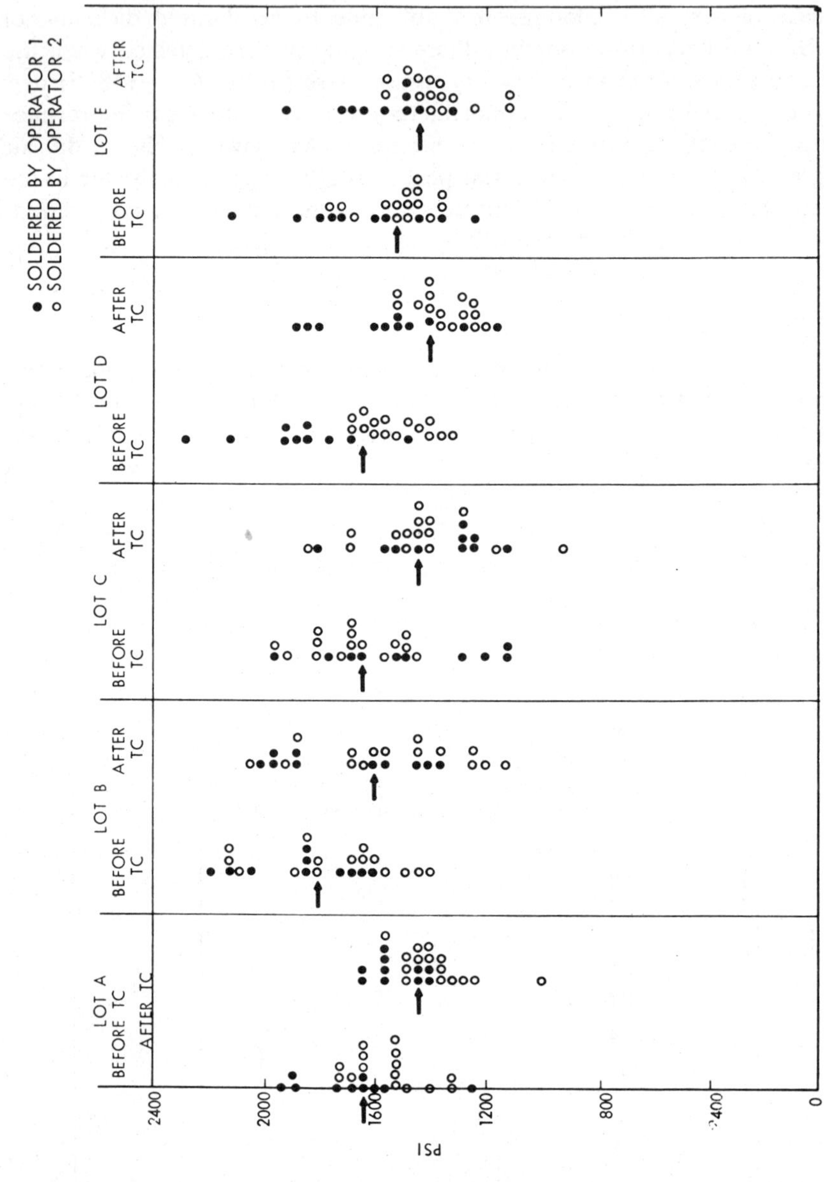

FIG. 7—Operator influence results.

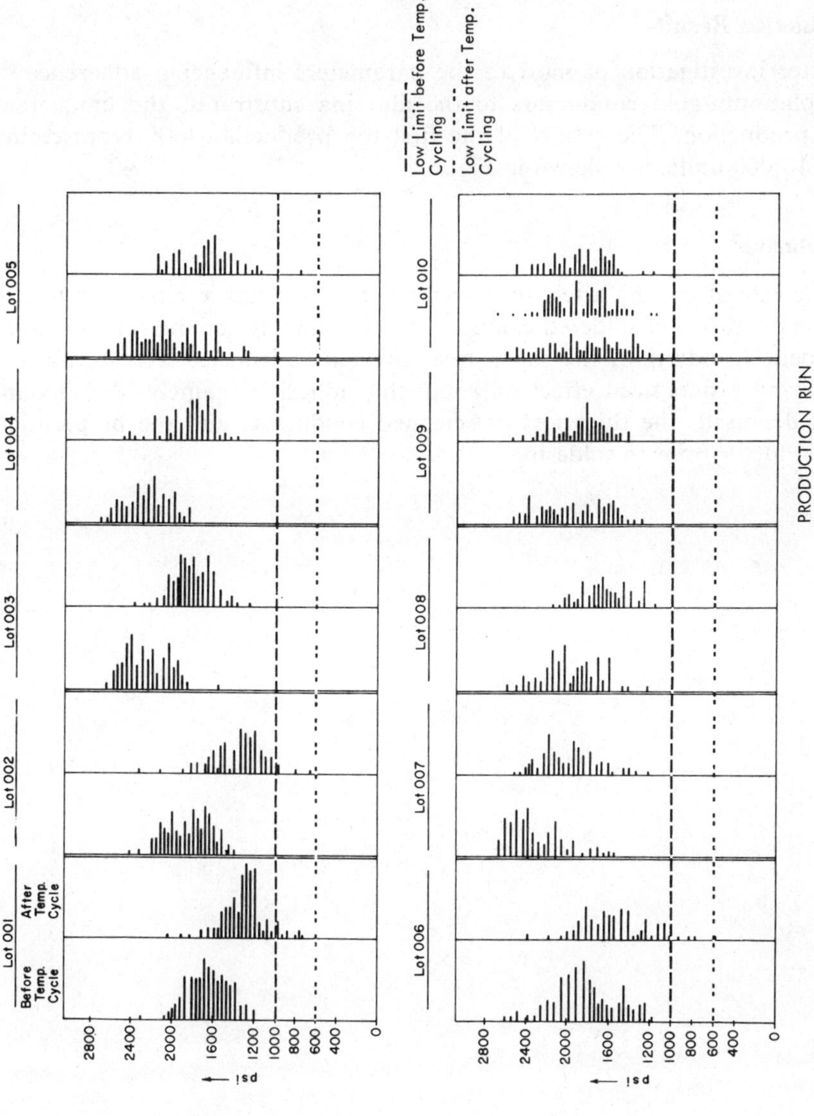

FIG. 8—*Production run.*

The conclusion was drawn that there is some difference between those two operators as far as soldering skill is concerned. Operator 1 average values were almost always higher. An analysis of the statistics might prove the difference significant.

Production Results

After investigation of most of the parameters influencing adherence of the platinum-gold conductors to the alumina substrates, the units went into production. The results of the first ten production lots, representing over 10 000 units, are shown in Fig. 8.

Conclusion

The rate of cooling after the soldering operation has a marked influence on the reliability of soldered components. A moderate cooling rate is recommended. In addition, the three areas previously reported were confirmed as having pronounced effect on conductor adhesion, namely, the amount of solder used, the thickness of screened conductor, and the preparation of conductor prior to soldering.

Adhesion Measurement of Deposits and Coatings

J. W. Dini[1] *and H. R. Johnson*[1]

Adhesion Testing of Deposit-Substrate Combinations

REFERENCE: Dini, J. W. and Johnson, H. R., "**Adhesion Testing of Deposit-Substrate Combinations,**" *Adhesion Measurement of Thin Films, Thick Films, and Bulk Coatings, ASTM STP 640,* K. L. Mittal, Ed., American Society for Testing and Materials, 1978, pp. 305–326.

ABSTRACT: For quantitative measurements of adhesion in coating-substrate combinations, the means most often used at Sandia Laboratories are ring shear, conical-head tension, and I-beam tension tests. Details presented for each test include specimen size, shape, and preparation, and the manner in which the test is performed. The pros and cons of each test are discussed in the light of substantial practical experience, and data for the various deposit-substrate combinations are included.

KEY WORDS: adhesion tests, adhesion, plating, tension tests, shear tests

The adhesion of electrodeposited metals to their substrates has always been a fundamental concern of both the electroplater and his customers. With the increasing number and diversity of basic engineering applications for plated coatings, the emphasis on adhesion has increased substantially in recent years.

One such engineering use of plated coatings at Sandia Laboratories is that of joining by plating, or electrochemical joining, an application which by definition requires the ultimate in adhesion. To verify that certain cleaning and plating cycles are indeed capable of providing the best in plated-bond adhesion, quantitative tests are essential, and over the years a number of different ones have been used to provide the necessary adhesion data. Among these, the three that offer the greatest versatility and reliability in measuring static bond strength are ring shear, conical-head tension, and I-beam tension tests. All have been discussed in detail in previously published work, and test data showing their usefulness have been given [*1*][2]. The purpose of the present paper is to provide an updated account of Sandia

[1]Members of technical staff, Sandia Laboratories, Livermore, Calif. 94550.
[2]The italic numbers in brackets refer to the list of references appended to this paper.

experience with these tests and to present new information for various deposit-substrate combinations.

Ring Shear Tests

Ring shear tests have been used at Sandia Laboratories since 1970 to provide quantitative data on the bond between plated coatings and a variety of substrates. The reason that this test is favored over others is simply that it can be done in a shorter time and is less costly.

For the test, a cylindrical rod is coated with separate rings of electrodeposit of predetermined width. The rod is forced through a hardened steel die having a hole whose diameter is greater than that of the rod but less than that of the rod and the coating. The bond strength A (in MN/m^2 or psi) is determined by the formula $A = W/\pi\, dt$, where d is the diameter of the rod, t the width of the deposit, and W the force required to cause failure in the specimen.

Zmihorski first used this technique for measuring the adhesion of chromium electrodeposits on steel. The method was subsequently elaborated upon by Mockus [*3*], who experimented with many different substrates and coatings. Zakirov [*4*] used the technique for testing the adhesion of thick iron deposits on steel substrates. Chen et al [*5*] used ring shear tests to evaluate various pretreatments for 4131 and 4340 steel prior to cobalt-alumina plating. Chakrapani et al [*6*], using ring shear tests to measure bond strength of copper-plated stainless steel, showed that test results were highly reproducible. Recently, Mittal [*22*] published a comprehensive review on adhesion measurement of electrodeposits. Over the past six years, the present authors have used ring shear tests for many different substrate-coating combinations (Table 1). Because interest in this test is increasing to such an extent, the American Society for Testing and Materials is currently developing a standard methodology for the procedure [*16*].

Test Procedure

Most of the work done at Sandia Laboratories has been with rods 12.7 mm (0.5 in.) in diameter and 305 to 381 mm (12 to 15 in.) long. Rods of this diameter are used because they are quite often stock items in many materials, and there is thus no need for machining prior to plating. Typically, five 25.4-mm (1-in.) segments are stopped off with plater's tape or surgical tubing, leaving alternate 25.4-mm segments available for plating. On occasions where thick, viscous solutions are used during the cleaning-pretreatment cycle and there is concern about trapping these solutions, the rods are not coated with stop-off but are plated all over so that separate specimens (usually five) can be machined from each, thereby offering the

TABLE 1—*Substrate-coating combinations evaluated by ring shear testing at Sandia Laboratories.*

Substrate	Coating	Reference
Aluminum 1100, 2024, 6061, 7075	electroless nickel	[*7*]
Aluminum 6061	nickel	[*7*]
Beryllium	nickel, copper	[*8*]
Beryllium copper	nickel, electroless nickel	[*9*]
Copper	copper	[*10*]
Kovar	nickel, gold	[*10*]
Stainless steel 303, 304, 321, 410, 430-Ti, 21-6-9, AM 363	gold	[*7,11*]
405, 410, 416	nickel	[*9*]
AM 363	nickel	[*7,12*]
AM 363, A 286, 303 Se	nickel-cobalt	[*10*]
303 Se	copper, nickel	[*10*]
17-4 PH	nickel	[*9*]
Steel-4340	nickel, electroless nickel	[*10*]
4130	nickel	[*9*]
maraging steel	nickel-cobalt	[*10*]
SA 106	nickel	[*9*]
Thorium	copper, nickel	[*9*]
Uranium-unalloyed	nickel	[*7,13*]
U-0.75 Ti, U-2.25 Nb	nickel	[*14,15*]
U-0.75 Ti	copper, electroless nickel	[*15*]

possibility of obtaining more than one item of data for each test condition. The rod is coated with a deposit about 1.52 mm (0.060 in.) thick, and then a ring 1.60 mm (0.0625 in.) thick is machined in each plated segment (Fig. 1). In cases where precious metal platings such as gold or silver are being evaluated, only enough of that metal, normally 0.076 to 0.102 mm (0.003 to 0.004 in.), is deposited to ensure a diameter greater than that of the opening in the die. The rest of the ring is then plated with either copper or nickel to a thickness of 1.52 mm (0.060 in.). The latter procedure not only reduces costs but, in the case of gold, also shortens the plating time since copper and nickel can be deposited at rates much higher than that of gold.

Optimum results are obtained when the diameter of the die is 0.051 to 0.102 mm (0.002 to 0.004 in.) greater than the diameter of the rod. As the die diameter increases beyond these clearances, the failure mode changes since the specimen is then subject to a mixed bending and shear load. The dies are usually made of 4340 steel, heat-treated to a hardness of Rockwell C58. This material or another high-strength steel should be used to avoid any possibility of distorting the die surface, particularly that near the shearing edge.

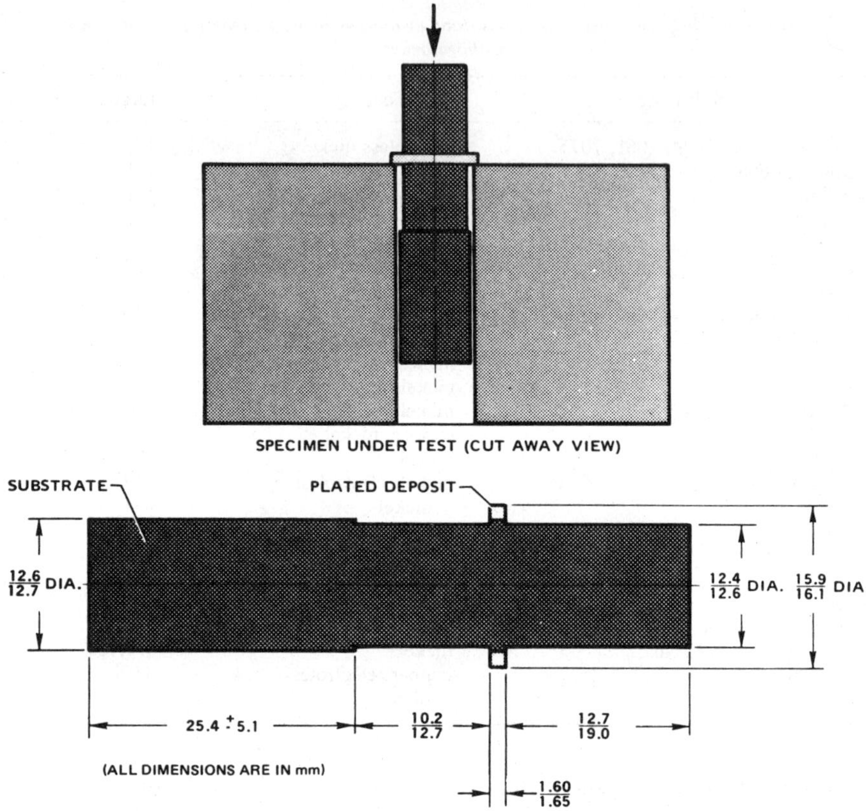

FIG. 1—*Ring shear test specimen and die.*

If neither 12.7-mm (0.5 in.) diameter rods nor material from which they can be machined are available, rods and dies of other sizes can be used. At Sandia, rods as small as 4.3 mm (0.170 in.) and as large as 19 mm (0.75 in.) have served. It should be pointed up, however, that the larger the rod diameter, the lower the ring shear values, probably because of an increase in bending moments with increasing rod diameter. This relationship is illustrated in Fig. 2 for nickel-plated 304 stainless steel, wherein all failures occurred in the nickel deposit rather than at the substrate-coating interface. Die clearance was 0.33 mm (0.013 in.).

A hand-operated laboratory press can be used for testing specimens; however, best results are obtained with a tension testing machine (an Instron, for example). The latter machine permits easier duplication of testing conditions. Crosshead speed during testing is normally 1.27 mm/min (0.050 in./min). Table 2 shows the effect of crosshead speed and die clearance on nickel-plated 17-4 stainless steel. With a die clearance of 0.1 mm (0.004 in.),

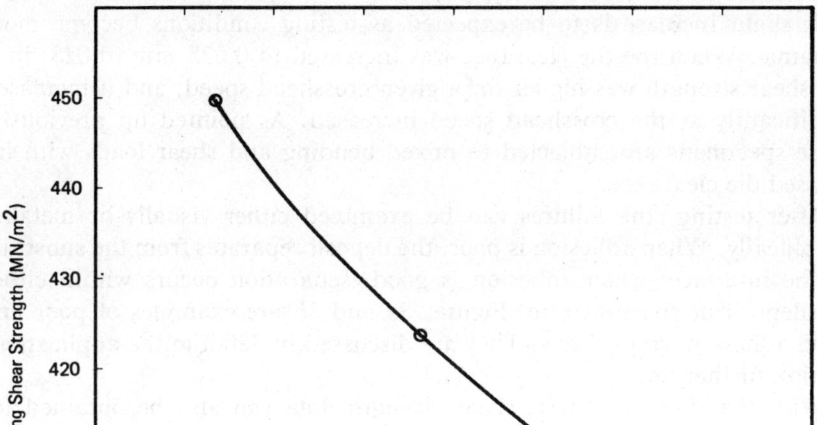

FIG. 2—*Influence of rod diameter on ring shear strength for nickel-plated 304 stainless steel.*

TABLE 2—*Influence of crosshead speed and die clearance on ring shear strength of nickel-plated 17-4 stainless steel.*[a]

Instron Crosshead Speed,		Die Clearance 0.1 mm (0.004 in.) Shear Strength[b]		Die Clearance 0.625 mm (0.025 in.) Shear Strength[b]	
mm/min	in./min	MN/m^2	psi	MN/m^2	psi
0.51	0.020	...	...	522	75 600
1.27	0.050	473	68 300	530	76 800
2.54	0.100	...	...	547	79 400
12.7	0.500	481	69 700	607	88 000
51.0	2.0	...	...	637	92 300
127.0	5.0	510	73 800	702	101 800

[a]Cleaning-activating cycle included cleaning, pickling in 30 percent (weight) HCl, anodically treating in 50 percent (weight) H_2SO_4 at 1070 A/m^2(100 A/ft^2) for 3 min, cathodically treating in Wood's nickel strike at 268 A/m^2(25 A/ft^2) for 3 min, then plating to final thickness in nickel sulfamate solution.

[b]All failures occurred in the nickel. Diameter of 17-4 stainless steel specimens was 12.7 mm (0.498 in.). Average of at least 3 specimens.

shear strengths increased very slightly with increasing crosshead speed. This slight increase is to be expected as testing conditions become more dynamic. When the die clearance was increased to 0.625 mm (0.025) in.), the shear strength was higher for a given crosshead speed, and it increased significantly as the crosshead speed increased. As pointed up previously, these specimens are subjected to mixed bending and shear loads with increased die clearance.

After testing, the failures can be examined either visually or metallographically. When adhesion is poor, the deposit separates from the substrate at the interface; when adhesion is good, separation occurs within either the deposit or the substrate. Figures *3a* and *3b* are examples of poor and good adhesion, respectively. They are discussed in detail in the applications section further on.

With the ring shear test, shear strength data can also be obtained for the substrate material. This procedure involves simply machining a ring in a rod of the material of interest and then testing it in the same manner as the coated material but with an appropriately sized die to accommodate it. Knowledge of the shear strength of the substrate material allows for quick assessment of the success or failure of the cleaning-activating process in question, or when the shear strength of the coated parts approaches that of the substrate material, good bonds are being achieved.

In a somewhat similar vein, approximate ring shear values for electrodeposited coatings can be obtained for a comparison base. The procedure involves plating a thick coating on a small-diameter rod and then machining rings entirely in the plated material. Testing then proceeds as before, only the resulting ring shear values are for the deposited material. This test was done for electrodeposited nickel, and the influence of heating on the shear strength of this material was also determined [*13*].

Typical Applications

A good example of the type of data obtainable with ring shear tests is provided by some recent work with beryllium. In this instance, an added value of the tests was that they provided a quick, relatively inexpensive screening method for checking various plating procedures. Later in the same program, conical-head tension tests were used for more detailed evaluation of those processes which the shear test results had shown to be among the more promising [*8*]. The data presented in Table 3 show that very poor adhesion, less than 60 MN/m^2 (8700 psi), was obtained when no zincate treatment was used and also when the pH of the zincate solution was at 9.3 or higher. (A zincate treatment consists of applying an adherent immersion zinc deposit before electrodepositing the material of interest for the intended application. For more information, see Ref *8*). Specimens given a zincate treatment in solutions ranging in pH from 3.0 to 7.7 ex-

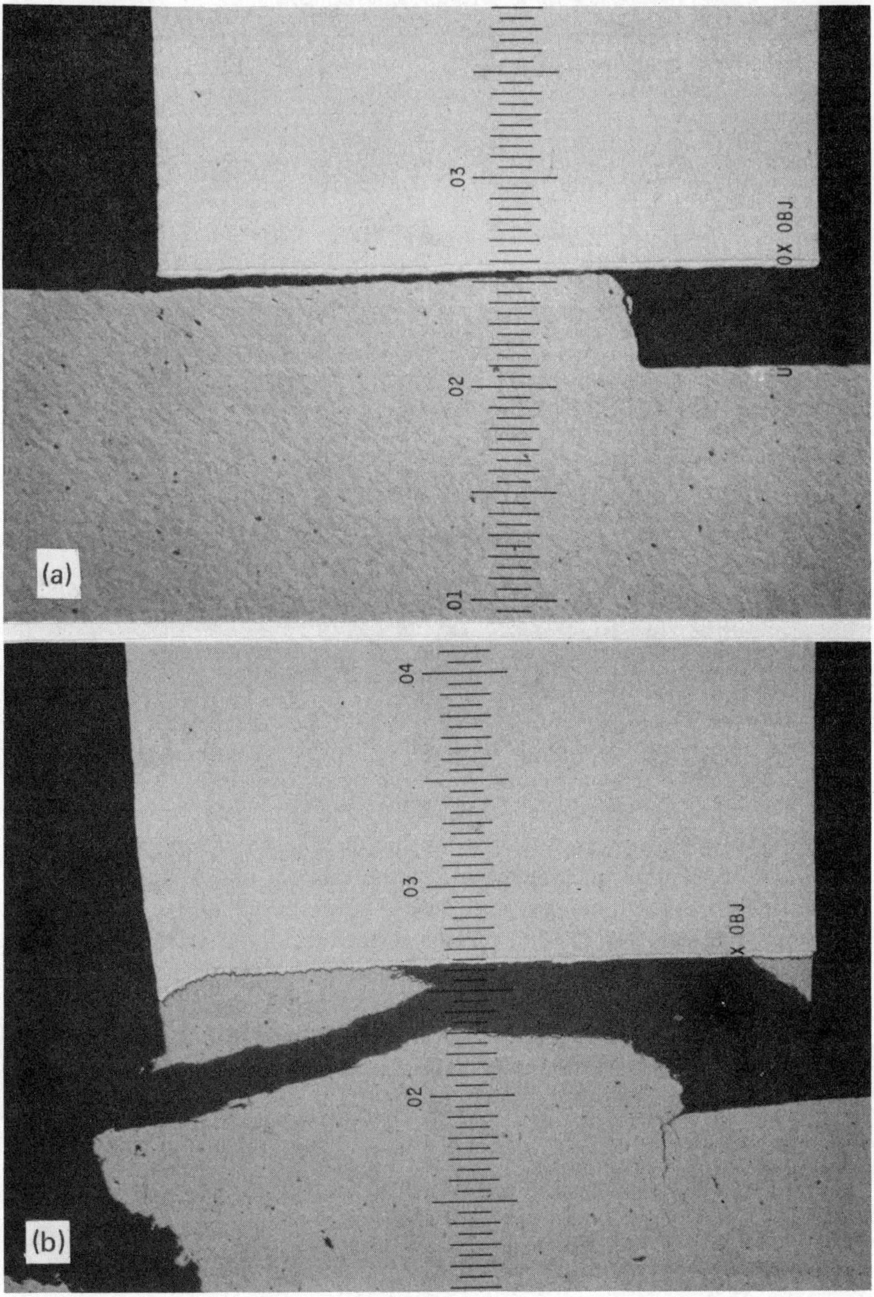

FIG. 3—*Cross sections of nickel-plated beryllium ring shear specimens* (×50). (a) *Poor adhesion, with interface separation at 25.5 MN/m² (3700 psi).* (b) *Good adhesion, with some separation in beryllium at 281 MN/m² (40 800 psi).*

TABLE 3—*Ring shear data for nickel-plated beryllium.*[a]

Treatment		Shear Strength MN/m²	psi
No zincate		0[c]	0
No zincate		51	7 400
Zincate[b]	pH 10.7	26	3 700
Zincate	pH 9.3	60	8 700
Zincate	pH 3.0	232	33 700
Zincate	pH 3.2	241	35 000
Zincate	pH 7.7	281	40 800

[a]Beryllium was S-200-E, 12.7 mm (0.5 in.) diameter rod. The nickel-plating solution contained 450 g/litre nickel sulfamate, 40 g/litre boric acid, and 1.0 g/litre nickel chloride. Current density was 268 A/m^2 (25 A/ft^2), pH 3.8 to 4.0, temperature 49°C (120°F), and anodes were SD nickel.

[b]Zincate treatment consists of applying an adherent immersion zinc deposit before the deposit of interest for the intended application (for more information, see Ref *8*).

[c]Specimens failed during machining before test.

hibited good adhesion. Shear strengths ranged from 232 to 281 MN/m^2 (33 700 to 40 800 psi). Figure 3*a* shows a specimen with poor adhesion, failing at 25.5 MN/m^2 (3700 psi) at the interface between the copper strike deposit and the beryllium, and showing no evidence of bonding between the plating and the beryllium. Figure 3*b* shows a specimen with excellent adhesion. This bond had a shear strength of 281 MN/m^2 (40 800 psi), failure actually occurring in a portion of the beryllium substrate.

Another application of ring shear testing at Sandia Laboratories was to aid in developing procedures for plating on a large casting of ASTM A-216 grade WCB material. Since none of the original material was available for test specimens and since it would have been too expensive to obtain an additional material lot, SA106 Grade B seamless pipe was used because its composition is very close to that of the casting material. Some of the specimens and the die used for testing are shown in Fig. 4. The test data in Table 4 show that best results were obtained with a pretreatment cycle consisting of anodic treatment in sulfuric acid followed by nickel striking prior to final nickel plating.

Conical-Head Tension Tests

With this test, the electrodeposit, the substrate, and the bond between the two are tested in tensile fashion, the bond being normal to the loading direction. Flat panels are plated on both sides with thick electrodeposit. Conical-head specimens are machined from the panels and then tested with standard tension testing procedures. Moeller and Schuler [*17*] first

FIG. 4—*SA 106 seamless pipe ring shear specimens and die used for testing.*

TABLE 4—*Ring shear data for nickel-plated SA 106 Grade B seamless pipe.*[a,b]

Cleaning/Activating Procedure	Ring Shear Strength,[c] MN/m^2	Ring Shear Strength,[c] psi
Clean,[d] anodic treat in 400 ml/litre sulfuric acid at 1080 A/m^2 for 3 min, sulfamate nickel strike[e] at 535 A/m^2 for 5 min, nickel plate	282	40 900
Clean, anodic treat in 100 ml/litre sulfuric acid at 1080 A/m^2 for 3 min, sulfamate nickel strike at 535 A/m^2 for 5 min, nickel plate	219	31 800
Clean, anodic treat in sulfamate nickel-strike at 535 A/m^2 for 3 min, cathodic treat in same solution at 268 A/m^2 for 5 min, nickel plate	232	33 700

[a]Composition (in weight-percent) of this material is 0.30 C, 1.0 Mn, 0.048 P, 0.056 S, and 0.10 Si.

[b]Shear strength of SA 106 steel in 383 MN/m^2 (55 800 psi).

[c]All reported values are the average of four tests.

[d]The cleaning cycle included degreasing in solvent and then anodically cleaning in hot alkaline solution.

[e]The sulfamate nickel strike contained 80 g/litre nickel (as nickel sulfamate) and 150 g/litre sulfamic acid; temperature was 49 °C (120 °F).

used this test to determine the bond strength and properties of copper and nickel deposits over a temperature range of −184 to +538 °C (−299 to +1000 °F). Helms used the procedure to evaluate various activation procedures for plating nickel on nickel [*18-20*] and also obtained reduction-in-area and tensile strength data for nickel deposits. Dini and Johnson have used this test for various substrate-deposit combinations (Table 5).

TABLE 5—*Substrate-coating combinations evaluated by conical-head testing at Sandia Laboratories.*

Substrate	Coating	Reference
Aluminum	nickel	[*1*]
Beryllium	copper, nickel	[*8*]
Nickel	nickel	[*18-20*]
Stainless steel—AM 363	copper	[*12*]
Mulberry	nickel	[*15*]
U-0.75 Ti	nickel	[*15*]

Test Procedure

Flat panels are plated on both sides with a deposit 2.5 to 5.0 mm (0.1 to 0.2 in.) thick. Since the thickness of the substrate panel is not critical, stock material is normally used, just as it is with shear tests. Most work at Sandia has been with substrates varying in thickness from 3.2 to 6.4 mm (0.125 to 0.250 in.). Some type of shielding, normally plastic, is used during plating to help maintain uniformity of deposition.

Conical-head specimens are machined to the requirements shown in Fig. 5. Dies, normally fabricated from 4340 steel and heat-treated to a hardness of Rockwell C58, are used to grip the specimens during testing (Figs. 5, 6). For testing at high temperatures, dies fabricated from either René 41 or Inconel 600 should be used. The comments made earlier about strain rate during testing of ring shear specimens apply also for conical-head testing; however, no specific data have been obtained on this matter. All conical-head testing at Sandia has been done at a crosshead speed of 1.27 mm/min (0.050 in./min).

After testing, visual examination is usually all that is needed to determine the location of the failure. Specimens with good adhesion fail in either the substrate or the deposit, thus providing a measure of the tensile strength of the failed member. Poor or mediocre adhesion is indicated by failure at the interface.

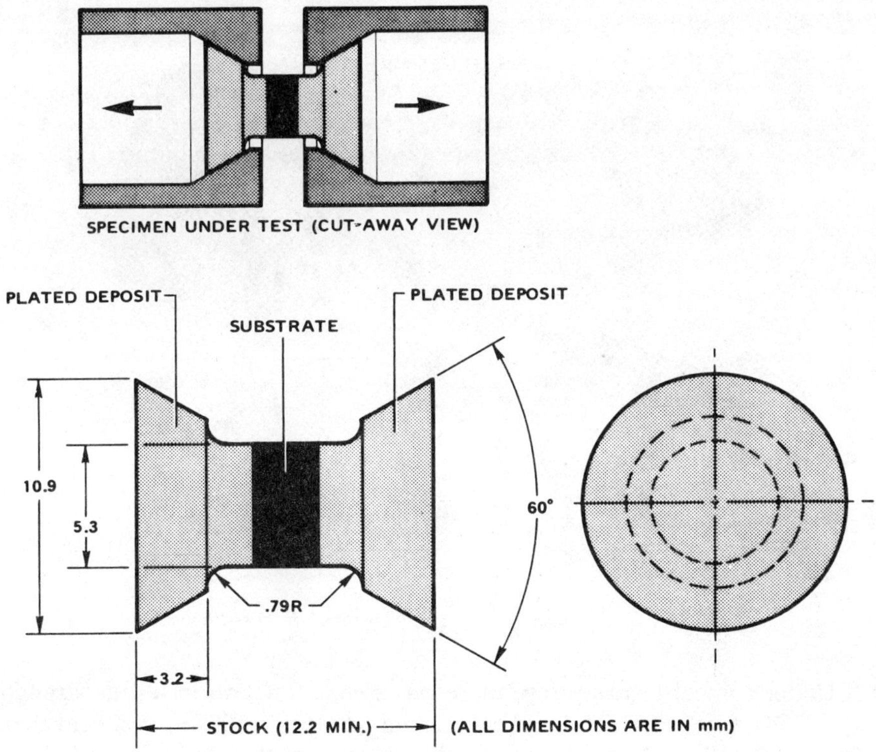

FIG. 5—*Conical-head tension specimen dimensions.*

Typical Applications

Conical-head tension tests have been used to evaluate procedures for plating on three aluminum alloys, 2024, 7075, and 7079. For initial treatment of the surfaces, the zinc immersion (zincate) method of preparation was used, followed by a copper strike deposited to thicknesses ranging from 0.0013 to 0.010 mm (0.00005 to 0.00038 in.). A thick deposit of nickel was the final step. To determine the effect of temperature on bond strength, some specimens of each alloy were heated to various temperatures before being tested at room temperature.

Data for the three alloys (Table 6) show that in all cases failure occurred in the aluminum for specimens not heated after plating. Strengths at failure agreed very well with handbook tensile-strength values for the various alloys (footnotes *g, h,* and *i*, Table 6). In the as-deposited condition, varying the copper strike intermediate layer from 0.0013 to 0.010 mm (0.00005 to 0.00038 in.) did not influence the bond strength; failure still occurred in the aluminum substrate material.

FIG. 6—*Conical-head tension specimens and dies used for testing.*

Under conical-head testing, there was some reduction in tensile strength of 7075 specimens which had been heated to 316 °C (600 °F) and held there for 2 h. The loss, however, was in the strength of the aluminum itself, not in the strength of the bond between it and the plating.

For 7079, heating also caused reductions in strength, and they were greater than those for the 7075. For example, the strength of the 7079 in the as-received (as-plated) condition was 579 MN/m^2 (84 000 psi). After plating and heating at 315 °C for 2 h this value was reduced to 334 MN/m^2 (48 600 psi). It is of interest that the specimens given a copper stike before nickel plating and heating failed at the interface between the initial deposit and the aluminum, while those not given a copper strike failed in the aluminum itself (Figs. 7*a, b*). However, strengths for both conditions were quite similar, 334 and 357 MN/m^2 (48 600 and 51 000 psi) (Table 6).

I-Beam Tests

I-beam tests are another method of tension testing substrate-coating combinations. The advantage of this test is that a piece of plated material of a given size will yield many more I-beam specimens than it will either of the other types discussed here. For example, at least 100 I-beam specimens can be obtained from one 152 by 152 mm (6 by 6 in.) plated substrate. Since the amount of machining required is also minimal, the cost per speci-

TABLE 6—*Conical-head tensile data for nickel-plated aluminum.*[a]

Aluminum	Treatment[b]	Tensile Strength		Location of Failure
		MN/m²	psi	
2024-T4	as-deposited[c]	445	64 500[f]	Al
2024-T4	heated at 149 °C for 75 min	428	62 100	Al
7075-T6	as-deposited[d]	572	83 000[g]	Al
7075-T6	heated at 149 °C for 1 h	572	82 900	Al
7075-T6	heated at 316 °C for 2 h	416	60 300	Al
7079-T6	as-deposited[e]	579	84 000[h]	Al
7079-T6	heated at 149 °C for 75 min	585	84 800	Al
7079-T6	heated at 149 °C for 24 h	593	86 000	Al
7079-T6	heated at 315 °C for 120 min	334	48 600	Al-Cu plate interface
7079-T6	heated at 315 °C for 120 min	357	51 800[i]	Al

[a]The aluminum was prepared for plating by the following process: clean, nitric pickle, zincate, nitric pickle, zincate, and copper strike; then final plating was done as outlined previously.
[b]All testing was done at room temperature.
[c]Thickness of copper deposit was 0.0013 mm (0.00005 in.).
[d]Thickness of copper deposit was 0.010 mm (0.00038 in.).
[e]Thickness of copper deposit was 0.010 mm (0.00038 in.).
[f]Handbook (Alcoa) value for tensile strength of 2024-T4, 432 MN/m^2 (64 000 psi).
[g]Handbook value for tensile strength of 7075-T6, 573 MN/m^2 (83 000 psi).
[h]Handbook value for tensile strength of 7079-T6, 538 MN/m^2 (78 000 psi).
[i]No copper deposit used before nickel plating.

men for I-beam testing is considerably lower than for either of the other types. I-beam tests are thus ideally suited for programs requiring many quantitative tests, for example, statistical evaluations of processes or evaluations of such things as temperature effects on bond strength. This test is a recent development and to date has been used only for stainless steel substrates [*1,21*].

Test Procedure

Parallel grooves 6.4 mm (0.25 in.) wide by 6.4 mm deep and spaced 1.57 mm (0.062 in.) apart are machined on one side of a 152 by 152 by 12.7 mm (6 by 6 by 0.5 in.) substrate. Aluminum strips 0.81 mm (0.032 in.) thick are then wedged into the grooves (Fig. 8). The substrate is cleaned, plated with a minimum of 2.54 mm (0.100 in.) of deposit, then machined (Fig. 9) at the experimenter's option. The I-beam specimens are then cut from this piece as shown in Fig. 10, and the aluminum is dissolved. Since final dimensions of the specimens are not critical, the cutting can be done

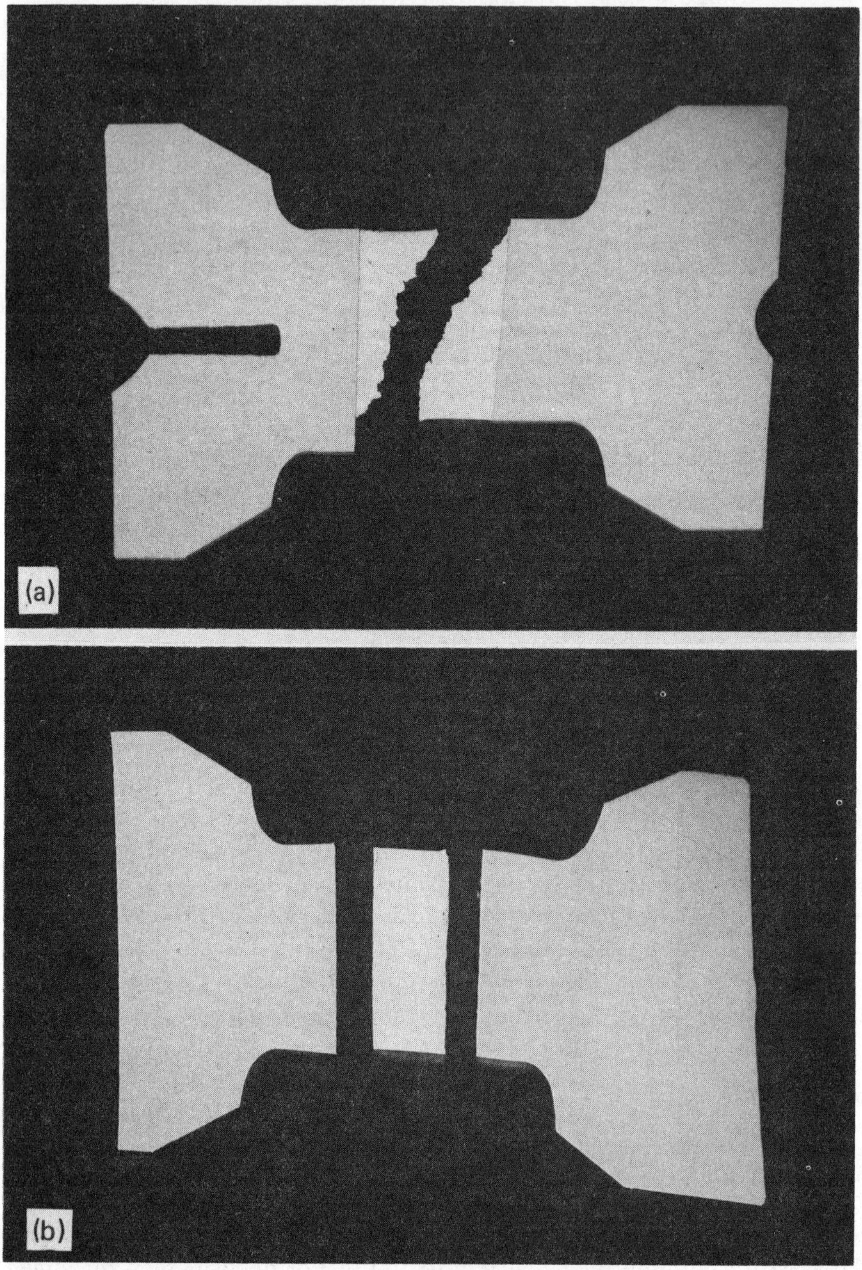

FIG. 7—*Cross sections of nickel-plated 7079 aluminum conical-head tension specimens after heating at 315°C (599°F) for 2 h and then testing at room temperature* (×*6*). (a) *No copper strike layer between aluminum and plating.* (b) *0.010 mm (0.38 mil) copper strike layer, failure at 334 MN/m²* *(48 600 psi).*

FIG. 8—*Early stage of substrate preparation for I-beam tension test specimens.*

FIG. 9—*I-beam panel after dissolution of aluminum strips.*

on a band saw or a cut-off wheel. Figure 11 shows typical I-beam specimens along with the grips used for testing them. The grips are normally made from 4340 steel heat-treated to a hardness of Rockwell C58. In cases where specimens are to be tested at elevated temperatures, the grips are made from René 41 or Inconel 600. Testing is done on a standard tension testing machine.

After the testing, visual examination is usually all that is needed to determine the location of failure. Specimens with good adhesion fail in either the substrate or the deposit, or else at the interface at a strength equal to that of the weaker component (Fig. 12). Poor or mediocre adhesion is indicated

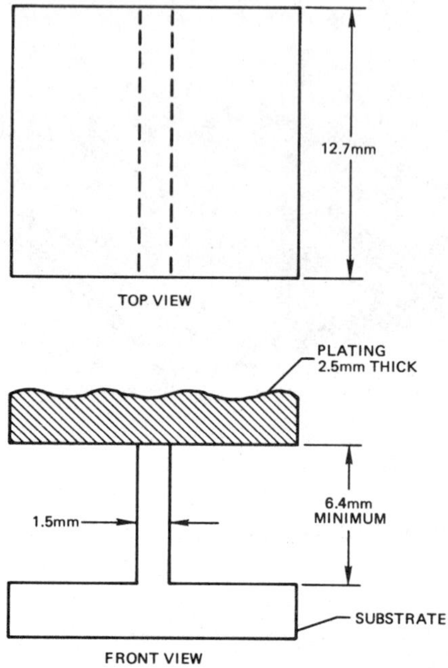

FIG. 10—*Dimensions of I-beam test specimens.*

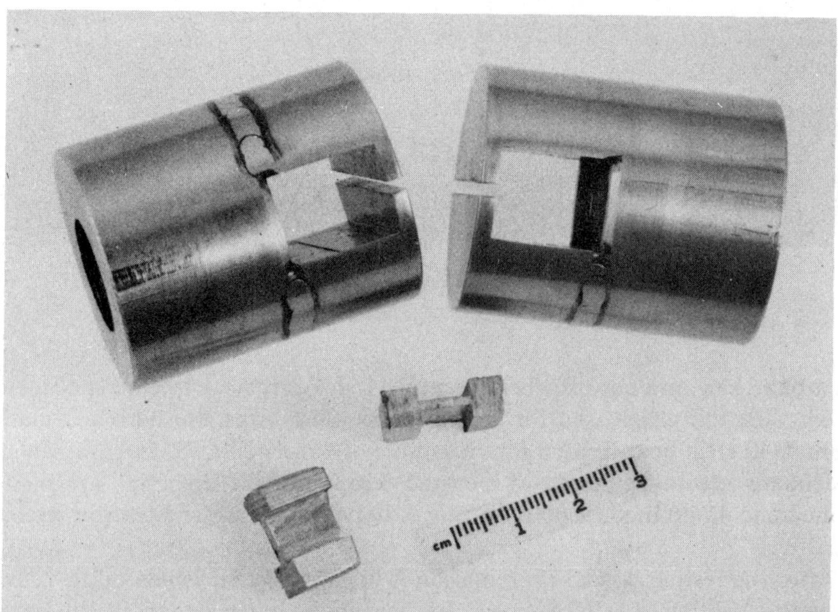

FIG. 11—*I-beam specimens and the grips used for testing.*

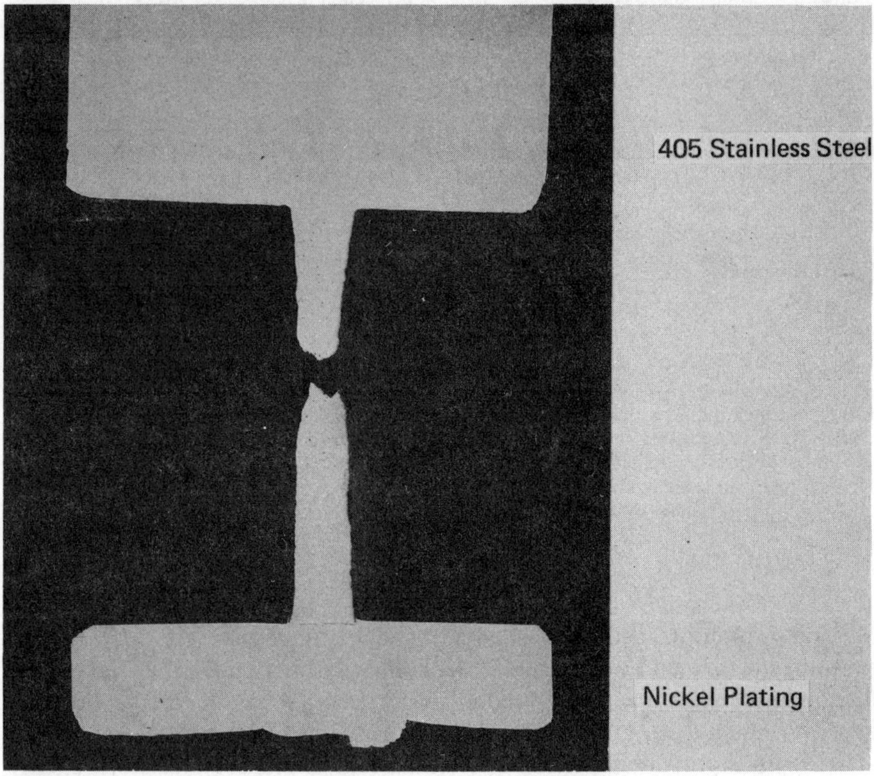

FIG. 12—*Nickel-plated stainless steel (405) I-beam specimen after testing (failure is in the stainless steel).*

by failure at the substrate-deposit interface at a strength less than that of the weaker component.

Typical Applications

I-beam tests have been used to evaluate the influence of temperature on the bond strength existing between electrodeposited nickel and 405 stainless steel. In one experiment, the stainless steel was given a copper strike before nickel plating and, in the other, no copper strike was used.

The I-beam room-temperature test data obtained are given in Table 7. In all cases for the panel which had been given a copper strike before plating with nickel, failure occurred in the copper or at the interface between the copper and the nickel plating. Specimen strengths were noticeably lower than those for the material which was not given the copper strike, and the strengths decreased progressively as a function of time at temperature.

TABLE 7—*Room-temperature I-beam test data for heated nickel-plated 405 stainless steel.*

Treatment	Time, h	Temperature, °C	I-Beam Tensile Strength,[c] MN/m²	I-Beam Tensile Strength,[c] psi
Copper strike[a]	control	...	283	41 000[d]
	30	316	338	49 000
	100	316	241	34 900
	30	538	22	3 200
	100	538	4	600
No copper strike[b]	control	...	414	50 000
	90	538	452	65 500
	720	538	452	65 500
	1510	538	455	65 900

[a]Substrate was cleaned, given a Wood's nickel strike, a 0.013 mm (0.0005 in.) thick copper strike, and then plated with 2.54 mm (0.1 in.) of nickel from a sulfamate solution.

[b]Substrate was cleaned, given a Wood's nickel strike, and then plated with 2.54 mm (0.1 in.) of nickel from a sulfamate solution.

[c]Each reported value is the average of five tests.

[d]All failures occurred in the copper or at the interface between the copper and the nickel plating.

[e]All failures occurred in the stainless steel.

Metallographic inspection revealed that these reductions were due to Kirkendall voids formed by diffusion of the copper strike deposit into the nickel. By contrast, there was no reduction in specimen strength for the material not given the copper strike. All failures occurred in the stainless steel, even after considerably longer times at elevated temperatures. This work clearly shows that if copper is eliminated from the plating cycle, parts intended for high-temperature applications will have considerably longer lives.

As an extension of this work, some specimens were tested at temperatures up to 649 °C (1200 °F). Because of the results given in Table 7, the specimens were prepared for plating without a copper-stike intermediate layer. When ready for testing, the specimens were resistance-heated to temperature and pulled within approximately one minute thereafter. As had happened with the previous work, specimens tested at room temperature failed in the stainless steel (Table 8), and the same thing happened with those tested at 204°C (399°F). The remainder of the specimens failed partly in the stainless steel and partly at the interface; however, in all cases the strength was comparable to that of 405 stainless steel at the particular temperature of interest. In fact, specimens tested at 538 and 649 °C (1000 and 1200 °F) were considerably stronger than the handbook values for stainless steel at these temperatures. The conclusion drawn from this work is that the high-temperature bond strength of nickel-plated 405 stainless steel is at least equal to the high-temperature strength of the steel itself.

The latter series of tests in particular shows the value of I-beam tests. Thirty specimens, all from the same panel, were tested to eliminate any

TABLE 8—*Influence of temperature on bond strength of nickel-plated 405 stainless steel.*[a]

Test Temperature		I-Beam Joint Strength[b]		Tensile Strength of 405 Stainless Steel[c]	
°C	°F	MN/m^2	psi	MN/m^2	psi
22	72	433	62 800	443	64 200
204	400	372	53 900	384	55 700
316	600	346	50 200	378	45 900
427	800	308	44 700	320	46 400
538	1000	268	38 900	199	28 800
649	1200	159	23 100	106	15 400

[a] Substrate was cleaned, given a Wood's nickel strike, and then plated with 2.54 mm (0.1 in.) of nickel from a sulfamate solution.

[b] Each reported value is the average of three tests.

[c] Data from Allegheny Stainless Type 405 Blue Sheet, Allegheny Ludlum Steel Corp., Pittsburgh, Pa.

doubt about one specimen seeing a different preparation than did another. In fact, upon completion of testing, additional untested specimens were available for further tests, had any been needed.

Discussion

From a practical standpoint, two important testing considerations are the cost per test for each technique and the time required to prepare and test the specimens. Table 9 is presented in an attempt to answer these questions. Many metals come in stock configurations that will satisfy substrate requirements for ring shear and conical-head tension specimens. This is not the case for I-beam substrates, which require about three hours of machining.

Plating thicknesses required for the various tests are 1.52 mm (0.060 in.) for ring shear, 2.54 mm (0.1 in.) for I-beam, and 5.1 mm (0.2 in.) for conical-head specimens. Therefore, if one assumes a constant plating rate, the I-beam and particularly the conical-head specimens take longer to plate than do ring shear specimens.

Final machining operations range from a simple band saw cut for I-beam specimens to rather precise machining operations for conical-head specimens. Since final dimensions of ring shear specimens are not critical, they can be machined quickly on a lathe. For all three types, the testing is straightforward, rapid, and requires only a tension testing machine.

The ring shear tests are used more often than the other two tests, primarily because the substrate material is easier to obtain and specimens cost the least to fabricate. The I-beam tests, a recent innovation, offer promise for use as an inexpensive screening test and will undoubtedly find

TABLE 9—*Sample preparation time.*[a]

Test	Substrate Dimensions	Time to Machine Substrate, min	Minimum Plating Thickness		Time to Machine One Substrate After Plating, min	Time to Test One Specimen, min
			mm	mils		
Ring shear		0[b]	1.52	60	20	5
Conical-head	3.2 or 6.4-mm (0.125 or 0.25-in.) thick plate	0[c]	5.10	200	120	10
I—beam	12.7 or 19.0-mm (0.5 or 0.75-in.) thick by 152 by 152-mm (6 by 6-in.) plate	180	2.54	100	2	5

[a] In all cases, it is assumed that the substrate material is stainless steel and the plating is either nickel or copper.
[b] It is assumed that 12.7-mm (0.5-in.) diameter rods are stock items.
[c] It is assumed that 3.2 or 6.4-mm (0.125 or 0.75-in.) thick plate is stock material.

increased use. The conical-head tension tests, although more expensive than the other two, fill a distinct need in providing tensile properties of the plated system.

Acknowledgments

Since a great many parts have been tested by means of the techniques described in this report, considerable help was required from a wide variety of Sandia staff working in plating, machining, testing, and analytical support. We thank all who have been actively engaged at one time or another in the work, particularly J. R. Helms and D. E. Brown in plating, W. Young in machining, A. Clark and S. Grisby in testing, and T. L. Bryant in metallurgy. Thanks too to Don Spencer for technical editing.

This work was supported by the United States Energy Research and Development Administration, Contract Number AT-(29-1)-789.

References

[*1*] Dini, J. W. and Johnson, H. R., *Proceedings*, Society of Vacuum Coaters 18th Annual Conference, 1975, pp. 27–41.
[*2*] Zmihorski, E., *Journal of the Electrodepositors Technical Society*, Vol. 23, 1947–48, p. 203.
[*3*] Mockus, E. S., University of California at San Diego, Aug. 1968, private communication.
[*4*] Zakirov, Sh. Z. *Chemical Abstracts*, Vol. 66, 1967, 101030C.
[*5*] Chen, E. S., Lakshiminarayanan, G. R., and Sautter, F. K., "The Adhesion of Electrodeposited Cobalt and Cobalt-Alumina on High-Carbon Steels," Electrochemical Society Fall Meeting, New York, Oct. 1974.
[*6*] Chakrapani, S., Venkatachalam, R., Subramanian, R., and Chenoi, B. A., *Proceedings*, 13th Electrochemical Seminar, Central Electrochemical Research Institute, Karaikudi, India, 1974, pp. 307–314.
[*7*] Dini, J. W., Johnson, H. R., and Helms, J. R., *Electroplating and Metal Finishing*, Vol. 25, March 1972, p. 5.
[*8*] Dini, J. W. and Johnson, H. R., *Plating and Surface Finishing*, Vol. 63, 1976, p. 41.
[*9*] Dini, J. W. and Johnson, H. R., *Surface Technology*, Vol. 5, 1977, p. 405.
[*10*] Dini, J. W. and Johnson, H. R., *Metal Finishing*, Vol. 72, Aug. 1975, p. 44.
[*11*] Dini, J. W. and Helms, J. R., *Plating*, Vol. 57, 1970, p. 906.
[*12*] Dini, J. W., Johnson, H. R., and Jacobson, R. S. in *Properties of Electrodeposits, Their Measurement and Significance*, The Electrochemical Society, Princeton, N. J., 1974, Chapter 18, p. 307.
[*13*] Dini, J. W., Johnson, H. R., and Helms, J. R., *Plating*, Vol. 61, 1974, p. 53.
[*14*] Johnson, H. R. and Dini, J. W., *Metal Finishing*, Vol. 74, March 1976, p. 37.
[*15*] Johnson, H. R. and Dini, J. W., "Adhesion of Electrodeposited Coatings on U-Ti and Mulberry," Sandia Laboratories, SAND76-8225, Livermore, Calif., May 1976.
[*16*] Polleys, R. W., Chairman, ASTM 808.10.05, Nov. 1975, private communication.
[*17*] Moeller, C. E. and Schuler, F. T., "Tensile Behavior of Electrodeposited Nickel and Copper Bond Interfaces," Rocketdyne Division of North American Rockwell Corporation, paper presented at ASM Metals Show, Cleveland, Ohio, Oct. 1972.
[*18*] Helms, J. R., *Metal Finishing*, Vol. 73, July 1975, p. 21.
[*19*] Helms, J. R., *Metal Finishing*, Vol. 74, July 1976, p. 36.
[*20*] Helms, J. R., "Anodic Activation Treatment of Nickel in Nickel Plating," Sandia Laboratories, SAND75-8056, Livermore, Calif., Dec. 1975.

[21] Dini, J. W. and Johnson, H. R., *Plating and Surface Finishing*, Vol. 64, Aug. 1977, p. 44.
[22] Mittal, K. L. in *Properties of Electrodeposits, Their Measurement and Significance*, The Electrochemical Society, Princeton, N. J., 1974, Chapter 17, p. 273.

DISCUSSION

K. L. Mittal[1] (*written discussion*)—I was quite impressed by the combination of techniques you described. I would like to find out if you used the surface analytical tools to look at the locus of failure, particularly interfacial failure. If yes, what did you find?

J. W. Dini and H. R. Johnson (*author's closure*)—We did not use surface analytical tools to look at the locus of failure. Quite often we would prepare metallographic mounts and use optical microscopy to examine the fracture zone.

[1]IBM Corporation, Hopewell Junction, N. Y. 12533.

C. A. Deckert[1]

Methods for Evaluating Adhesion of Photoresist Materials to Semiconductor Devices

REFERENCE: Deckert, C. A., "**Methods for Evaluating Adhesion of Photoresist Materials to Semiconductor Devices,**" *Adhesion Measurement of Thin Films, Thick Films, and Bulk Coatings, ASTM STP 640,* K. L. Mittal, Ed., American Society for Testing and Materials, 1978, pp. 327–341.

ABSTRACT: A technique has been developed for making comparative evaluations of the adhesion of photoresists to various types of substrates. The technique is based on the use of a fluoride-containing etchant which exaggerates the rate of undercutting of photoresist on silicon dioxide (SiO_2) surfaces and thus permits direct microscopic measurement of the relative rates of undercutting that result with various types of SiO_2 surfaces and with various types of photoresist materials. Using the special etchant, the benefits of adhesion promoters such as hexamethyldisilazane are readily demonstrated. An empirical equation is given which approximately describes the shape of the undercut oxide edges, thus allowing a numerical measure of the relative adhesion (that is, resistance to etchant undercutting) of photoresist/SiO_2 composites.

A variation of this test can be used to detect nonuniform adhesion across a wafer surface. A photomask with a very-fine-geometry, highly recurrent pattern is used to define the photoresist, after which the wafer is etched. By using an optical microscope, the resultant dielectric/silicon substrate can be inspected very rapidly for localized adhesion variation because of the distinct color difference between areas which have or have not been undercut.

KEY WORDS: photoresist, adhesion, undercutting, semiconductors (materials), fluoride etchants, adhesion promoter, hexamethyldisilazane, evaluation

The methods of adhesion measurement described herein were developed for a specific application, namely, adhesion of thin coatings of photoresist materials to thin-film silicon dioxide (SiO_2) on silicon wafers. The methods are really accelerated usage tests rather than ways of measuring classical adhesion strength. Nevertheless we believe that this type of usage test has considerable merit and may, in fact, be more useful to a product manufacturer than a more conventional test. It appears that the methods them-

[1]Member of the technical staff, RCA Laboratories, Princeton, N. J. 08540.

selves could be adapted to study various other types of bonds, and the basic requirements will be outlined later.

The photolithographic process, wherein complex patterns are delineated on semiconductor devices, is one of the principal factors in limiting the yields realized in silicon wafer processing. During the photolithographic process, certain light-sensitive organic polymers, known as photoresists, are used to mask particular areas of a dielectric/silicon substrate while the dielectric is etched away in the unmasked areas. Since the photoresist film is the only protective coating for the dielectric layer during etching, good adhesion between the photoresist material and the dielectric film, particularly at the edges of the delineated pattern, is of prime importance. Many photolithographic failures can be traced directly to poor adhesion at the photoresist/dielectric interface. In the worst case, poor adhesion can cause sudden lifting of entire photoresist areas from the wafer, so that the supposedly masked dielectric is completely etched away. The most common mechanism of photoresist adhesive failure, however, involves attack by the etchant at the resist/dielectric interface around the periphery of the pattern, so that the resist is gradually "undercut" [*1,2*].[2] In this way new portions of dielectric are constantly being exposed to the etchant and the edges of the delineated dielectric become sloped rather than exhibiting a relatively sharp edge.

The adhesion between commercially available photoresist polymers and the surface layers of silicon device wafers is essentially determined by van der Waal's forces rather than by chemical bonding, and thus adhesion is influenced by the exact nature of the surface to be coated with photoresist, the relative humidity of the room where the resist is applied, the chemical composition of the particular type and lot of photoresist being used, and the exact conditions of application, thermal pre-exposure curing, exposure, development, drying, and post-development thermal treatment of the resist.

It is important to note that in the majority of cases, photoresist adhesion is indeed adequate for patterning dielectric films on silicon, but silicon wafer processing involves so many photolithographic steps that only a small failure rate per step can lower device yields very significantly. Since any particular dielectric patterning step has a high probability of success, it is generally not feasible to use the failure rates from the etching process itself as a feedback mechanism for maximizing photoresist/dielectric adhesion.

A simple method of measuring photoresist adhesion, at least qualitatively, was desired. Of the conventional adhesion tests, few are suitable for thin-film work [*3*], and we are aware of no such test which determines edge adhesion without first altering the specimen in some way. Furthermore, service conditions for photoresist/dielectric bonds involve tempera-

[2]The italic numbers in brackets refer to the list of references appended to this paper.

ture cycling, exposure to corrosive environments, etc., so that factors other than adhesion are involved. Thus, as has been suggested elsewhere [4], adhesion tests which simulate usage conditions can be at least as significant as more conventional tests. We describe here simple versatile usage tests for evaluating photoresist film adhesion under simulated usage conditions.

Undercutting Tests

As mentioned previously, photoresist adhesion failure shows up primarily when etchant solutions undercut the resist at the photoresist/dielectric interface. This behavior is depicted in Fig. 1. The etchant normally dissolves only dielectric which is not masked by photopolymer, but if the etchant successfully breaks down the adhesive bond holding the resist to

FIG. 1—*Etchant undercutting at the photoresist dielectric interface.*

the dielectric, then newly exposed areas of dielectric are also etched, causing a sloped edge to be produced. In order to distinguish among poor, marginally acceptable, and good photoresist adhesion, it seemed logical to develop a method of exaggerating this undercutting effect. In this way, the commonly observed phenomenon of photoresist undercutting could be turned to the advantage of the investigator, by pointing out those process variables which affect photoresist adhesion most strongly.

In such a test, rates of undercutting would correlate roughly inversely with adhesion strength, and these rates could be measured rather easily by observing the etched dielectric/silicon substrates under an optical microscope. As shown in Fig. 2, the dielectric layer usually varies from 2000 to 10 000 Å in thickness. In this study, thermally grown SiO_2 was the dielectric used. Thin, uniform films of SiO_2, as well as those of other commonly

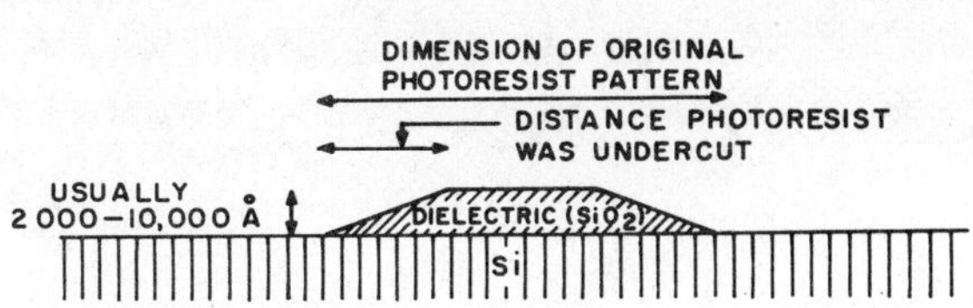

FIG. 2—*Profile of dielectric pattern resulting from photoresist undercutting.*

used dielectrics, appear brightly colored because of light interference. From zero to 10 000 Å thickness, five orders of interference colors can be seen, so that the thickness of a thin SiO_2 film can be determined by visual inspection if the order is known. Also, and pertinent to the measurement of undercutting rates, a sloped edge on a patterned SiO_2/silicon substrate is easily characterized as to thickness and lateral dimension by use of an optical microscope. Figure 3 shows a photograph of a patterned SiO_2 area whose edges are rather gradually sloped due to etchant attack at the photoresist/SiO_2 interface. Figure 4, on the other hand, shows a typical specimen etched for 5 min in a buffered hydrofluoric acid (HF) (BHF) solution. (The BHF used was that manufactured by Transene Co.) The sharply

FIG. 3—*SiO_2* (dark area) *pattern edge is sloped due to photoresist undercutting during etching.*

defined pattern edge indicates that essentially no undercutting has occurred. Most etched wafers yield this type of result; therefore, an exaggerated undercutting rate must be artificially induced in order to discern differences in adhesion.

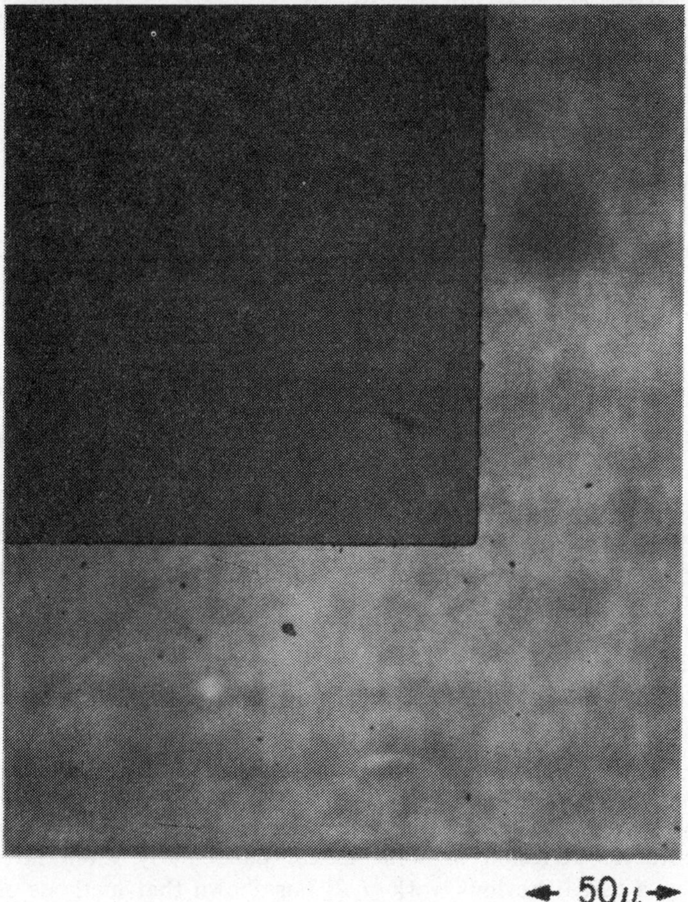

FIG. 4—*Typical SiO_2 area delineated with BHF etchant.*

In order to produce reproducibly a very highly sloped SiO_2 pattern, it was necessary to determine which components of the etchant foster undercutting. Previous studies have indicated [*5*] that low pH etchants lead to high rates of undercutting; we have found that fluoride-containing species are also involved. Thus, by varying the pH and the total fluoride content, it is possible to formulate etchants which can undercut the photoresist layer to any desired extent.

A mixture consisting of BHF and 49 percent HF in the volume ratio 3:1 was found to exaggerate the effect of etchant undercutting suitably. Three photoresists, two negative and one positive, were tested under a variety of processing conditions: (*a*) the advantage of pretreatment by an organosilane adhesion promoter, hexamethyldisilazane (HMDS), was investigated; and (*b*) adhesion to freshly prepared ($\leq$ ½ h old) thermally grown SiO_2 was compared with that to SiO_2 aged overnight in room air (30 to 40 percent relative humidity).

The silicon substrates were steam oxidized at 1000°C (1832°F) to produce about 6000 Å of SiO_2, and were then processed with photoresist as recommended by the manufacturer, except for the process variables mentioned. Three different photoresists, chosen at random from among those widely used in the semiconductor industry, were used: Kodak's KTFR and Hunt's Waycoat IC resist (both negative-working) and Shipley's AZ-1350B (positive-working). The substrates were then etched in the BHF/HF mixture at 25°C (77°F) for 1 min, stripped of photoresist, and examined microscopically. The results of the undercutting tests are shown in Figs. 5–8.

It is seen that all the specimens etched in the mixed etchant exhibit substantially larger undercutting than the specimens etched in ordinary BHF, as typified in Fig. 4. In Figs. 5–8, side-by-side comparison is made of undercutting observed with and without the organosilane adhesion promoter, HMDS; in every case the HMDS result was either superior or equal to that obtained without HMDS. The use of organosilane reagents for improving photoresist adhesion to SiO_2 is well known in the semiconductor industry [*6*]; hence, the demonstration of better performance in the undercutting test described here is indicative that the test does indeed correlate with adhesion strength.

Most interestingly, by using the undercutting test, the effect of other process variables, whose effect would not be known *a priori*, can be evaluated. For instance, the use of freshly steam-grown SiO_2 appears to be detrimental to adhesion in some cases, particularly when no adhesion promoter is used. Previous work [*1,2*] has shown that methods of preparing and pretreating SiO_2 surfaces affect photoresist adhesion significantly. Such effects have not been clearly elucidated and could now be studied more quantitatively.

It would be of considerable interest if the rate at which a photoresist/dielectric composite were undercut could be expressed numerically. Since the slopes of the etched oxide areas are not linear, a simple first-order dependence of undercutting rate on time cannot be hypothesized. However, the following logarithmic equation was found to fit the oxide profiles reasonably well

$$k_u = \frac{r_0 \ln(x + 1)}{y - y_s} \tag{1}$$

FIG. 5—*KTFR photoresist used. SiO_2 was aged overnight before resist application.* (a) *without HMDS;* (b) *with HMDS. Etched with BHF/HF for 1 min at 25°C (77°F).*

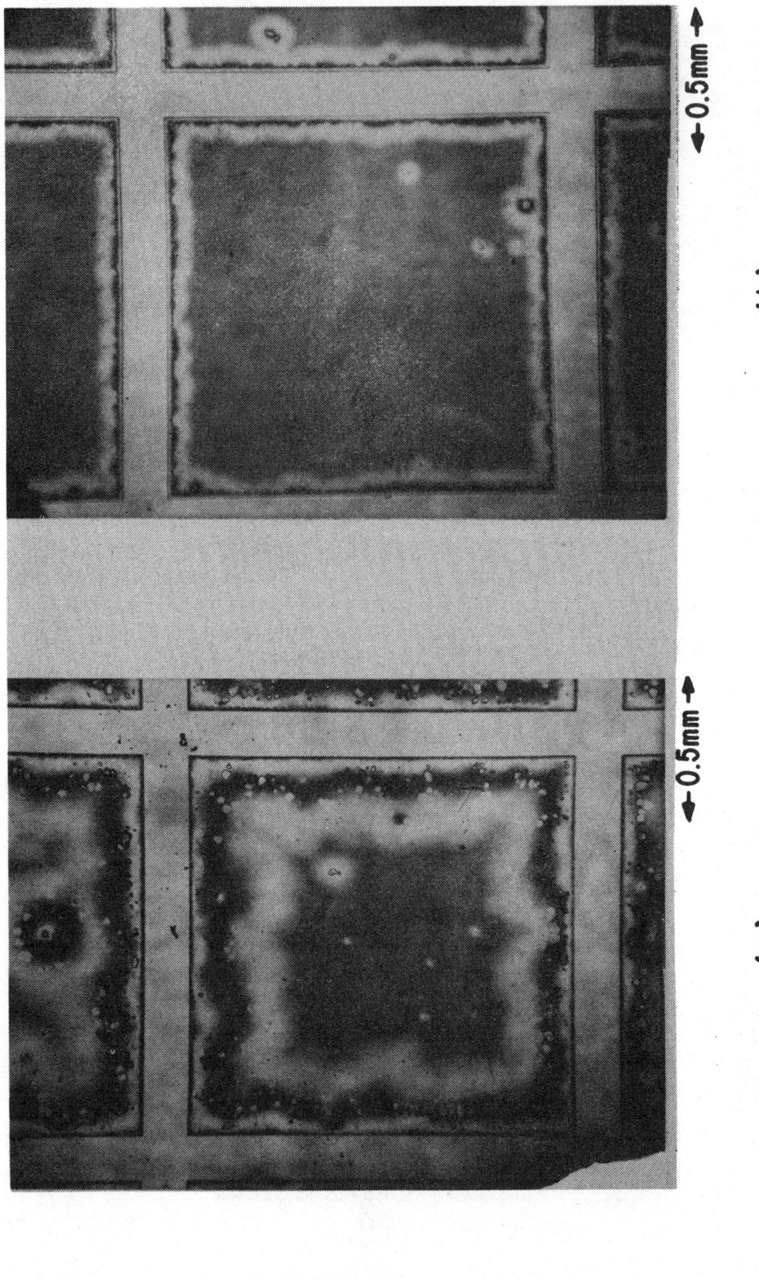

FIG. 6—*KTFR photoresist used. Thermally grown SiO_2 was <½ h old.* (a) *without HMDS;* (b) *with HMDS. Etched with BHF/HF for 1 min at 25°C (77°F).*

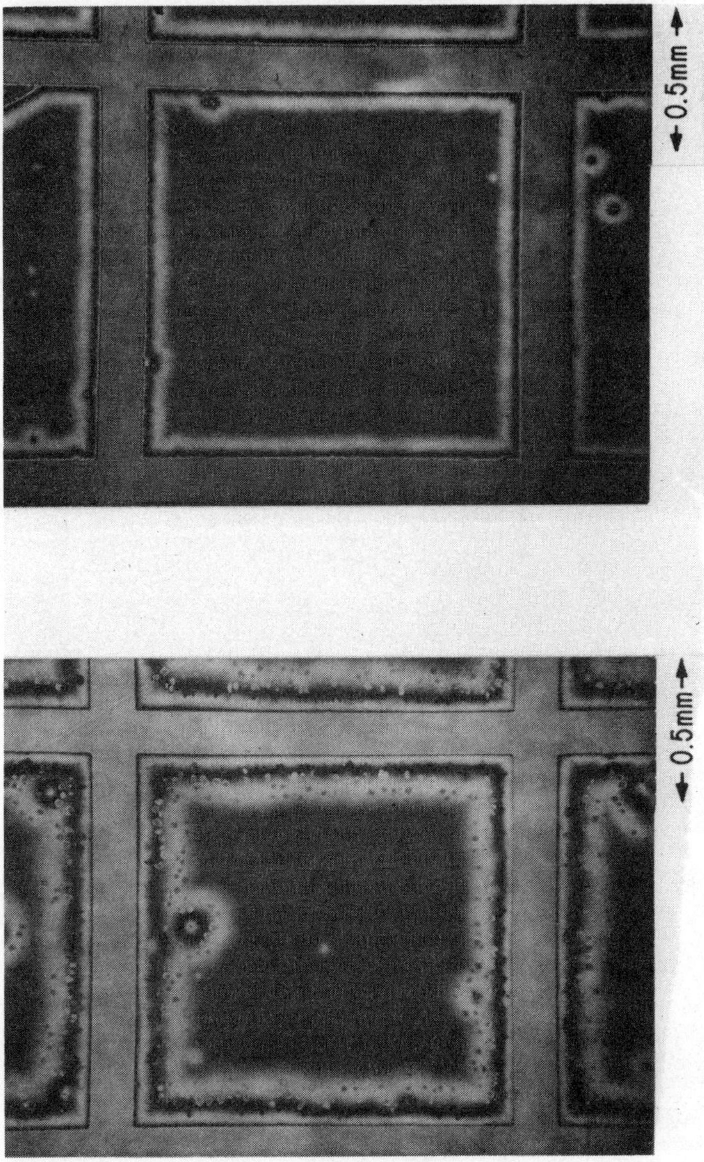

FIG. 7—*Waycoat IC negative photoresist used.* SiO_2 *was aged overnight before resist application.* (a) *without HMDS;* (b) *with HMDS. Etched with BHF/HF for 1 min at 25°C (77°F).*

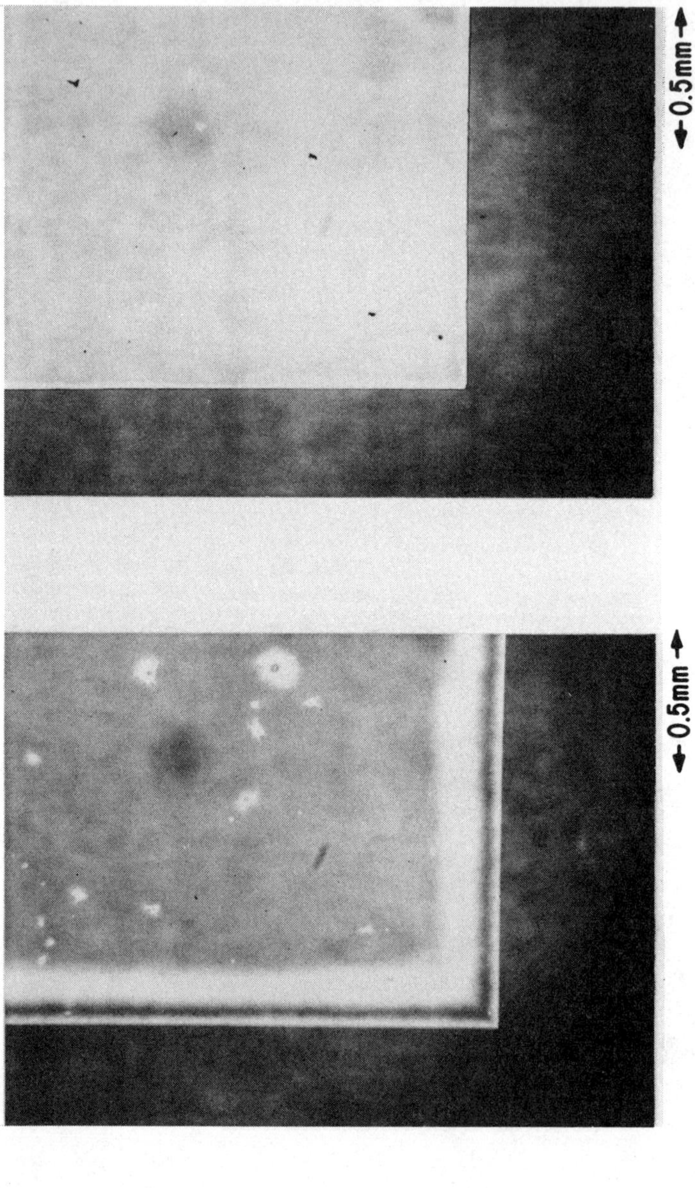

FIG. 8—*AZ1350B positive photoresist used. SiO_2 was aged overnight before resist application.* (a) *without HMDS;* (b) *with HMDS. Etched with BHF/HF for 1 min at 25°C (77°F).*

where

k_u = undercutting constant, with units ln(distance)/time,
r_0 = oxide etch rate of etchant solution being used,
x = lateral distance undercut to a chosen spot on pattern,
y = thickness of SiO_2 at that lateral distance, and
y_s = thickness of oxide film in unmasked areas between patterns.

These variables are also defined in Fig. 9. In order to use this equation, it is necessary that the unmasked areas not be etched all the way through to the silicon substrate (y_s must be ≥ 0). Since oxide thickness can typically be measured to only about ± 200 Å accuracy by optical examination, the error in measuring a 2000 Å step might be as much as ± 20 percent. However, the magnitude of the error can be made quite small by evaluating k_u at several points on the sloped edge and averaging the results.

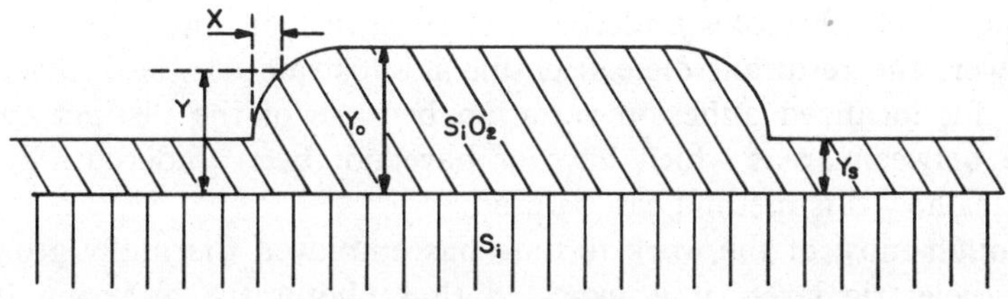

FIG. 9—*Schematic diagram of a patterned SiO_2/silicon specimen after being subjected to the undercutting test and stripped of photoresist. The parameters of Eqs 1 and 2 are illustrated.*

Now, $r_0 = (y_0 - y_s)/t$, where t = etch time; thus, in cases where $y = y_0$ = original thickness of oxide, Eq 1 is simplified to

$$k_u = \frac{\ln(x + 1)}{t} \quad (2)$$

where t = etch time.

Thus a large value for the undercutting constant, k_u, corresponds to poor photoresist/dielectric adhesion. A relative measure of the adhesion strength, or resistance to the undercutting test, might be given by $1/k_u$. Values of k_u and $1/k_u$ for the specimens shown in Figs. 5–8 are given in Table 1.

The test just described assumes that photoresist adhesion is uniformly good or bad across the wafer surface. In cases where adhesion is suspected to be nonuniform, a variation of this test has been found useful. The wafer is coated with photoresist using a very-fine-geometry, highly recurrent photomask pattern, and is then etched either with a standard etchant or

TABLE 1—*Values of* k_u *and* $1/k_u$ *obtained in undercutting tests.*

Specimen	HMDS	k_u, ln(μm)/min	$1/k_u$, relative adhesion
KTFR/aged SiO_2	−	4.0	0.25
	+	3.4	0.29
KTFR/new SiO_2	−	5.4	0.18
	+	4.3	0.23
Waycoat IC/aged SiO_2	−	4.9	0.20
	+	3.7	0.27
AZ-1350B/aged SiO_2	−	6.3	0.16
	+	3.5	0.28
Ideal specimen[a]	...	~0.3	~3

[a]Based on theoretical specimen having perfect photoresist adhesion, in which lateral distance undercut equals thickness of SiO_2 etched away (~0.4 μm in a test of 1-min duration).

with one which promotes undercutting. By using an optical microscope at low power, the resultant dielectric/silicon substrate can be inspected very rapidly for localized adhesion variation because of the distinct color difference between areas which have or have not been undercut. Figure 10 illustrates how this method can be used.

Although most of the work to date has employed thermally grown SiO_2 as the dielectric layer, it is expected that photoresist adhesion to other dielectrics, such as chemically vapor-deposited (CVD) SiO_2, Si_3N_4, phosphosilicate and borosilicate glasses, can be monitored by similar methods.

Conclusion

The results given here are not meant to provide conclusive answers to the question of how photoresist processes can be optimized, but rather to demonstrate the potential applicability of undercutting tests in assessing relative resistances to etchant undercutting (that is, photoresist adhesion) of photoresist/dielectric systems, and in effecting optimization of photoresist processes.

It is important to emphasize that the results of the undercutting test may not correlate well with actual adhesion strengths of photoresist/dielectric bonds as would be measured in more conventional tests. Resistance to etchant undercutting certainly involves factors such as chemical reactivity and hydrophobicity which are probably unrelated or related in a different manner to classical adhesion strength. However, since the tests described simulate actual usage conditions for photoresist/SiO_2 composites, the relative photoresist adhesion, as measured in these tests, is likely to be a more generally useful parameter than a "true" adhesion strength measurement.

FIG. 10—*Color differences in delineated SiO_2 areas readily indicate localized poor photoresist adhesion.*

In certain photolithographic processes a tapered edge is desired [*7,8*], and scanning electron microscopy (SEM) analysis is used to monitor the slope. Use of the undercutting test should increase understanding of the factors which must be controlled to achieve a specific taper.

The ability to obtain a measure of the adhesion of photoresist films to semiconductor device wafers would be of considerable value as an in-line control of the properties of the interface between the semiconductor device wafer to be patterned by photoengraving techniques and the photoresist film (after the exposure, development and baking steps have been performed). Specific problems which could be detected would include contaminated or otherwise unsuitable wafer surfaces and improper spin-on, exposure, development, or drying conditions. The availability of tests for measuring photoresist adhesion is expected to be extremely useful in developing improved materials and processes, such as optimized means for preparing surfaces before applications of photoresists, including use of adhesion promoters, and improved photoresist formulations. It is expected that tests of this sort, although not quantitative, are potentially useful in increasing device yields and lowering costs.

As was mentioned earlier, it appears that the undercutting adhesion test method might be adapted for measuring adhesion of other materials. The basic requirements are as follows. One substrate, in the form of a uniform thin film, is bound firmly to a supporting material; this thin-film substrate must be nonreflective and must exhibit interference colors corresponding in a known way with the film thickness. In our studies this thin film is the SiO_2 and the backing is a single crystal silicon wafer. The other substrate material, corresponding to the photoresist film in this study, need not necessarily be a thin film; it must simply be amenable to patterning so that the underlying substrate is exposed. The solution in which the composite specimen is tested must contain components both to attack the interface and to etch the thin-film substrate. The ratio of attacking agent to etching agent would then be adjusted to produce the most clear-cut test result. The overlying film must, of course, be capable of easy removal after the test has been carried out, and the results could be analyzed in a manner similar to that described in the foregoing.

Acknowledgments

The author wishes to thank G. L. Schnable and W. Kern for the many helpful discussions which took place throughout this study. Their critical comments on the manuscript, as well as those of A. W. Fisher and H. Hook, are also gratefully acknowledged.

References

[1] Bergh, A. A., *Journal of the Electrochemical Society,* Vol. 112, 1965, p. 457.
[2] Lussow, R. O., *Journal of the Electrochemical Society,* Vol. 115, 1968, p. 660.
[3] Weaver, C., *Journal of Vacuum Science and Technology,* Vol. 12, 1975, p. 18.
[4] Chapman, B. N., *Journal of Vacuum Science and Technology,* Vol. 11, 1974, p. 106.
[5] Kern, W., Vossen, J. L., and Schnable, G. L., *Proceedings,* 11th Annual Reliability Physics Conference, Institute of Electrical and Electronic Engineers, 1973, p. 214.
[6] Collins, R. H. and Decerse, F. T., U.S. Patent 3,549, 368, 1970.
[7] Haken, R. A., Baker, I. M., and Beynon, J. D. E., *Thin Solid Films,* Vol. 18, 1973, p. S3.
[8] Ham, E. J. and Soden, R. R., U.S. Patent 3,839,111, 1974.

DISCUSSION

Carl Dahlquist[1] (*written discussion*)—Does the silane treatment reduce the spreading pressure of the etchant solution?

C. A. Deckert (*author's closure*)—We have not studied spreading pressure *per se* on these surfaces; however, we have found that the contact angle made by water on an SiO_2 surface is increased markedly if the SiO_2 has been treated with silane.

K. L. Mittal[2] (*written discussion*)—Have you compared the results of your adhesion test with other adhesion measurement techniques? I feel that the blister method for measuring adhesion should be looked into for adhesion of photoresist materials.

C. A. Deckert (*author's closure*)—As you know, there are very few adhesion measurement techniques which can be used on thin films without altering the specimen in some way. We have not yet found an alternative technique with which to compare our adhesion test, but the blister method which you mention might be a good possibility.

J. J. Bikerman[3] (*written discussion*)—I was delighted to hear you mentioning "frozen" stresses, or tensile stresses, in your photoresist films. Did you, or anyone in your laboratory, try to correlate the magnitude of these stresses with the observed undercutting?

C. A. Deckert (*author's closure*)—Although we have determined that photoresist films on SiO_2 surfaces are in a state of tensile stress, these results are only semiquantitative, at best. Thus we did not think it feasible to compare magnitudes of stresses of photoresist films on surfaces which had undergone different pretreatments.

[1] 3M Company, St. Paul, Minn. 55101.
[2] IBM Corporation, Hopewell Junction, N. Y. 12533.
[3] Department of Chemical Engineering, Case Western Reserve University, Cleveland, Ohio 44106.

Mineo Masuoka[1] *and Kazumune Nakao*[1]

Effect of Aspect Ratio on Tensile Bond Strength for Butt Joint of Internal Fracture—Theoretical and Experimental Analysis

REFERENCE: Masuoka, Mineo and Nakao, Kazumune, **"Effect of Aspect Ratio on Tensile Bond Strength for Butt Joint of Internal Fracture—Theoretical and Experimental Analysis,"** *Adhesion Measurement of Thin Films, Thick Films, and Bulk Coatings, ASTM STP 640,* K. L. Mittal, Ed., American Society for Testing and Materials, 1978, pp. 342-361.

ABSTRACT: A general formula for the tensile bond strength of a butt joint in the case of internal fracture is given in terms of a simplified model constructed from measurable parameters. The stress distribution under large plastic deformation is obtained from the stress equilibrium on the horizontal symmetry plane of the adhesive film. The yield condition of Mises or Tresca was employed. The theoretical formula shows good agreement with experimental results. The contraction ratio of lateral contractive strain to longitudinal tensile strain at yielding of the adhesive is given as a function of the aspect ratio and is independent of the adhesive materials used. The empirical relation between the bond strength and the contraction ratio is expressed by a single curve and also is independent of the adhesive materials used. The theoretical and empirical equations obtained indicate that the bond strength for the butt joint of internal fracture depends on both the yield strength and the aspect ratio of the adhesive.

KEY WORDS: butt joints, adhesion, thin films, internal fracture, tensile bond strength, stress analysis, aspect ratio, yield strength, contraction ratio

Much effort has been expended in analyzing the fracture behavior of butt joints from such viewpoints as the finite-element method [*1*],[2] three-dimensional elasticity [*2–4*], and fracture mechanics [*5*]; however, much remains to be done. It seems to be especially important that the stress distribution, combined with fracture criterion, be consistent with the

[1]Bachelor of science and doctor of science, respectively, Osaka Prefectural Industrial Research Institute, Enokojima Kamino-cho, Nishiku, Osaka 550, Japan.

[2]The italic numbers in brackets refer to the list of references appended to this paper.

observed locus of failure and that the related fracture parameters be experimentally detected for evaluating practical joint strength.

Thin adhesive films are those whose bonded area to force-free (lateral) surface ratio is usually larger than 100. This geometric ratio, sometimes called the aspect ratio of the adhesive film, has been adopted as a measure of the effect restricting the deformation of the film under loading [*6*]. The thin adhesive film cannot contract laterally as a result of the geometric restriction. Therefore, the restrictive effect is approximated by a triaxial stress state [*7,8*].

The relationship between the tensile bond strength and the aspect ratio has been reported for circular-shaped butt joint by the use of linear stress-strain behavior [*9*]

$$\sigma_b = (2EW)^{1/2}(1 + a^2/2h^2)(1 + 3k^2a^2/h^2)^{-1/2} \tag{1}$$

where

E = Young's modulus of adhesive film,
W = total strain energy density in edge region,
a/h = aspect ratio of adhesive film, and
k = unknown factor of stress concentration effect.

The terms of Young's modulus and the energy density of the adhesive in Eq 1 have been reported to be mostly affected by the aspect ratio [*2,10*]. It is difficult to explain the fracture behavior of the joints by the use of the linear stress-strain relation.

In this paper, in order to evaluate practical joint strength, the effect of the restriction on the tensile bond strength was examined by using a simplified model which was constructed from the stress equilibrium. The stress distribution obtained was related to the fracture criterion. An adhesive of square cross section was employed in this study because practical joints are usually square, rectangular, or elliptical. The effect of the anisotropic shape of the cross section on joint strength is very interesting and will be reported in a forthcoming paper by the author.

Stress Analysis

The following analysis was carried out to show the relation between the bond strength, the aspect ratio, and the fracture condition. The dimensions of the adhesive and the stress and strain are shown in Fig. 1. The equilibrium condition in the z-direction at Point z on the horizontal symmetry plane of the adhesive film, as shown in Fig. 2, is expressed by [*11*]

$$-\sigma_z r d\theta + (\sigma_z + d\sigma_z)(r + dz)d\theta - \sigma_x dz d\theta = 0 \tag{2}$$

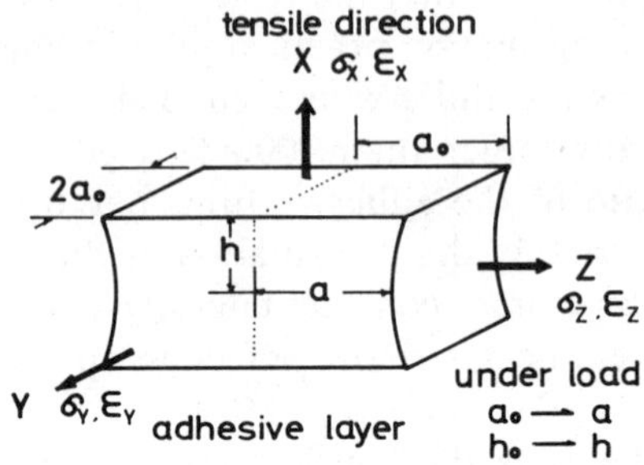

FIG. 1—*Dimension of adhesive and definition of stress* (σ) *and strain* (ϵ) *on horizontal symmetry plane of adhesive film.*

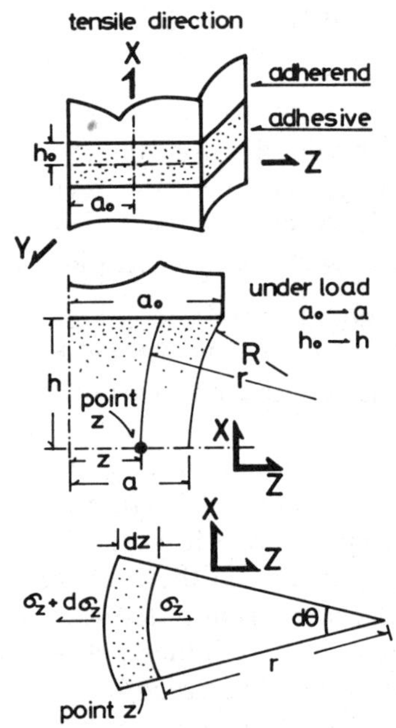

FIG. 2—*Stress trajectory* (middle) *and all forces acting at Point* z *under loading* (bottom); r *and* R *are the radii of the stress trajectory, respectively.*

The last term in Eq 2 is derived as follows. Denoting principal stress by σ_1 on dz and approximating as $\sigma_1 = \sigma_x$, the shear component is expressed as $2\sigma_1 dz \sin(d\theta/2) \simeq \sigma_1 dz d\theta = \sigma_x dz d\theta$. When dz/r is very much smaller than unity, therefore, the equilibrium is expressed in the following form

$$d\sigma_z = (1/r)(\sigma_x - \sigma_z)dz \tag{3}$$

where r is the radius of curvature of the stress trajectory at Point z. The principal stress trajectory is approximated by a circle as shown in Fig. 2. Taking into account the constency of the effective stress on the symmetry plane, $(\sigma_x - \sigma_z)$ must be constant σ_0. On the basis of geometric analysis, the relation between r and R may be represented by

$$1/r = z/aR \tag{4}$$

where a is the distance from the center to $z = a$ on the symmetry plane and R is the radius of curvature of the trajectory at the edge. Substituting Eq 4 into Eq 3 gives

$$d\sigma_z = (z/aR)\sigma_0 dz \tag{5}$$

Tensile and shear stresses at Point z are given by integrating Eq 5 from z to a

$$\sigma_z = \sigma_0(a^2 - z^2)/2aR \tag{6}$$

and

$$\sigma_x = \sigma_0[1 + (a^2 - z^2)/2aR] \tag{7}$$

Equations 6 and 7 give the stress distributions on the symmetry plane in the lateral and the tensile directions, respectively. The following gives σ_0 from Tresca's and Mises's yield conditions:

Maximum shearing stress condition (Tresca)

$$\sigma_x - \sigma_z = \sigma_t \tag{8}$$

Maximum shearing strain energy condition (Mises)

$$(\sigma_x - \sigma_y)^2 + (\sigma_y - \sigma_z)^2 + (\sigma_z - \sigma_x)^2 = 2\sigma_t^2 \tag{9}$$

where σ_t is the yield strength of adhesive material obtained from the uniaxial tension of a dumbbell.

Equation 8 gives immediately $\sigma_0 = \sigma_t$. Assuming the triaxial stress state ($\epsilon_y = \epsilon_z = 0$; then $\sigma_y = \sigma_z$), $\sigma_0 = \sigma_t$ is also obtained from Eq 9.

Integrating Eq 7 over the bonded area, and denoting σ_b as the tensile bond strength of unit cross-sectional area, the reduced bond strength (σ_b/σ_t) is given by the equation

$$(\sigma_b/\sigma_t) = 1 + (1/4)(a_b/R_b)$$

$$a_b = a_0(1 - \epsilon_{zb}) \simeq a_0, \qquad \epsilon_{zb} \ll 0 \tag{10}$$

$$= 1 + \frac{1}{4(R_b/a_0)}$$

where the subscript b means yielding of the adhesive.

The ratio R_b/a_0 is estimated from the geometric relation of the adhesive and may be approximated by

$$(R_b/a_0) = 1/2\,(h_0/a_0)^2\,(1 + \epsilon_{xb})^2/\epsilon_{zb} \tag{11}$$

where

ϵ_{xb} = longitudinal tensile strain at yielding of the adhesive and
ϵ_{zb} = lateral contractive strain at yielding of the adhesive.

Substituting Eq 11 in Eq 10 gives the tensile bond strength for the butt joint in the case of internal fracture

$$(\sigma_b/\sigma_t) = 1 + 1/2\,(a_0/h_0)^2\,\epsilon_{zb}/(1+\epsilon_{xb})^2$$

$$= 1 + 1/2\,(a_0/h_b)^2\,\epsilon_{zb} \tag{12}$$

where

$$h_b = h_0(1 + \epsilon_{xb})$$

Equation 12 is related to measurable parameters of the yield strength of the material, the tensile and lateral strains at yielding of the adhesive, and the aspect ratio. This equation is a newer expression applied to the square-shaped butt joint, equivalent to the formula derived by Davidenkov and Spiridonova for the necking of a circular bar under tension [*15*]. This relation may be independent of the form of stress-strain curve of the adhesive film if the values of the tensile and lateral strains are determined. However, the discrepancy of the yield conditions employed was not reflected in Eq 12. It may be due to the adoption of a square cross section.

Equation 12 indicates that the plot of (σ_b/σ_t) versus $(a_0/h_b)^2\,\epsilon_{zb}$ should give a straight line which is independent of the adhesive material used and that the tensile bond strength of the internal fracture was affected mainly by both the square of the aspect ratio and the yield strength of the adhesive material.

Experimental

Surface Treatment of Adherend

The aluminum adherend (E = 40 kg/mm^2) with a square bonding area (18 by 18 mm) was polished by No. 400 emery paper, degreased in distilled acetone and toluene with ultrasonic cleaner, boiled for 30 min in water treated by ion-exchange resin, and then dried under vacuum for 4 h. These treatments of the adherend were carried out to cause internal fracture within the adhesive film.

Adhesive

Four kinds of commercial ethylene-vinylacetate copolymer (EVA) were used as adhesives. The EVA sheets molded were degreased in distilled acetone with ultrasonic cleaner and dried under vacuum for 4 h before making the bond. The tensile modulus and yield strength of the EVAs were obtained by uniaxial tension of dumbbells. Stress-strain relations and the definition of the yield strength (σ_t) are shown in Fig. 3. The compositions and the mechanical properties of EVAs are given in Table 1. These values were averaged over ten specimens. The rigidity of the EVAs was calculated by the relation $E = 3G$.

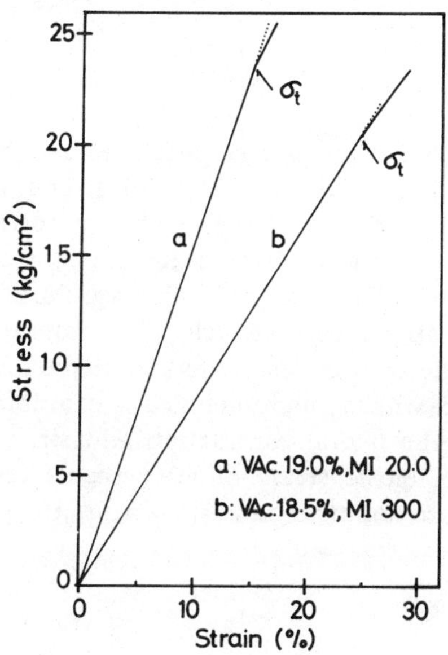

FIG. 3—*Stress-strain relations from uniaxial tension test on EVA dumbbells. Arrows show yielding points. Stress is employed as yield strength* (σ_t) *of adhesive materials used.*

TABLE 1—*Compositions and mechanical properties of ethylene-vinylacetate copolymers used as adhesives.*

Vinylacetate Content of EVA, weight %	Melt Index Number	Tensile Modulus,[a] E, kg/cm^2	Rigidity,[b] G, kg/cm^2	Yield Strength,[a] σ_t, kg/cm^2
19.0	20.0	154.0	51.3	23.7
18.5	300.0	81.8	27.3	20.4
31.1	24.8	29.7	9.9	10.7
29.0	300.0	28.1	9.4	9.2

[a]These were obtained from uniaxial tension test of dumbbells.
[b]These were calculated from $E = 3G$.

Bond Formation

The butt joint was formed by hot-melt bonding at 140°C (284°F) for 10 min and allowed to cool through the use of a device to assure parallelism between two bonded surfaces. The bonded specimen was aged at 30°C (86°F) in a vacuum desiccator for 5 days prior to tension in order to relax a residual stress resulting from thermal shrinkage of the adhesive film [*8*]. In order to estimate exactly the lateral contractive strain, parallel lines (0.20 or 0.15 mm in width) were scribed on a lateral surface of the joint at intervals of 1.0 mm after aging as shown in Fig. 4. The line interval and width were used as scale. The adhesive thickness was in the range from 0.8 to 3.5 mm.

Tension Test

A universal joint and some special attachments were used to make a straight pull of the joint. The tensile bond strength (σ_b) and the yield strength (σ_t) of the EVAs were measured by an Instron tension tester at 25°C (77°F) at a strain rate of 0.6 to 1.25 min^{-1}; σ_b was almost independent of the strain rate. The change in the profile of the adhesive under loading was photographed continuously. The lateral and tensile strains at yielding of the adhesive were estimated from the photograph by the use of projector provided with a micrometer. An example of such estimation is shown in Fig. 4. The lateral contraction and strain were 0.50 mm and 5.56 percent, at the tensile strain of 40 percent, respectively. The data shown in all figures are not average.

Results and Discussion

Internal Fracture

The apparent load-elongation curves for the adhesives are shown in Fig. 5. The arrows indicate the conditions for internal crack or cavity at

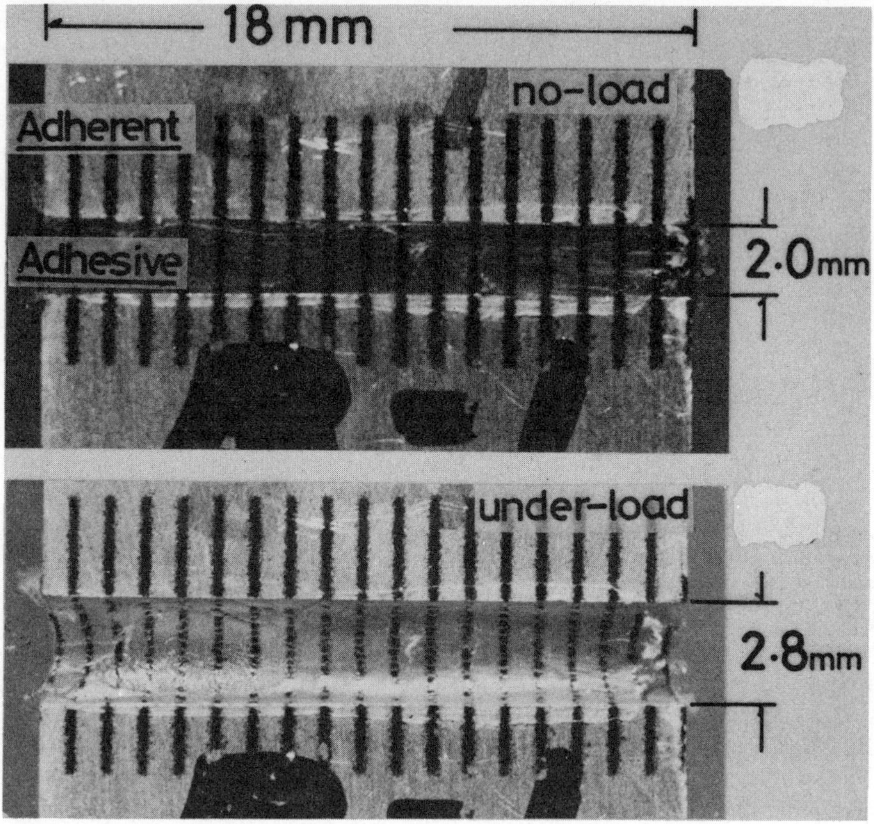

FIG. 4—*Example for estimation of tensile strain* (ϵ_x) *and lateral strain* (ϵ_z) *of adhesive film for aluminum-EVA-aluminum butt joint. Tensile strain* (ϵ_x) = *0.40; lateral contraction* = *0.50 mm; lateral strain* (ϵ_z) = *0.055*$_6$; *contraction ratio* ($\nu = -\epsilon_z/\epsilon_x$) = *0.14.*

the center region of the symmetry plane (see Figs. 6 and 8), where the stress and the strain were employed as the tensile bond strength and the yielding tensile strain, respectively. The adhesive elongated considerably prior to gross failure. The tensile bond strength is equivalent to the cracking stress reported by Gent and Lindley for a circular-disk bonded specimen [*3*]. The cracking stress has been related empirically by the following equation to Young's modulus of the adhesive material used for a specimen 3 mm thick

$$S = 0.55\,E + 0.5 \tag{13}$$

where S is the cracking stress, kg/cm^2 and E is Young's modulus of material.

From Eq 13, the calculated bond strength of the EVA adhesives 3 mm thick is 85.3, 45.2, 16.8, and 16.0 kg/cm^2, respectively, using the tensile

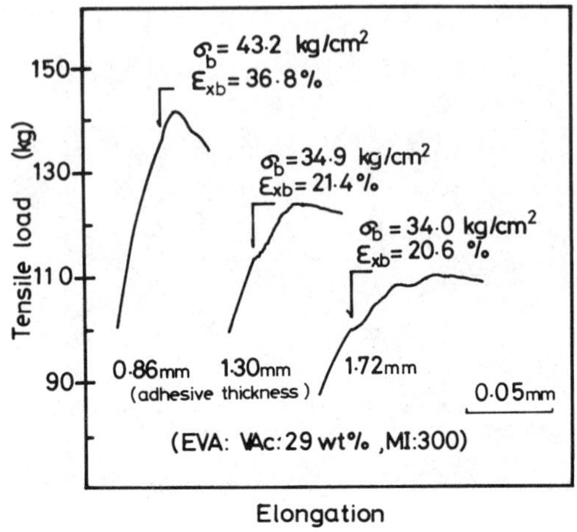

FIG. 5—*Load-elongation curves from uniaxial tension on bonded square-shaped specimens. Arrows indicate conditions of internal fracture, where partial yielding of the adhesives occurred at their center region, as shown in Fig. 6. Stress (applied tensile load divided by cross-sectional area) is denoted as tensile bond strength* (σ_b).

modulus given in Table 1. These calculated bond strengths agree quite well with the results shown in Fig. 11.

The shape of internal voids in the adhesive is shown schematically in Fig. 6 (*top*). The fractographs are given in Fig. 6 (*bottom*). The adhesive was swollen in toluene at 30°C (86°F) for one day and cut along the symmetry plane with a sharp cutter after the tension test. The swollen specimen was allowed to dry and then photographed. Two different shapes of the internal voids were observed with the composition of EVA, however, they were independent of both the adhesive thickness and the melt index number of EVA. It was impossible to deduce the inherent flaw size and number from these photographs.

Stress Distribution and Locus of Failure

In order to confirm the validity of the analysis, the calculated stress distribution curve was compared with the observed locus of failure. From Eqs 6 and 7, the stress distribution curves on the symmetry plane were calculated by using experimental values of R_b; these are shown in Fig. 7. Distributions for both σ_x and σ_z had a maximum at the center ($z = 0$) and a minimum at the edge ($z = a$). This indicates that the internal voids are initiated at the center of the symmetry plane and grow toward the edge, and that the stress required to initiate the internal voids increases with the decrease in the adhesive thickness. These suggestions can be confirmed by the experimental results.

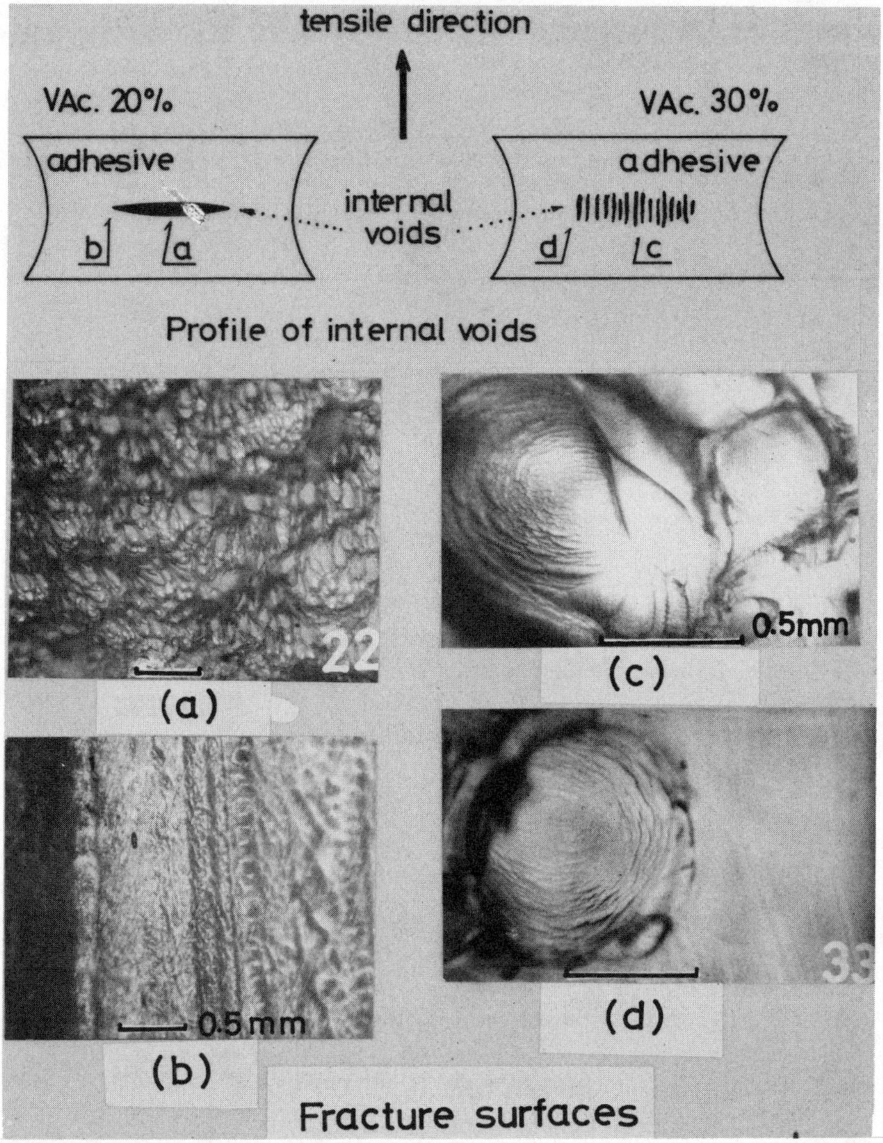

FIG. 6—*Schematic representations of internal voids* (top) *and fractographs of crack and cavity* (bottom): (a) *center region of crack;* (b) *front of crack;* (c) *center of cavity region;* (d) *edge of cavity region.*

The observed locus of failure is given in Fig. 8 (*right*), which was obtained by the same method as Fig. 6. White parts in the center region indicate the internal crack or cavity. These suggest that the partial yield-

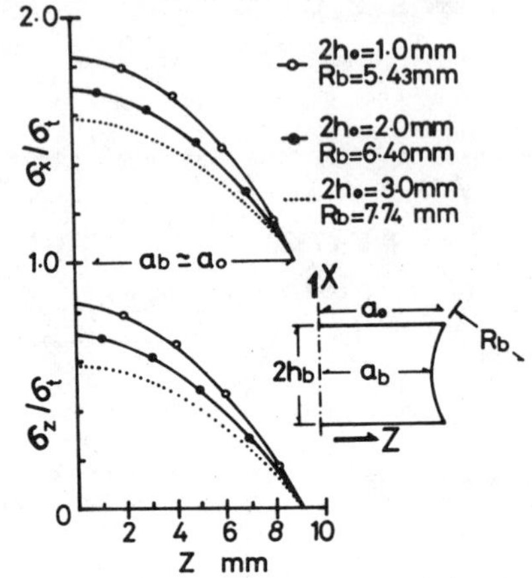

FIG. 7—*Stress distribution curves for* σ_x *and* σ_z *on symmetry plane at yielding of adhesive, calculated from Eqs 6 and 7 using experimental values of* R_b.

ing of the adhesive film is initiated at the center on the symmetry plane. The calculated distribution of σ_z decreases monotonically as the function of z^2 as shown in Fig. 7. On the other hand, the shear stress would be uniform for $z/h_0 \leq 8.5$, as shown in Fig. 8 (*left*). The uniform shear stress region agrees very well with the result of the finite-element calculation [*1*].

Comparison Between Theory and Experiment

In order to verify Eq 12, the dependence of the tensile bond strength and the tensile and lateral strains at yielding of the adhesive on the adhesive thickness was determined.

The tensile strain decreases with increasing thickness, as shown in Fig. 9. The tensile strains differ with the adhesive materials used. On the other hand, the lateral strain increases and is independent of the adhesive materials used, which had a rigidity of 9.4 to 51.3 kg/cm², as shown in Fig. 10. This suggests that the lateral strain is affected mainly by the geometry of the adhesive film rather than by the mechanical properties of the materials.

The tensile and the reduced tensile bond strengths decrease with increasing thickness, as shown in Fig. 11, which agrees with the prediction of the stress distribution as shown in Fig. 7.

From Eq 12, the plot of (σ_b/σ_t) versus $(a_0/h_b)^2\ \epsilon_{zb}$ is shown in Fig. 12. This plot gives a single straight line which is independent of the adhesive

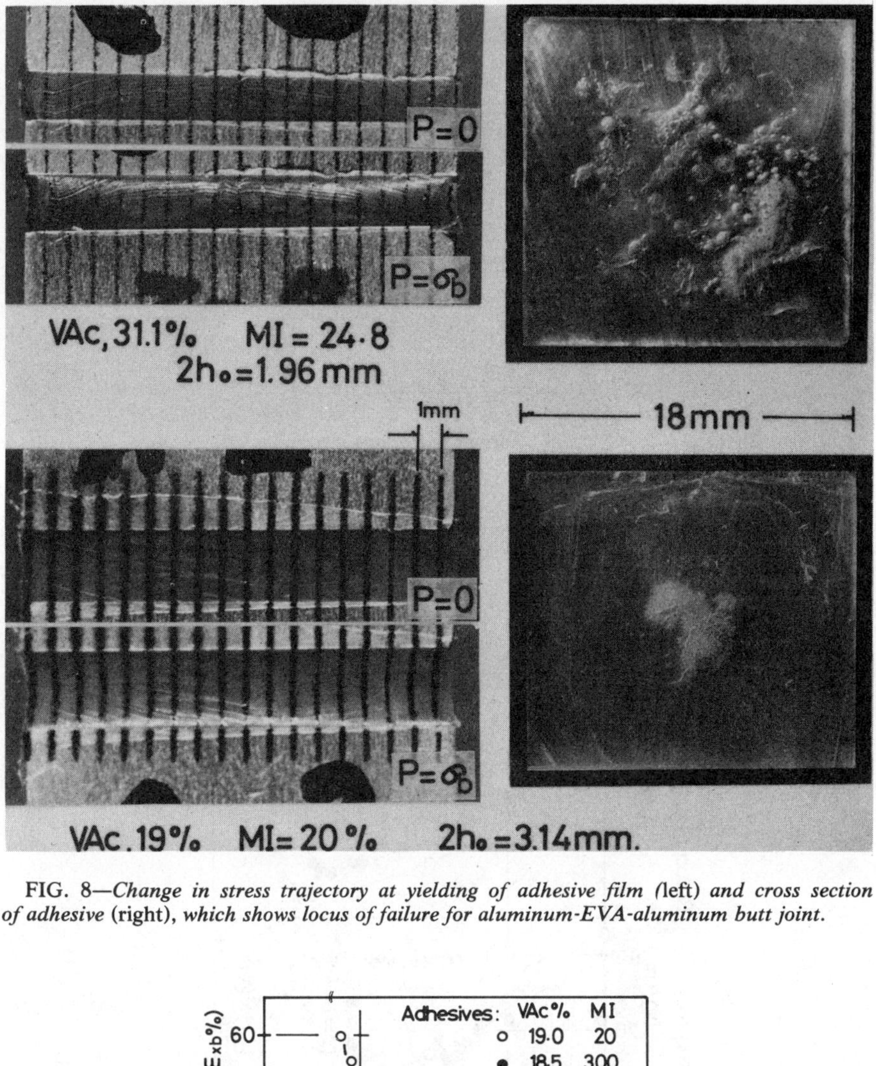

FIG. 8—*Change in stress trajectory at yielding of adhesive film (*left*) and cross section of adhesive* (right), *which shows locus of failure for aluminum-EVA-aluminum butt joint.*

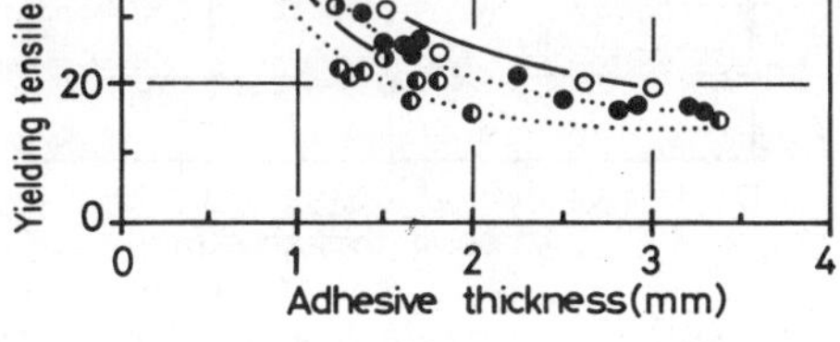

FIG. 9—*Dependence of yielding tensile strain* (ϵ_{xb}) *on adhesive thickness for aluminum-EVA-aluminum butt joint.*

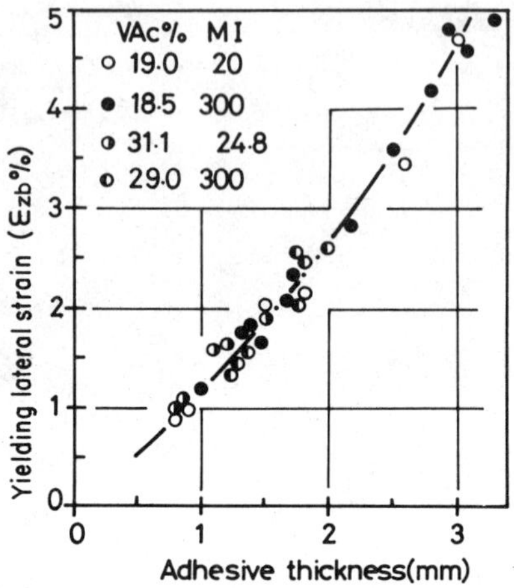

FIG. 10—*Dependence of yielding lateral strain* (ϵ_{zb}) *on adhesive thickness for aluminum-EVA-aluminum butt joint.*

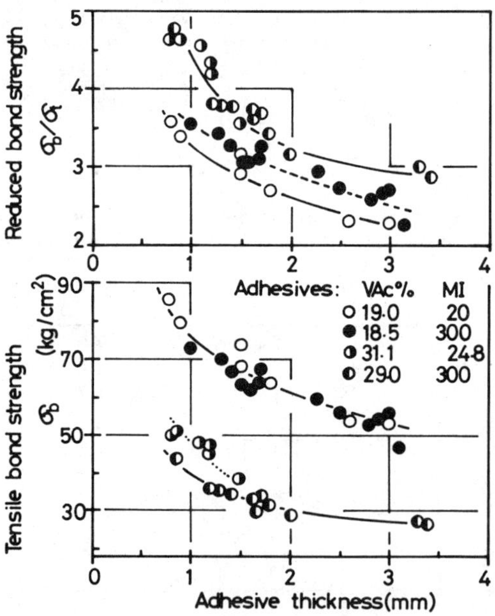

FIG. 11—*Dependence of tensile bond strength* (σ_b) *and reduced tensile bond strength* (σ_b/σ_t) *on adhesive thickness for aluminum-EVA-aluminum butt joint.* σ_b = *tensile bond strength;* σ_t = *yield strength of adhesive material used.*

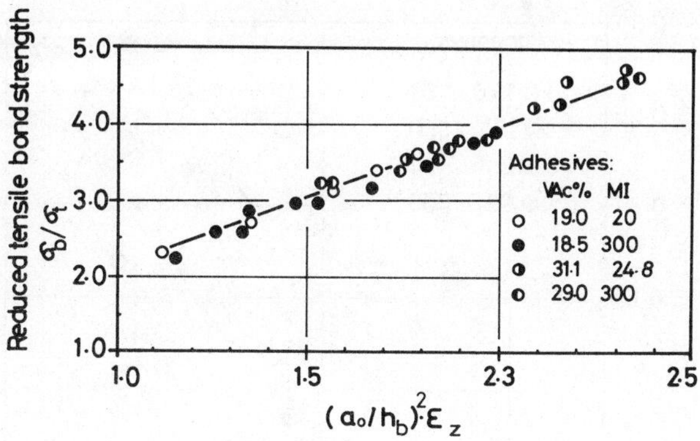

FIG. 12—*Plot of* (σ_b/σ_t) *versus* $(a_0/h_b)^2\epsilon_{zb}$ *for aluminum-EVA-aluminum butt joint according to Eq 12.*

material used. The empirical relationship between (σ_b/σ_t) and $(a_0/h_b)^2$ ϵ_{zb} is represented by

$$(\sigma_b/\sigma_t) = 0.35 + 1.8\,(a_0/h_b)^2\,\epsilon_{zb} \tag{14}$$

It should be noted that both Eq 12, which was theoretically derived, and the empirical relation given in Eq 14 show a linear relationship between the reduced tensile bond strength and the product of the square of the aspect ratio and the lateral strain. However, Eq 14 differs from Eq 12 in the slope and the intercept of the linear line. The intercept of unity in Eq 12 results because of the relation $\sigma_x - \sigma_z = \sigma_0$; therefore, the discrepancy of the intercept between Eqs 12 and 14 cannot be explained at present. The difference in the slopes of both equations suggests that the approximations used for Eqs 4 and 11 may be insufficient. However, the theoretical relationship given in Eq 12 agrees quite well with the experimental results.

Contraction Ratio of Adhesive

A ratio of a lateral contractive strain to a longitudinal tensile strain at yielding of the adhesive is defined as the contraction ratio ($\nu_b = -\epsilon_{zb}/\epsilon_{xb}$) in this paper, and is calculated from the results shown in Figs. 9 and 10. The contraction ratio, which would be measured within the elastic limit, corresponds to Poisson's ratio. The dependence of the contraction ratio on the adhesive thickness is shown in Fig. 13. The contraction ratio increases monotonically with increasing adhesive thickness from 0.01 to 0.28 and depends only on the thickness. The contraction ratio is also independent of the mechanical properties of the material used. Poisson's ratios

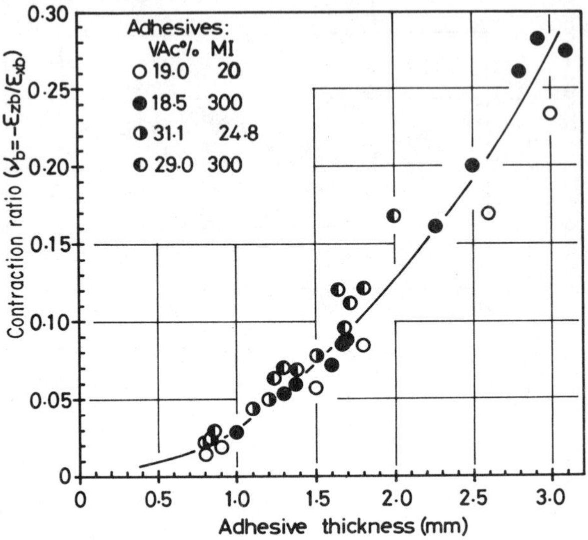

FIG. 13—*Dependence of contraction ratio* ($\nu_b = -\epsilon_{zb}/\epsilon_{xb}$) *on adhesive thickness for aluminum-EVA-aluminum butt joint. Solid line expresses equation:* $\nu_b = 11.9\,(a_0/h_0)^{-2.9}$.

reported for some polymeric substances and the adhesive layers are summarized in Table 2. Poisson's ratios for linear and crosslinked polymers are more than 0.33. Table 2 shows that the contraction and Poisson's ratios of the adhesive are smaller than the ordinary Poisson's ratio and that Poisson's ratio of the adhesive is smaller than the contraction ratio [*8*]. The small contraction and Poisson's ratios of the adhesive film are distinctly attributed to the geometric restriction. It has been pointed up that a Poisson's ratio of less than 0.33 would equally indicate a "superelastic solid" [*12*]; however, it is not clear whether the adhesive film corresponds to a superelastic solid or not.

Denoting the longitudinal strain by ϵ_x, the lateral strain is written as $-\nu\epsilon_x$ under the condition of constant volume. The Poisson's ratio (ν) is given by

$$\nu = (1/\epsilon_x)[1 - 1/(1 + \epsilon_x)^{1/2}] \qquad (15)$$

It was found from this equation that the Poisson's ratio is a function of the tensile strain. The value of ν falls rapidly as the strain increases and is already 0.47 for $\epsilon_x = 0.01$ and 0.40 for $\epsilon_x = 0.35$. The dependence of the contraction and the Poisson's ratios on the tensile strains (ϵ_x and ϵ_{xb}) is shown in Fig. 14. The Poisson's ration calculated from Eq 15 was considerably larger than the contraction ratio. The difference between Poisson's ratios and the contraction ratios would correspond to the increase in the

TABLE 2—*Poisson's ratios of polymeric materials and adhesive films.*

Materials	Strain Range	Poisson's Ratio	Author and Reference
Polyurethane rubber	large	0.5	T. L. Smith[a]
Polystyrene	small	0.336	D. S. Hughes[a]
Polystyrene (P = 15000 psi)		0.353	
Polypropylene (−50 to 40°C) (−58 to 104°F)	small	0.33 to 0.35	Z. Rigbi[a]
Filled polyvinylchloride	small	0.5	
PVC 0 to 50 volume percent glass beads	large	0.2 to 0.5	T. L. Smith[a]
Cured epoxy resin (Epikote 828-FB3)	small	0.377	S. Kawabata [*14*]
Adhesive Joint:			
Steel-nylon 12 cross lap joint ruptured at interface	small at break	0.005 0.04[b]	M. Masuoka [*8*]
Aluminum-EVA butt joint	at break	0.01 to 0.28	the present work

[a]These values are from J. A. Rinde's paper [*13*].
[b]These are defined as the contraction ratios of the adhesive to distinguish them from Poisson's ratio.

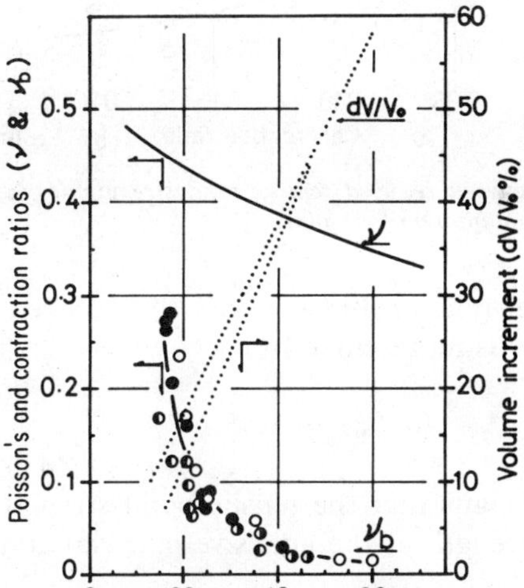

FIG. 14—*Dependence of both Poisson's ratio and contraction ratio* (ν *and* ν_b)*, and volume increment* (dV/V_0) *of adhesive, on tensile strains* (ϵ_x *and* ϵ_{xb})*. Solid line is plot of* ν *versus* ϵ_x *calculated by Eq 15; circles are plot of* ν_b *versus* ϵ_{xb}*; dotted line is plot of* (dV/V_0) *versus* ϵ_{xb} *calculated by Eq 16.*

adhesive volume until yielding of the joint. The average volume increment of the adhesive film (dV/V_0) is adequately approximated by

$$(dV/V_0) = \epsilon_{xb}(1 - 2\nu_b) \qquad (16)$$

The volume increment calculated from Eq 16 is given by the dotted line in Fig. 14. The volume increment increases linearly from 7 percent to 57 percent with the increase in the yielding tensile strain. That is, the volume increment increases with decreasing contraction ratio or the adhesive thickness. The smaller value of the contraction ratio and the larger increment of the adhesive volume until yielding of the joint seem to be the important characteristics of the adhesive deformation under the triaxial stress state. These characteristics result from the restriction of the adhesive geometry.

The dependence of the reduced tensile bond strength on the contraction ratio is shown in Fig. 15. The reduced bond strength increases with decreasing contraction ratio and is independent of the adhesive materials

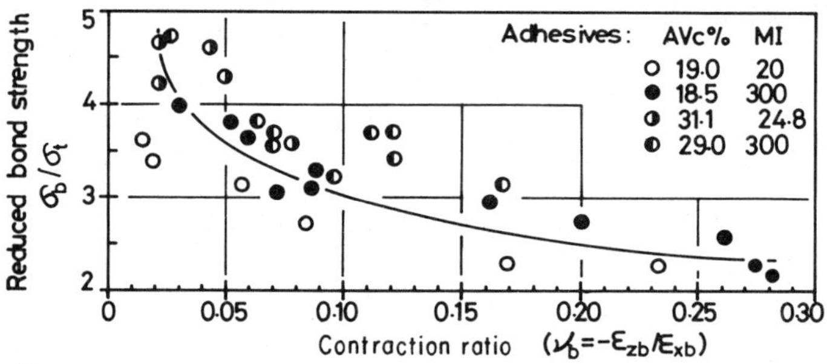

FIG. 15—*Relation between reduced tensile bond strength* (σ_b/σ_t) *and contraction ratio* (ν_b) *for aluminum-EVA-aluminum butt joint.*

used. The empirical relationship between the reduced bond strength and the contraction ratio is represented by

$$(\sigma_b/\sigma_t) = 1.55\nu_b^{-0.23} \quad (17)$$

This relation also means that the tensile bond strength is affected mainly by both the yield strength of the adhesive material and the aspect ratio of the adhesive film.

Conclusion

The theoretical expression for the tensile bond strength, which was derived from a simplified model, agrees very well with experimental results for the square-shaped butt joint. The characteristics of the fracture behavior of adhesive film are clarified fairly well by the methods applied in this study. Most importantly, measurement of the contraction ratio of the adhesive is very useful for determining the restrictive effect on the bond strength. For further development of the present study, it would be necessary to employ an adhesive film of anisotropic cross section, such as rectangle or ellipse, and the thermosetting systems.

References

[1] Harrison, N. L. and Harrison, W. J., *Journal of Adhesion*, Vol. 3, 1972, pp. 195–212.

[2] Tanaka, T. and Taniyama, K., *Proceedings*, 19th Congress on Materials Research, The Society of Materials Science, Japan, 1976, pp. 179–185.

[3] Gent, A. N. and Lindley, P. B., *Proceedings of the Royal Society*, London, Series A, Vol. 249, 1959, pp. 195–205.

[4] Gent, A. N., Henry, R. L., and Roxbury, M. L., *Transactions*, American Society of Mechanical Engineers, Series E, *Journal of Applied Mechanics*, Vol. 41, Dec. 1974, pp. 855–859.

[5] Trantina, G. G., *Journal of Composite Materials*, Vol. 6, April 1972, pp. 192–207.

[6] Gent, A. N. and Meinecke, E. A., *Polymer Engineering and Science*, Vol. 1, No. 1, Jan. 1970, pp. 48–53.

[7] Smith, T. L., *Polymer Science Symposium*, No. 32, 1971, pp. 269–282.

[8] Masuoka, M. and Nakao, K., *Journal of the Adhesion Society of Japan*, Vol. 12, No. 2, Feb. 1975, pp. 41–48.

[9] Gent, A. N., *Journal of Polymer Science*, Part A-2, Vol. 9, 1971, pp. 283–294.

[10] Masuoka, M. and Nakao, K., *Journal of the Adhesion Society of Japan*, Vol. 13, No. 3, March 1977, pp. 82–88.

[11] Hill, R., *The Mathematical Theory of Plasticity*, Clarendon Press, Oxford, 1950, Chapter 10, pp. 258–277. (This book was published in Japan by arrangement with the Clarendon Press of Oxford and translated into Japanese by H. Wasizu, Y. Yamada, and H. Kudho.)

[12] Rigbi, Z., *Applied Polymer Symposia*, No. 5, 1967, pp. 1–8.

[13] Rinde, J. A., *Journal of Applied Polymer Science*, Vol. 14, 1970, pp. 1913–1926.

[14] Kawabata, S., Sagami, H., and Hozeki, Y., Preprints of the 25th Conference of the Society of Materials Science of Japan, May 1976, pp. 53–54.

[15] Davidenkov, N. N. and Spiridonova, N. I., *Proceedings*, American Society for Testing and Materials, Vol. 46, 1946, pp. 1147–1158.

DISCUSSION

J. J. Bikermann[1] (*written discussion*)—I was delighted to hear a discussion of the stress pattern in adhesive joints. Three points remain unclear to me. (1) Is the radius of curvature (R) along the profile of the adhesive-air interface supposed to be constant along the whole profile? (2) According to your slides, the observed Poisson's ratio (ν) increases rapidly when the thickness t of the adhesive film rises; if your curve is extrapolated a little, results above the bulk value for this ratio (0.4) are obtained. I would expect the ν to approach this value asymptotically. (3) You compare the observed increase of the breaking stress on lowering the thickness t with the

[1]Department of Chemical Engineering, Case Western Reserve University, Cleveland, Ohio 44106.

increase calculated from the stress changes caused by the changes in t. But the tensile (or shear) strength of a solid decreases when the volume of the specimen increases also independently of the stress distribution; this effect is caused by the lesser probability of a particularly bad flaw in a smaller specimen. Apparently, your theory disregards this important effect.

Mineo Masuoka and Kazumune Nakao (*authors' closure*)—Your questions are highly significant to our treatment. (1) The curvature along the whole profile was constant for the thinner adhesive—less than about 2 mm. In the thicker region, the discrepancy of curvatures between the supposed circle and the real profile was very slight and would not be significant for our treatment because the experimental parameter needed is the lateral contract strain (ϵ_{zb}) on the horizontal symmetry plane. (2) The contraction ratio of the adhesive film, which was defined in the present paper, corresponds to Poisson's ratio at yielding. Therefore, the ratio will approach gradually an inherent value of the material used when the aspect ratio (a_0/h_0) becomes smaller than unity with increasing adhesive thickness (h_0), and then the deformation of the adhesive will be free from the lateral restriction. (3) In general, the strength of a solid material depends on the inherent flaw size and distribution in a specimen. For example, the relationship between the strength of a material (σ) and the volume of a specimen (V) was expressed by Griffith as $\sigma = V^{-1/m}$, where m is a constant. According to this equation, the increase in the tensile bond strength with decreasing thickness, as shown in Fig. 11, can be explained qualitatively. However, a similar increase in the tensile bond strength of the interfacial fractured joint was observed by the authors. That is, the nature of the bond strength would be equal to the work of deformation of the adhesive film up to breaking the joint. Therefore, the lateral constraint effect caused by the adhesive geometry on the deformation would be more significant for the bond strength than the volume effect caused by the flaw distribution in the adhesive film.

L. H. Sharpe[2] (*written discussion*)—Two comments should be made. Implicit in this treatment is the assumption of a mode of failure—that is, by yield. The treatment, therefore, cannot apply to joints in which failure does not occur by yield. The point also should be made that the material parameters, specifically in this case Poisson's ratio, do not change because of constraints due to confinement in a joint. The shape of the deformed adhesive is due to the particular stress distribution in the adhesive and not to any change in the Poisson's ratio of the material.

[2]Engineering and Development Division, Bell Laboratories, Murray Hill, N.J. 07974.

Mineo Masuoka and Kazumune Nakao (*authors' closure*)—Your comments are of great importance from the viewpoint of our treatment. (1) It is necessary to make clear the nature of the bond strength that the stress distribution combined with the fracture or yield criterion would coincide with the observed locus of failure. Therefore, the application of our treatment is significant when the partial yield occurs at the center region in the adhesive. (2) The Poisson's ratio of a material means a material parameter when it is measured under the condition of free deformation. Under the triaxial stress state, the adhesive film cannot contract laterally; therefore, the apparent contraction ratio has a smaller value than the inherent value. It should be noted that the observed contraction ratio as a function of the aspect ratio is one of the important characteristics of adhesive deformation under the triaxial stress state. As pointed up by you, the shape of the deformed adhesive is certainly due to the stress state and the stress distribution in the adhesive.

N. I. Egorenkov[1] *and V. A. Belyi*[1]

Measuring the Temperature Dependence of the Strength of Metal-Polymer Joints

REFERENCE: Egorenkov, N. I. and Belyi, V. A., "**Measuring the Temperature Dependence of the Strength of Metal-Polymer Joints,**" *Adhesion Measurement of Thin Films, Thick Films, and Bulk Coatings, ASTM STP 640,* K. L. Mittal, Ed., American Society for Testing and Materials, 1978, pp. 362–368.

ABSTRACT: The method of measuring the temperature dependence of adhesion characterized by peel strength, which is based on the application of the temperature gradient in the plane of adhesive contact, is discussed. The foil is peeled in the direction of the temperature gradient. In this case the adhesiogram is the temperature dependence of peel strength. The method determines in a simple way the dependence of the strength of adhesive metal-polymer joints on the test temperature for a wide temperature range. The gradient method has been employed to study the effect of the test temperature (250 to 390 K) on the adhesion of amorphous and crystalline polymers to aluminum. The temperature dependence of adhesion in the case of amorphous polymers to the metal is described by a curve with the maximum at 15 to 35 K higher than the polymer glass transition temperature. The strength of adhesion in the case of crystalline polymers is lower than that of amorphous ones, and decreases with increasing test temperature.

KEY WORDS: adhesion, metal-polymer joints, peel strength, temperature dependence, measurement, temperature gradients, adhesive strength, crystalline polymers, amorphous polymers

The adhesion of polymers to metals that is characterized by the stress to failure of the adhesive joint is usually determined at normal, that is, room, temperature. More valuable information about the strength of an adhesive joint can be obtained, however, when its temperature dependence is known. Test temperatures influence greatly the adhesion of polymers to metals [*1–3*].[2] The most simple and widely used method of measuring adhesion is the peel method [*2*]. The temperature dependence of adhesion is usually deter-

[1] Head of laboratory and professor, respectively, Institute of Mechanics of Metal-Polymer Systems of the Byelorussian Academy of Sciences, Gomel, U.S.S.R.

[2] The italic numbers in brackets refer to the list of references appended to this paper.

mined by peeling the film under isothermal conditions. Such a method is labor- and time-consuming and requires a lot of specimens to measure the temperature dependence of adhesion.

While determining the temperature dependence of adhesion, only the temperature in the peeling area is important. Thus, if the temperature varies in the peeling area as the front line of failure spreads forward, then the adhesiogram is the temperature dependence of the adhesion. This condition is provided by several means. First, it is possible to vary the temperature of the specimen when peeling is done, by heating or cooling the chamber in which the tests are carried out; also, by heating or cooling the specimen itself, the substrate, or the coating (for example, by passing electric current through the metal, by using high-frequency current or contact heaters, etc.). Second, it is possible to create temperature gradients in the specimen in the plane of the adhesive contact along the front line where failure spreads. In the case of rigid bulk substrates such as rods and plates, it is possible to create a temperature gradient by heating or cooling the ends of the substrate [*4,5*].

By placing a heater at one end of the metallic substrate and a cooler at its other end, it is possible to govern the temperature gradient over wide ranges in the plane of an adhesive contact. It is advisable to establish a steady temperature gradient. In this case the measurement of the temperature in the specimen is simplified.

Experimental Techniques

Figure 1 is a schematic of the device used to measure temperature dependences of adhesion of polymers to metals by the gradient method. The device had been designed as an attachment to the ZM-10 instrument (as produced in the German Democratic Republic) that is being used for studying mechanical properties of materials. Referring to the figure, a hollow metallic rod (square prism) made of aluminium or stainless steel, *1*, serves as the basic element of the device. In the cavity at one end of the rod there is an electric heater, *2*; at the opposite end there is a cooler, *3*. Cooling is provided either by water or by liquid nitrogen. The gradient rod is mounted onto the base, *4*.

To estimate the temperature dependence of the adhesion of a polymer to the metal, specimens (foil-polymer-foil) were placed onto the surface of the gradient rod and pressed with the frame along the perimeter. Specimens, *5* in Fig. 1 were aluminium foils 200 μm thick jointed with polymer films 200 μm thick. Th foils were polished beforehand with abrasive paper and washed with acetone. The specimens were obtained by melting the polymer between the aluminium foils over 300 s at the following temperatures: polyvinylbutyral (PVB) at 520 K, polyethyleneterephthalate (PETP) at 550 K, poly-3,3-bis(chloromethyl)-oxacyclobutane at 470 K. PETP and

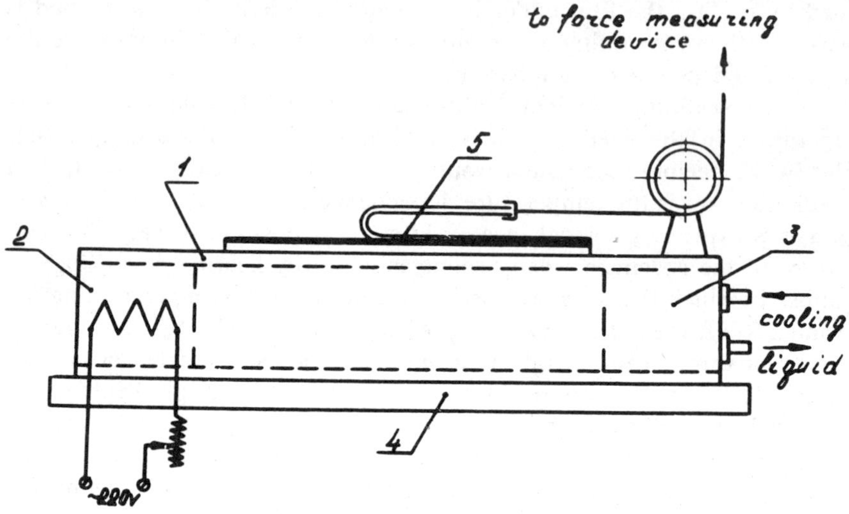

FIG. 1—*Diagram of apparatus for measuring temperature dependence of adhesion* (*explanations are given in the text*).

poly-3,3-bis(chloromethyl)-oxacyclobutane were turned into the amorphous state in adhesive joints by cooling the specimens in ice water. Crystalline polymers were obtained by cooling the specimens at room temperature in air. Specimens containing PVB were also cooled in air. The degree of crystallinity in the polymer films was estimated by X-raying and infrared spectroscopy. The adhesion was measured while the 10-mm-wide foil was being peeled from the polymer film at an angle of 180 deg with a velocity of about 0.2 mm/s. With the help of the frame, the specimens were pressed onto the surface of the rod having a temperature gradient of 10 deg/cm. Mathematical treatment of the experimental data showed that for the 10-mm-wide specimens the variation factor, under the test conditions, does not exceed 7 to 8 percent. The investigations indicated that by increasing the width of the film being peeled it is possible to improve the accuracy of adhesion measurements using the peel method. For instance, in the case of PVB-based specimens at test temperatures of 290 to 295 K, the variation factor decreases from 7.1 to 2.1 percent when the film width is increased from 10 to 50 mm. During peel tests, the temperature measurements were carried out with thermocouples whose joints were within the system of foil-polymer-foil, which had an area of 100 mm^2 along the temperature gradient (the interval was 10 mm). The deviation of the temperature during operation of the gradient device under steady-state conditions was not higher than ± 1.5 K. Comparison of the results obtained by the gradient and nongradient methods was made beforehand. For the nongradient method the specimens were pressed with the frame against the solid of constant surface temperature, and after the specimen reached the same temperature peel tests were carried out. The discrepancies in the values of adhesion obtained with the gradient and nongradient methods were within the experimental error (peel method). The peel temperature of the polymers was determined dilatometrically with a thermomechanical instrument (Rigaku, Japan).

Results and Discussion

Figures 2 through 4 show data on the effect of the test temperature on the adhesion of PETP, PVB, and poly-3,3-bis(chloromethyl)-oxacyclobutane to the metal. For PVB in the amorphous state, the temperature dependence of adhesion is described by a curve with a maximum in the temperature range of 340 to 380 K (Fig. 2, Curve 1). At lower temperatures the adhesion seems not to vary with variations in the temperature. The adhesion starts to increase at temperatures of 340 to 345 K. On the dilatometric curve (Fig. 2, Curve 2) there is a break characteristic of a relaxation transition (glass transition). Maximum adhesion was achieved at 370 K, which exceeds the glass transition temperature by 25 to 30 K. Here the adhesion reaches the value of 4 kN/m.

Similar relations were observed for the adhesive joints made of materials based on crystalline polymers which were obtained in the amorphous state as a result of rapid cooling from the melt. For instance, in the case of PETP in the amorphous state the temperature dependence of adhesion

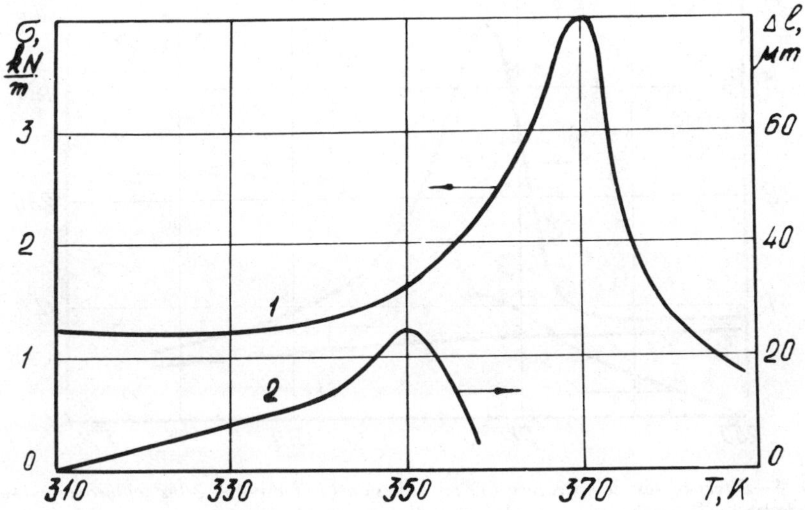

FIG. 2—*Effect of test temperature* (T) *on ahesion* (σ) *for* (1) *polyvinylbutyral to aluminium;* (2) *dilatometrical curve for polyvinylbutyral.*

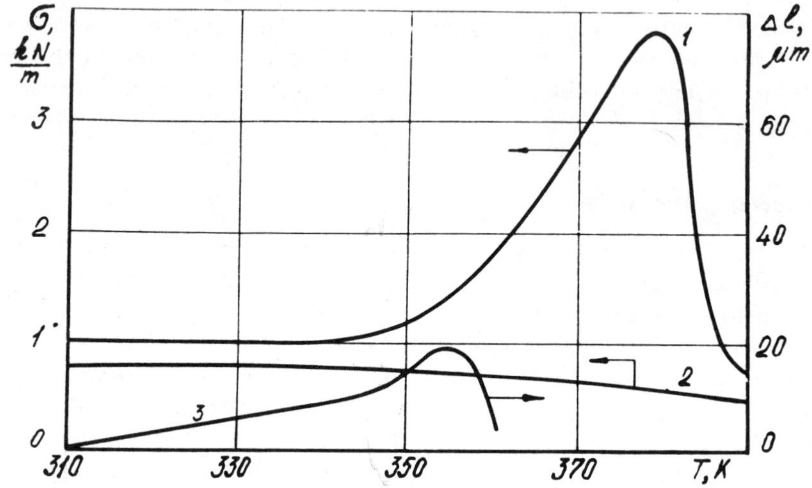

FIG. 3—*Effect of test temperature* (T) *on adhesion* (σ) *for polyethyleneterephthalate to aluminium:* (1) *amorphous state;* (2) *crystalline state;* (3) *dilatometrical curve for polyethyleneterephthalate.*

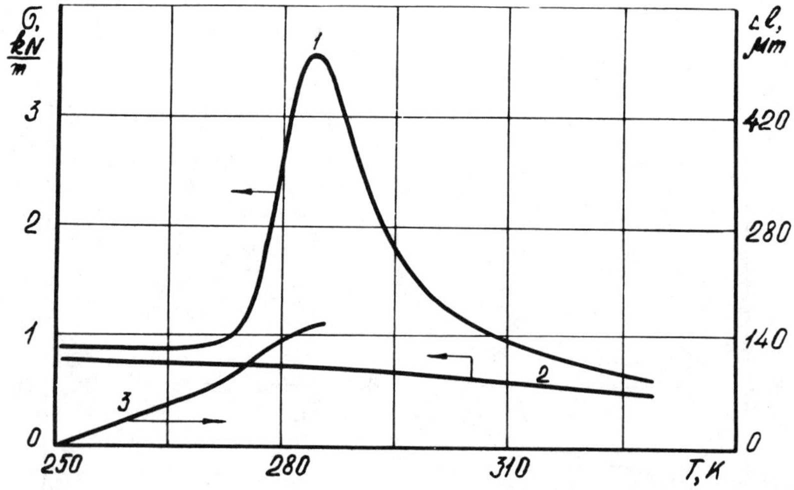

FIG. 4—*Effect of test temperature* (T) *on adhesion* (σ) *for poly-3,3-bis (chloromethyl)-oxacyclobutane:* (1) *amorphous state;* (2) *crystalline state;* (3) *dilatometrical curve for poly-3,3-bis (chloromethyl)-oxacyclobutane.*

is described by a curve with a maximum found at 375 to 380 K, which is higher than the glass transition temperature, that is, 345 to 350 K (Fig. 3, Curve 3), by 25 to 35 K. Here the adhesion reaches the value of 3.8 kN/m. As in the case of PVB, increase in adhesion is observed before the temperature reaches the glass transition point. At lower temperatures the adhesion is, in fact, not dependent on the test temperature. Unlike the amorphous PETP, the adhesion of crystalline PETP decreases with increasing temperature (there is no maximum on the curve). X-ray analysis gives the degree of crystallinity for PETP not higher than 25 to 30 percent. Although the crystallized PETP contains almost 70 percent of nonordered macromolecules (amorphous phase), the character of its temperature dependence is completely different from that of the wholly amorphous PETP.

In the case of poly-3,3-bis(chloromethyl)-oxacyclobutane the glass transition temperature is 270 K, that is, below room temperature (Fig. 4, Curve 3). As in the case of amorphous PETP and PVB, adhesion of poly-3,3-bis(chloromethyl)-oxacyclobutane increases only on reaching the glass transition temperature. Maximum adhesion is observed near room temperature (285 to 290 K) (Fig. 4, Curve 1), that is, 15 to 20 K higher than the glass transition temperature. As a result, the amorphous poly-3,3-bis(chloromethyl)-oxycyclobutane possesses high adhesion to the metal (3 to 4 kN/m) at room temperature. Under these conditions, however, spontaneous crystallization of the polymer takes place, because its glass transition temperature is found near room temperature. At room temperature the process of crystallization is completed in 4 to 8 h [*6*]. As a result of crystalization the adhesion decreases [*6*]. There is a slight dependence of the adhesion of crystallized poly-3,3-bis(chloromethyl)-oxacyclobutane to the metal on the test temperature. There is no maximum on the curve (Fig. 4, Curve 2) though the degree of crystallinity of the polymer does not exceed 25 to 27 percent (according to infrared spectroscopy).

From the comparison of data, one can see that the adhesion of crystallized polymers to metals is lower than that of the amorphous ones for a wide temperature range. Maximum values of adhesion for the polymers in the amorphous state are 3.5 to 4 times higher than those for the crystallized ones. Presumably, this is associated with the conditions of polymer deformation and stress concentration in the area of adhesive contact failure. Low values of adhesion for the crystallized specimens are not associated with chemical decomposition of the polymer in the process of their cooling in air. If the crystallized specimens are melted and cooled rapidly in water, the adhesion increases and, in fact, does not differ from that of specimens which were cooled in water.

Conclusions

The gradient method allows a simple way of obtaining information on the character of the dependence of adhesion of polymers to metals on the test temperature over wide temperature ranges. The investigations showed that the adhesion characterized by the stress of peeling-off is higher for polymers in the amorphous state than for those in the crystalline state. Maximum differences (several times) are achieved at temperatures exceeding the glass transition temperature of polymers by 15 to 35 K.

Acknowledgment

The work reported herein was done at the Institute of Mechanics of Metal-Polymer Systems of the Byelorussian Academy of Sciences, Gomel, U.S.S.R.

References

[1] Belyi, V. A., Egorenkov, N. I., and Pleskachevskii, Yu. M., Adhesion of Polymers to Metals, Nauka i Tekhnika, Minsk, 1971, p. 263.
[2] Voyutskii, S. S., *Autohesion and Adhesion of High Polymers,* Rostekhizdat, Moscow, 1960, pp. 158, 198.
[3] Bikerman, J. J., *The Science of Adhesive Joints,* Academic Press, New York, 1968, p. 309.
[4] Egorenkov, N. I., *Mekhanika Polimerov,* No.6, 1975, p. 1104.
[5] Egorenkov, N. I. and Snezhkov, Eu. L., *Zavodskaya Laboratoria,* No.2, 1975, p. 249.
[6] Egorenkov, N. I. and Egorenkov, A. I., *Vysokomolekulyarnye Soyedinenia,* Vol. 1755, No. 6, 1975, p. 465.

O. J. Klingenmaier[1] *and S. M. Dobrash*[1]

Peel Test for Determining the Adhesion of Electrodeposits on Metallic Substrates

REFERENCE: Klingenmaier, O. J. and Dobrash, S. M., "**Peel Test for Determining the Adhesion of Electrodeposits on Metallic Substrates,**" *Adhesion Measurement of Thin Films, Thick Films, and Bulk Coatings, ASTM STP 640,* K. L. Mittal, Ed., American Society for Testing and Materials, 1978, pp. 369-390.

ABSTRACT: The applicability of the Jacquet peel test for determining the adhesion of plated coatings on metallic substrates was investigated. A procedure was developed to permit testing of various plated coatings without the need for elaborate equipment, and conditions were optimized for determining the peel strength on different metallic substrates. Use of the peel test was demonstrated on a production part. The influence of some pre- or post-plating treatments on adhesion was examined using the peel test. A relationship was established between the tensile strength of the substrate and the possible maximum peel strength.

KEY WORDS: adhesion, peel tests, electroplating, plating on aluminum, mechanical tests

The usefulness of electrodeposited coatings is determined to a large extent by the degree of adhesion to the substrate. In decorative applications, resistance to corrosion is enhanced if the coating is adherent because corrosion through a scratch or an imperfection in the deposit will not be able to spread as readily between the coating and the substrate. In engineering applications, the full potential for which the deposit was intended may be realized and premature failure may be avoided when optimum adhesion is obtained.

Unfortunately, unless a coating is practically nonadherent, there is no easy way to determine the degree of adhesion. Various methods for determining adhesion which have been reviewed elsewhere [*1-5*][2] can be classified as either qualitative or quantitative.

The simpler and more commonly used procedures for determining adhesion, such as hitting the edge of a plated part with a blunt instrument,

[1] Senior research chemist and senior research technician, respectively, Electrochemistry Department, Research Laboratories, General Motors Corporation, Warren, Mich. 48090.

[2] The italic numbers in brackets refer to the list of references appended to this paper.

sawing against the coating, or bending until rupture occurs, are entirely qualitative methods and are, for the most part, subjective. At best, such tests indicate that adhesion was satisfactory or unsatisfactory under the particular test conditions. Such a test may be suitable as a go or no-go gage in some instances where adhesion is not the primary consideration, but it definitely would not be satisfactory in many engineering applications where good adhesion is of utmost importance.

Most of the quantitative methods which have been proposed for determining adhesion are similar, in part, to the qualitative tests in that they result in destruction of the plated coating and the part. Quantitative tests have not been widely used because of the difficulty or length of time involved in preparing specimens, the experience required to perform the test and in interpreting the results, and the elaborate equipment necessary for testing.

The peel test was one of the early quantitative tests proposed for determining adhesion. Jacquet first used this method in 1934 to measure the adhesion of copper on nickel [*6*]. He electrodeposited strips of copper, 5 mm wide and 300 μm thick, on suitably masked nickel substrates. Nonadherent tabs to initiate peeling were obtained at one end of the nickel surface by treating this area with a casein peptone solution before depositing the copper. The force required to peel the copper normal to the surface was used as a measure of adhesion.

In 1950, Brenner and Morgan [*7*] attempted to adapt the Jacquet method to test commercially plated coatings. They were not satisfied with the method because the force of peeling varied with the thickness of the overlay. Later, Linford and Venkateswarlu [*8*] used a modified Jacquet test to determine the influence of copper oxide films on the adhesion of nickel. An acid copper bath was used to build up the peel strips to a thickness of 2 mm. Nonadherent tabs for peeling were obtained by treating a portion of the panel with colloidal graphite. The force required for peeling was dependent primarily on the thickness of the copper oxide films.

In 1962, Wittrock and Swanson [*9*] proposed a modified Jacquet test for determining the peel strength of plated aluminum. The test was subsequently used by Such and Wyszynski [*10*] to obtain an improvement in a zincate solution for treating aluminum before plating. In their procedure, strips of deposit, 25 mm wide and 250 μm thick, were peeled manually from the substrate using a spring balance to measure adhesion. To obtain a tab for peeling, the aluminum was not completely immersed in the zincate solution, thereby producing a nonadherent deposit on the untreated area. Peel strengths as high as 236 N/cm (135 lb/in.) were reported.

A Jacquet-type peel test has been used extensively since 1965 to determine the peel strength of plated plastics [*11,12*]. For this test, an overlay of copper, 37 to 50 μm thick, is deposited on the plated plastic. Strips of metal, 25 mm wide, are then peeled normal to the surface. Peel strengths around 26 N/cm (15 lb/in.) are commonly obtained. Since adhesion is low, a tab to initiate peeling can be easily obtained by prying the deposit loose from the substrate.

Other than these investigations, there has not been much significant use of the peel test for determining adhesion. Lack of additional interest in using this type of test with plated metallic substrates has probably resulted from a number of reasons, including the following:

1. Insufficient information available on optimum conditions for testing.
2. Difficulties encountered in initiating tabs for peeling.
3. Time and care necessary to prepare specimens.
4. Special equipment which may be required to prepare and test specimens.

Regardless of these possible problems, the peel test appeared to be a good method to obtain an objective measure of the degree of adhesion.

It should be emphasized that the peel test, and likewise most other quantitative tests, can be used to measure adhesion only when the adhesive strength of the electrodeposit on the substrate is less than the cohesive strength of the deposit or substrate. In such instances rupture will occur solely or at least predominantly at the interface between deposit and substrate. When rupture no longer occurs at the interface but in a boundary layer near the interface, optimum adhesion has been obtained and the peel strength is no longer a measure of adhesion but is a measure of the cohesive strength of the layer undergoing rupture. Thus the applicability of the peel procedure is extended not only to determine adhesion but also to evaluate those factors that influence the properties of the deposit and the immediate surface of the substrate.

The purpose of this study was to determine the optimum conditions for peel testing, to simplify the procedure as much as possible, and to investigate the influence of some pre- and post-plating treatments on the peel strength of plated substrates.

Experimental

Substrate Materials

Test panels were 7.5 by 10 cm in size and from 2.5 to 6.25 mm thick. Initial tests to establish optimum conditions were conducted using a 6061 T6 wrought aluminum. Other materials used in later tests were 3003 H14, 5052 H93, and 7046 T63 wrought aluminum, F 132 cast aluminum, CDA 110 copper, AISI 1015 steel, and ASTM Type 20 grey cast iron. Aluminum pistons were used to show applicability of the peel test to a production part.

Surface Preparation—Mechanical

Surfaces of panels were belt-polished through successive stages of 120,

180, and 220 grits, and then buffed. In several specific instances, panels were plated in an as-polished or vapor-blasted condition. Aluminum pistons were machine ground.

Surface Preparation—Chemical

All surfaces were degreased. Aluminum alloys were cleaned for 1 min in a proprietary-type acid cleaner at 62°C (144°F) using ultrasonic agitation and were etched lightly in an alkaline solution at 72°C [162°F]. Etching residues were removed by treating for 30 s in a solution containing 50, 25, and 25 volume percent of nitric (70 weight percent), sulfuric (96 weight percent), and hydrofluoric (48 weight percent) acids. The acid solution was also used to strip the first zinc layer from the aluminum substrate during a double zincate treatment.

Two techniques, the zincating process and the phosphoric acid anodizing process [*13*] were used to prepare aluminum surfaces for electrodeposition. In the zincating process the aluminum was immersed for 45 s in a concentrated solution of sodium zincate[3] at 20 to 25°C (68 to 77°F) to form a displacement coating of zinc on the surface. In most instances, this step was repeated after the first layer of zinc was stripped to obtain what is commonly called a double zincate treatment. Copper or nickel was then plated on the zinc-coated aluminum using live electrical contact during entry in the solution. In the anodizing method a porous anodic coating was formed on the aluminum, as a base for plating, by anodizing at 15 to 50 V for 10 min in a 36 weight percent phosphoric acid solution at 38°C (100°F). A nickel strike was plated on the anodic coating. Live electrical contact was not required here.

Substrate materials other than aluminum were cleaned cathodically for 1 to 2 min in a proprietary alkaline cleaner and activated for 5 to 15 s in a 4 weight percent sulfuric acid before plating.

Plating Baths

Copper and nickel strikes on treated aluminum surfaces were plated from a low pH cyanide solution[4] and a neutral nickel solution,[5] respectively. Chromium was plated from a 100/1 ratio chromic acid bath.[6] Iron was deposited from a ferrous chloride electrolyte.[7] Nickel strikes on substrates other than aluminum were plated from a sulfamate bath.[8]

[3] 600 g/litre Alumetex Immersion Salts (Proprietary).

[4] Low pH bath: CuCn, 30 g/litre; free KCN, 7.5 g/litre; Rocheltex, 5 volume percent pH 10.2; 40°C; 2 A/dm^2; 3 min.

[5] Neutral nickel bath: U. S. Patent 3,417,005; 12/17/68; General Motors Corp.

[6] Chromic acid bath: CrO_3, 240 g/litre; H_2SO_4, 2.4 g/litre; 55°C; 30 A/dm^2.

[7] Ferrous chloride bath: U. S. Patent 3,404,074; 10/1/68; General Motors Corp.

[8] Nickel sulfamate bath: Ni $(NH_2SO_3)_2$, 325 g/litre; H_3BO_3, 40 g/litre; $NiCl_2 \cdot 6H_2O$, 4.3 g/litre; Anti-pit agent, 3.8 g/litre; pH 3.8; 55°C.

The nickel overlay used for peel testing also was deposited from the same type of sulfamate bath. The nickel was plated at a current density of 2 A/dm^2 for 30 min, then at 6 to 8 A/dm^2 until the desired thickness was obtained. To obtain a thickness of 160 μm required about 2 h.

Open-ended rectangular cells approximating a linear conductor [*14*] were used to obtain a uniform current density over the surface of panels during the deposition of the nickel overlay. Depth of throw in the plating cells was at least half the width of the panels. Shielded racks were used to plate pistons. Thickness variation of the nickel overlay was less than 5 percent.

Peel Test Procedure

The procedure used to prepare specimens for the peel test is presented in Table 1. Surfaces were usually prepared for the test after being plated with a strike deposit, generally less than 5 μm in thickness. In several

TABLE 1—*Procedure for peel test.*

1. Deposit a thin adherent flash of nickel from a sulfamate bath (1 to 2 min at 2A/dm^2) over the freshly plated surface.
2. Rinse and dry.
3. Passivate a 20-mm-wide strip along one edge of the nickel-plated surface perpendicular to the desired direction of peeling by brushing on a 4 g/litre sodium dichromate solution that has been adjusted to a pH of 1.5 to 2.0 with hydrochloric acid.
4. After 15 to 30 s rinse the passivated area, taking care that the passivating solution does not contact the nontreated area.
5. Immerse in 10 weight percent sulfamic acid for about 1 min.
6. Plate with 140 to 180 μm of nickel from a sulfamate bath. Add 40 μm of nickel for each 70 N/cm (40 lb/in.) [in peel strength expected in excess of 280 N/cm (160 lb/in.)]. Use a current density of 2 to 3 A/dm^2 for the first half hour. Complete plating at a current density of 6 to 8 A/dm^2.
7. Prepare 25-mm-wide strips for the peel test by sawing through the deposit to the base metal.
8. File or grind the edge that was treated with the passivating solution to loosen the nickel tab for peeling. Start a peeled strip manually by carefully gripping the tab with an arc-joint pliers and prying back to initiate rupture.
9. Pull the nickel tabs in a direction normal to the surface at a constant speed of approximately 75 mm/min using a tension tester or other available equipment having a 100-kg scale with 1-kg divisions.

instances, to show that the test could be applied to thicker or multilayered deposits, or to both, cast iron panels were plated with a combined thickness of 30 μm of iron and chromium before being plated with the nickel overlay. To obtain adherent coatings on chromium-plated surfaces, such deposits were plated first with a high-acid nickel strike[9] immediately after rinsing, then with the sulfamate nickel. To demonstrate that the procedure could

[9] High-acid strike: $HiCl_2 \cdot 6H_2O$, 240 g/litre; HCl, 55 to 65 g/litre; 25°C; 2 A/dm^2; 2 min.

be used to determine adhesion on a plated part, aluminum pistons were removed from a plating line after being plated with copper and hard iron, then were prepared for the peel test.

An important step in the test procedure was preparation of a small peel tab (about 1 cm long at one edge of the specimen) to attach the jaws of the peeling device. In many instances on plated substrates such as aluminum, where the peel strength was less than 175 N/cm (100 lb/in.), a peel tab could be initiated by forcing the edge of the overlay away from the substrate with an arc-joint pliers after the plated edge had been ground away and a corner of the deposit had been pried loose with a hammer and chisel. However, this method was time-consuming, and a quicker or more reliable technique was desired to produce a tab for peeling.

Of various procedures investigated to produce a nonadherent tab through electrodeposition, the best method found was to deposit a thin adherent coating of sulfamate nickel on the plated specimen, lightly passivate a small portion of the nickel surface with a dilute dichromate solution, then resume plating in the nickel sulfamate bath until the desired overlay thickness was obtained. After the peel cuts were made and the treated edge was ground to loosen the nickel overlay, the tab could be bent back to initiate rupture. If the overlay was sufficiently thick and good adhesion had been obtained with the various deposits, rupture would occur below the initial strike.

The pH and concentration of the sodium dichromate passivating solution were critical in the limits described in Table 1. Excessive passivation produced by higher pH or concentration frequently caused the nickel overlay to break or tear easily at the treated and nontreated borders. Another method of obtaining a nonadherent overlay—painting the tab area with a conductive solution (colloidal graphite or a silver preparation)—also frequently resulted in broken peel tabs.

Saw cuts for peel strips were best made using tungsten carbide cutting wheels.[10] A device was constructed in which as many as four parallel cuts, 10 or 25 mm apart, could be simultaneously made on a panel. Satisfactory cuts were also made with a band saw using a fine-toothed steel blade. In the latter instance, a metal straight edge was clamped to the face of the panel to guide the blade into a straight cut and the panel was pressed vertically against the open blade. If the cuts were rough or jagged and not clean or sharp, the peel strip tore readily. Peel strips on aluminum pistons were prepared circumferentially on the piston skirt, one on each side. A hacksaw was used to prepare cuts on pistons.

An Instron tension tester was used in early tests to determine peel strengths. Panels were held in a parallelogram-type fixture similar to that used for

[10] Tungsten carbide wheels, 42 teeth, 50 mm diameter, 1.14 mm thickness.

plated plastics [12]. The fixture pivoted in a horizontal plane to keep the direction of pull normal to the plated surface.

A special fixture was built to hold a piston and to permit it to rotate on its axis during peeling. A hole was drilled in the head of the piston to accommodate the shaft of the fixture. A notched plate, which rotated with the piston, was used to support the skirt and to keep it axially normal to the applied force.

To demonstrate that elaborate equipment was not required to determine the peel strength, specimens were also tested using a hand-operated winch to provide the force for peeling and a spring balance to measure peel strength. Figure 1 shows a piston and a panel being tested in this manner.

From 5 to 15 peel determinations were made for each of the various preplate or plating conditions investigated in this study. At least three peel determinations were made for each of the post-plating conditions (influence of heat). In many instances both sides of panels were prepared for the peel test; usually two peel strips were readied per side. The maximum, minimum, and the mean (midpoint) values of peel strength were recorded for each peel strip. The range (overall high and low) of peel strength encountered for each series of peel determinations and the average of the mean values are presented in this paper. Although a peel width of 25 mm was used in most instances, peel strength is presented in newtons per centimetre (N/cm) to avoid the need for decimals or additional digits in the numerator or denominator of the term.

Results and Discussion

Overlay Deposit

Nickel plated from a sulfamate bath was chosen for the overlay used for peeling over other possible deposits because of its optimum properties and convenience.

The tensile and yield strengths of sulfamate nickel are adequate[11] to permit its use in determining relatively high peel strengths without requiring excessive thickness. The deposit has good ductility and toughness to resist fracture during peel preparation and testing. Moreover, sulfamate nickel can be deposited with low residual stress and its surface is easily passivated to aid in forming a nonadherent tab to initiate peeling. Nickel sulfamate plating solutions are readily available, simple and convenient to operate,

[11]Tensile strength, 765 MPa (111 ksi); yield strength, 540 MPa (78.5 ksi); elongation, 7 percent; internal stress <41 MPa (6 ksi) tensile. Sample, C. H. and Knapp, B. B. in *Symposium on Electroforming—Applications, Uses, and Properties of Electroformed Metals, ASTM STP 318,* American Society for Testing and Materials, 1962, p. 32; Electroforming with Nickel, INCO.

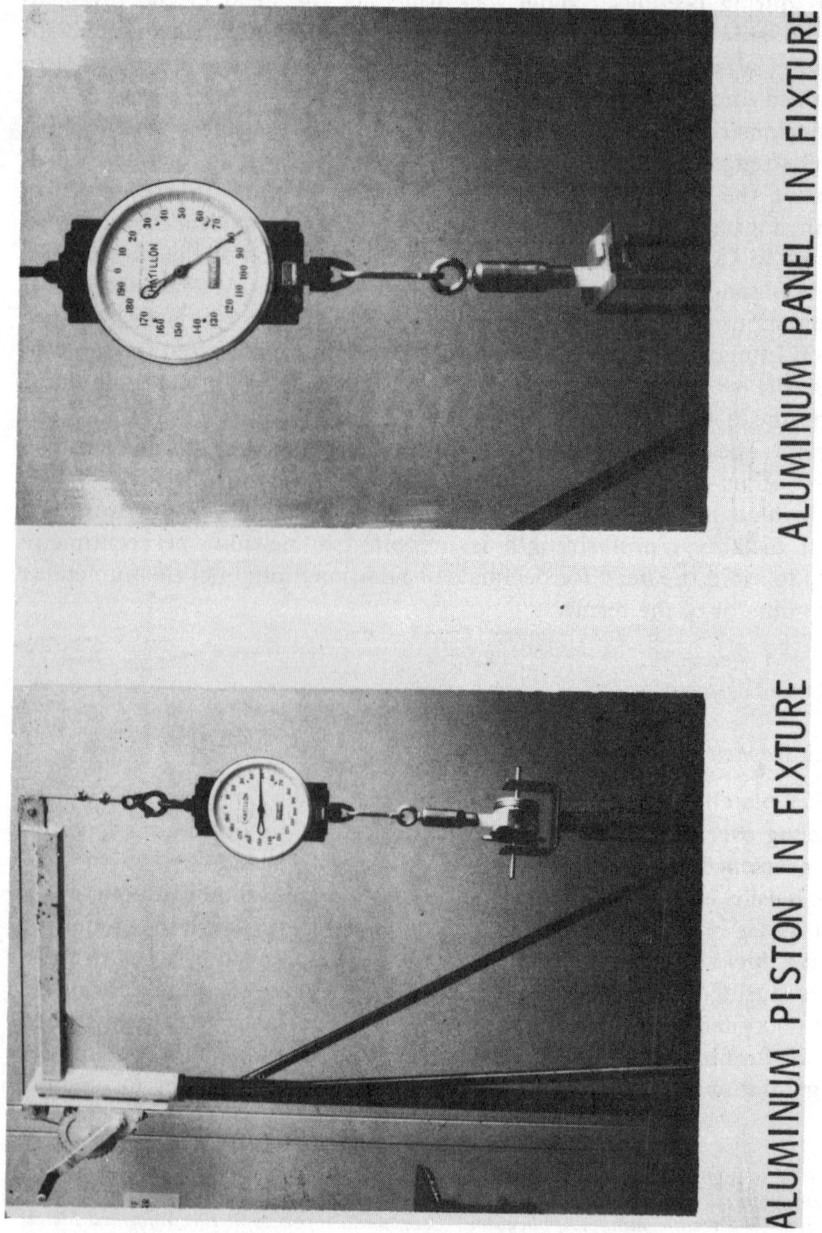

FIG. 1—*Simplified test apparatus for determining peel strength (scale, lb/in.).*

and can be used at relatively fast plating rates (up to 125 μm/h), thus decreasing specimen preparation time.

Thickness and Width of Peel Strips

The influence of thickness of the nickel overlay and the width of the strip on the peel strength of plated aluminum is presented in Fig. 2.

Results indicated that the peel strength of plated aluminum was not influenced appreciably by the thickness of the overlay if it was kept in the range of 140 to 180 μm. A deposit thickness of 160 μm appeared to be optimum for plated aluminum and could be used to peel plated substrates with peel strengths up to 280 N/cm (160 lb/in.) without tearing. Nickel overlays thinner than 140 μm broke in many instances at the edge of the peel tab area or tore during peeling, particularly if the peel strength exceeded 100 N/cm (57 lb/in.). Overlays thicker than 180 μm produced abnormally high values of adhesion for plated aluminum. For example, an overlay thickness of 220 μm resulted in peel strength of about 14 N/cm (8 lb./in.) greater than that obtained with a thickness of 160 μm (Fig. 2).

The added influence of excessive thickness of the overlay on the peel strength has been observed previously with plated plastics. Such apparently higher peel strengths with thicker overlays are produced by the greater force required to bend a thick overlay. Also, a large radius of curvature may be obtained with a thicker overlay which increases the area of the material undergoing failure at any one time, thus requiring greater applied force.

When extremely high peel strengths were encountered, such as with nickel-plated copper or steel, proportionately thicker overlays were required

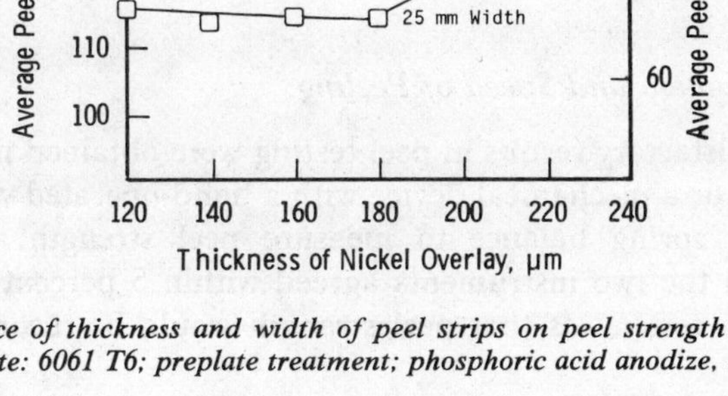

FIG. 2—*Influence of thickness and width of peel strips on peel strength of nickel-plated aluminum. Substrate: 6061 T6; preplate treatment; phosphoric acid anodize, 30 V for 10 min.*

to prevent the strips from breaking during peeling. To obtain satisfactory peel results, the overlay thickness was increased about 40 μm for each 70 N/cm (40 lb/in.) expected in excess of 280 N/cm (160 lb/in.).

The peel strength of plated plastics is generally determined using 25-mm-wide strips [*11,12*]. Similar widths were used in recent investigations involving plated metallic substrates [*9,10*]. A width of 25 mm was determined to be an optimum size in this study. Narrower widths consistently produced higher peel strengths. As shown in Fig. 2, 10-mm-wide strips resulted in peel strengths that were about 10 N/cm (6 lb/in.) greater than that obtained with the 25-mm width. When peel strips were both narrow and excessively thick, the apparent peel strength was increased by as much as 30 percent.

Angle of Peeling

To obtain reproducible results during peeling, not only was it important to maintain close control over the dimensions of the peel strip but also over the angle of peeling. As indicated by previous investigators [*6–12*] and verified in our study, optimum results are obtained by applying the peel force normal to the surface. A peel angle of 90 deg also is recommended in ASTM Measurement of Peel Strength of Metal-Plated Plastics (B 533-70). Conveniently, a 90 deg angle is readily obtained by use of a parallelogram type of fixture for holding panels or by rotating circular parts on their axis during peeling.

Significant changes in the apparent peel strength should be expected if the correct peel angle is not maintained. In our study, for example, average peel strengths of 133 and 88 N/cm (76 and 50 lb/in.) were obtained when nickel-plated and anodized aluminum was peeled at angles of 75 and 105 deg, respectively. Peeling the same strips normal to surface produced an average peel strength of 105 N/cm (60 lb/in.). The higher peel strength obtained at a 75-deg angle was produced by the peel force applied partially against the direction of peel and thus possibly causing a larger area to undergo rupture at any one time—similar to the effect encountered by an abnormally thick overlay. Conversely, lower peel strength obtained at a 105 deg angle was produced by the peel force applied partially in the peeling direction and thus possibly causing a smaller area to undergo rupture.

Type of Instrument and Speed of Peeling

Equally satisfactory results in peel testing were obtained using an Instron tension tester or a mechanical device with a hand-operated winch to provide force, and a spring balance to measure peel strength. Peeling results obtained with the two instruments agreed within 5 percent. An advantage of the Instron tester was that peel strength would be recorded on a chart for future reference, in contrast to the manually operated device, where peel strength was presented momentarily on the balance scale and accuracy in measurement was dependent partly on the operator's attentiveness and care in applying force and in observing the scale. Despite the apparent disadvantage, the manually operated device was preferred for testing not only because of convenience and availability, but also because specimens could be peeled faster.

The rate at which force was applied for peeling was not critical in the range measured; a linear rate of strip removal from 25 to 75 mm/min was satisfactory with the Instron tester and from 25 to 150 mm/min with the manually operated device. Speeds of 75 to 150 mm/min were generally used with the latter.

Of greater importance than speed was the application of a steady force so that the specimen could pivot gradually and smoothly in the direction opposite to that of peeling and thus maintain a peel angle as close to 90 deg as possible. Use of a uniform peel rate also insured that the failure mechanism during peeling remained constant and that it was primarily one of material distortion—not one of rapid crack propagation, which could have occurred when nonuniform peeling was used.

Peel Strengths of Copper and Nickel Strikes on Zincate-Treated Aluminum

A double zincate treatment is often used for treating aluminum before plating as it provides for more thorough and complete coverage with a dense, tight film of zinc.

Tests conducted with plated aluminum alloy 3003, Table 2, indicated that approximately 20 and 100 percent increases in the peel strengths of copper and nickel strikes, respectively, could be obtained if a double

TABLE 2—*Peel strength of copper or nickel strike on wrought aluminum alloy 3003.*

Strike Deposit	Preplate (Post) Treatment	Peel Strength, N/cm (lb/in.) Range	Peel Strength, N/cm (lb/in.) Average Mean
Copper	single zincate	69 to 91	81 (46)
	single zincate (150°C, 1 h)	105 to 146	128 (73)
	double zincate	79 to 115	98 (56)
	double zincate (150°C, 1 h)	173 to 184	180 (103)
Nickel	single zincate	43 to 255	103 (59)
	single zincate (150°C, 1 h)	138 to 245	208 (119)
	double zincate	173 to 245	196 (112)
	double zincate (150°C, 1 h)	207 to 314	257 (147)

zincate treatment was used in place of a single treatment. With a double zincate treatment, separation during peeling occurred primarily in the aluminum substrate compared with a single treatment, where separation was produced in the zinc layer or at the deposit-zinc interface. Higher peel strengths were usually obtained with a nickel strike rather than with copper. Heating plated panels for 1 h at 150°C (302°F) produced additional improvements in the peel strengths ranging from 30 to 100 percent regardless of whether a single or double zincate treatment was used.

Better adhesion also was produced on other aluminum alloys when a nickel strike was used in place of copper. As shown in Table 3, at least 30 percent higher peel strengths were obtained with nickel-plated wrought 6061, 5052, and cast F132 aluminum alloys than with similar substrates plated with copper. Copper-plated cast aluminum pistons had slightly better (10 percent) peel strength than similarly plated panels. Less porosity was encountered with the pistons.

The influence of heating for 1 h at 150°C (302°F) on adhesion was very pronounced for the wrought alloys, resulting in 50 to 100 percent improvements in peel strengths, with the nickel-plated ones again showing the greatest improvement. Heating the copper-plated cast alloy for 1 h at 150°C (302°F) improved adhesion to a level obtained with the as-plated nickel. The peel strength of nickel-plated cast alloy was not further improved by heating. Prolonged heating at a high temperature [12 weeks at 175°C (347°F)] was not detrimental to the adhesion of plated 6061 or F132 aluminum alloys.

The substantial difference obtained in peel strength between copper and nickel strikes on wrought aluminum suggests that the physical properties of the strike deposit may be influencing the apparent values obtained for adhesion.

In a mathematical development of the Jacquet peel test used for plated plastics, Saubestre et al [*11*] concluded that Young's modulus of elasticity for the deposited film is one of five factors influencing adhesion, others being the thickness of the electrodeposited metal and the thickness, Young's modulus, and tensile strength of the plastic-yielding film (substrate). It was further indicated that nickel deposits on plastic substrates should produce higher peel strengths than copper deposits because Young's modulus for nickel is greater than for copper.

Although the thickness of the strike deposits on aluminum substrates in this investigation amounted to only 2 percent of the thickness of the nickel overlay, it is possible that Young's modulus of the initial deposit could produce a greater influence on peel strength than Young's modulus for the entire overlay. This would explain the lower peel strength consistently obtained with a copper strike on aluminum.

Differences in peel strength between copper and nickel strikes could also be caused, in part, by the plating characteristics of the strike baths which

TABLE 3—*Peel strength of copper or nickel strike on other aluminum alloys using a double zincate treatment.*

Alloy	Strike	Post Treatment	Peel Strength, N/cm (lb/in.) Range	Average Mean
6061	copper	...	39 to 126	89 (51)
		150°C, 1 h	91 to 210	144 (82)
		150°C, 24 h	131 to 218	149 (85)
		175°C, 12 week	140 to 149	144 (82)
	nickel	...	70 to 210	107 (61)
		150°C, 1 h	196 to 220	210 (120)
		150°C, 24 h	198 to 250	228 (130)
		175°C, 12 week	226 to 265	245 (140)
5052	copper	...	61 to 92	77 (44)
		150°C, 1 h	96 to 149	123 (70)
	nickel	...	96 to 176	142 (81)
		150°C, 1 h	227 to 297	280 (160)
F132		...	60 to 100	84 (48)
(Pistons)	copper	...	44 to 105	77 (44)
(Panels)	copper	150°C, 1 h	105 to 109	107 (61)
		150°C, 24 h	105 to 114	109 (62)
		175°C, 12 week	79 to 105	91 (52)
	nickel	...	70 to 143	105 (60)
		150°C, 1 h	94 to 128	101 (58)
		150°C, 24 h	101 to 128	112 (64)
		175°C, 12 week	105 to 123	114 (65)

could adversely affect the zincate film on aluminum. The nickel strike bath has a lower pH (neutral) than the copper bath, shows better efficiency for depositing the metal, and does not readily form immersion deposits. Thus, the zincate film could be attacked to a lesser degree by the nickel strike bath, producing less "debris" on the surface before coverage is obtained. This would result in better metal-to-metal bonding on plating, with possible further improvement obtained through heating. Obviously, additional investigations are required in this area to determine more fully what is evidently a complex phenomenon.

Peel Strength of Plated Anodized Aluminum

The influence of an anodizing preplate treatment on the peel strength of several aluminum alloys is compared in Table 4 with that obtained with a double zincate treatment. More uniform peeling results were obtained through anodizing as indicated by the narrower range of peel strength values encountered for the various alloys. Individual pulls on anodized surfaces usually varied by less than ±5 percent from the mean peel strength compared with variations of as much as ±10 percent for zincated surfaces.

TABLE 4—*Influence of type of preplate treatment on peel strength of nickel-plated aluminum.*

Alloy	Preplate Treatment	Peel Strength, N/cm (lb/in.) Range	Average Mean
3003	double zincate	173 to 245	196 (112)
	phosphoric acid anodize, 15 V	183 to 227	198 (113)
6061	double zincate	70 to 210	107 (61)
	phosphoric acid anodize, 15 V	44 to 61	51 (30)
	phosphoric acid anodize, 20 V	70 to 88	79 (45)
	phosphoric acid anodize, 30 V	96 to 130	105 (60)
	phosphoric acid anodize, 40 V	131 to 149	140 (80)
7046	double zincate	96 to 210	156 (89)
	phosphoric acid anodize, 20 V	no adhesion	
	phosphoric acid anodize, 30 V	17 to 35	26 (15)
	phosphoric acid anodize, 40 V	70 to 87	79 (45)
	phosphoric acid anodize, 50 V	70 to 123	96 (57)

Different anodizing voltages were required for the three alloys given in Table 4 to obtain satisfactory adhesion. A substantially lower voltage (15 V) which was used to anodize the 3003 alloy produced peel strengths significantly higher than were obtained with the other alloys.

As presented in Table 4, the magnitude of the voltage used to anodize the 6061 and 7046 alloys determined the degree of peel strength. Although peel strengths comparable to that reached by zincating were obtained with the 6061 alloy by anodizing at 30 to 40 V, similar results were not obtained with the 7046 alloy even by anodizing as high as 50 V. The 7046 alloy has a much higher alloying content which may be contributing to the lower adhesion obtained through anodizing.

In contrast to the 3003 alloy where the ruptured zone on peeling occurred primarily in the aluminum substrate, plate 6061 and 7046 alloys showed separation largely in the anodic coating. Penetration of the nickel strike into the anodic coating was indicated in all instances.

Peel Strength of Chromium Deposits on Aluminum

The peel strength of electrodeposited chromium on zincate-treated aluminum using several procedures is presented in Table 5.

Relatively low peel strengths were obtained when the strike deposit of chromium was plated from a cold solution (Procedures *a* and *b*, Table 5). Use of a cold bath is described in ASTM Recommended Practice for Preparation of and Electroplating on Aluminum Alloys by the Zincate Process (B 253–73). During peeling, rupture occurred entirely within the chromium deposit, possibly in the initial layer, which is likely to have lower cohesive

TABLE 5—*Peel strength of electrodeposited chromium on various aluminum alloys using a double zincate treatment.*

Alloy	Plating Procedure	Post Treatment	Peel Strength, N/cm (lb/in.) Range	Average Mean
3003	(*a*) 24°C for 15 min, then	...	37 to 53	43 (25)
	transfer to second bath	260°C, 1 h	77 to 95	86 (50)
	at 55°C			
	(*b*) 24°C for 15 min, then	...	34 to 43	39 (23)
	raised to 55°C	260°C, 1 h	181 to 198	190 (110)
	(*c*) 55°C for 1 h	...	114 to 162	131 (75)
		260°C, 1 h	163 to 254	226 (129)
6061	55°C for 1 h	...	114 to 167	147 (84)
		260°C, 1 h	192 to 318	251 (144)
F132	55°C for 1 h	...	79 to 96	87 (50)
		260°C, 1 h	91 to 96	92 (53)

strength since it was plated from a cold solution. Extremely high hardness and poor wear resistance have been reported for such deposits [*15*].

Heating chromium-plated aluminum for 1 h at 260°C (500°F) doubled the peel strength in instances where two baths were used (Procedure *a*) and quadrupled the peel strength when the same bath was used for both temperatures (Procedure *b*). An improvement in cohesive strength of the deposits is indicated in both instances, possibly the result of dehydrogenation of the chromium.

On the other hand, when chromium was deposited from a high-temperature bath directly on zincate-treated aluminum, as shown in Procedure c, Table 5, the peel strength was approximately three times that obtained with the low-temperature procedures. Heating again produced a further improvement in peel strength.

Peel tests using the high-temperature chromium strike bath were repeated on two other aluminum alloys, wrought 6061 and cast F132. Excellent peel strengths were produced which compared favorably with those obtained with nickel-plated aluminum.

To obtain optimum adhesion results in depositing chromium on aluminum, it was important to use a live contact with a strike current density of at least 30 A/dm^2. The presence of a fluorochemical wetting agent in the plating bath produced poor adhesion regardless of the temperature of the solution.

Scanning electron micrographs (SEMs), Fig. 3, of the back of strips peeled from chromium-plated aluminum indicated that rupture occurred partially in the chromium and partially in the aluminum. At the left is presented the back of a partially peeled strip which was tested immediately after plating; at the right, the back side of another portion of the same

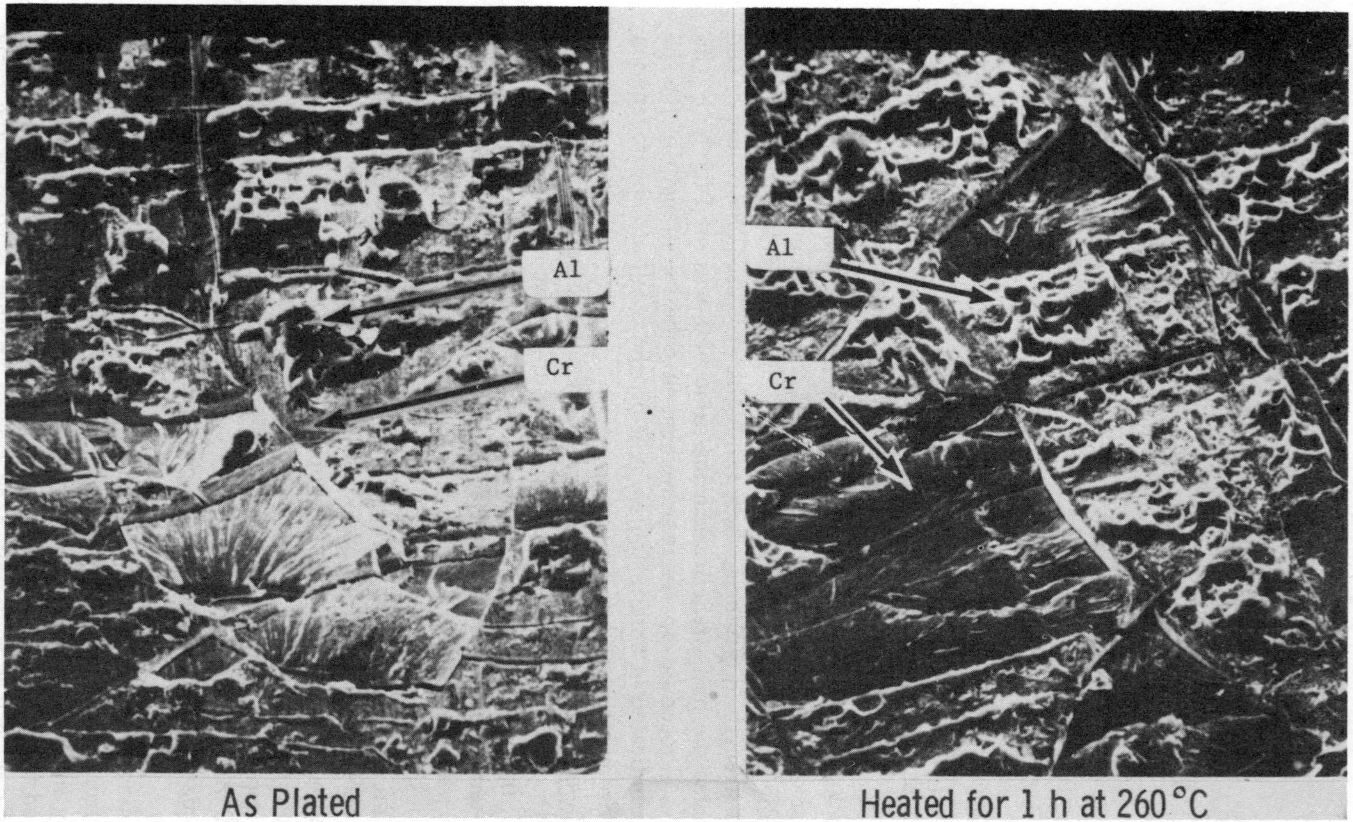

FIG. 3—*Back of plated strip after peeling chromium deposit from 6061 aluminum alloy (SEM ×300, 45 deg). Peel strength: as plated, 108 N/cm (62 lb/in.): heated, 215 N/cm (123 lb/in.).*

strip after heating. Notice areas where aluminum is torn from the substrate and is left adherent on the back of the chromium, other areas where there is no aluminum present, and still others where the chromium has ruptured. Although both pictures show several sections where chromium has ruptured, the right-hand specimen had a larger overall area covered with aluminum. The peel strength here was twice as great as before heating.

Peel Strength of Chromium Deposits on Cast Iron

The effect of various preplate procedures on the peel strength of chromium-plated steel or cast iron substrates is presented in Table 6.

An average peel strength of 158 N/cm (90 lb/in.) was obtained (Table 6) for chromium-plated steel in which a more or less conventional method was used to prepare the surface for plating. Adhesion was considered excellent and no other procedures were investigated to improve the bond strength. However, it would appear that an additional improvement in as-plated adhesion could be obtained by perhaps minor changes in treatment since the peel strength more than doubled after heating for 1 h at 260°C (500°F).

Unlike results with steel, difficulty can be experienced in attempting to obtain good adhesion of chromium on cast iron. The etching procedure used for steel cannot be used very readily with cast iron because of graphite which remains on the surface after etching and interferes with adhesion. As shown in Procedure *a*, Table 6, etching anodically for 3 to 5 s resulted in an average peel strength of 79 N/cm (45 lb/in.). Slight improvements in adhesion (30 to 50 percent) were obtained by heating the plated case iron for 1 to 3 h at 260°C (500°F). When etching times longer than 5 s were used to activate the cast iron, Procedure *b*, Table 6, little or no adhesion was obtained.

By replacing anodic etching with dilute acid activation, Procedure *c*, Table 6, approximately a 40 percent improvement was obtained in the peel strength of chromium-plated cast iron. To obtain optimum adhesion, it was essential to rinse quickly after acid activation and to plate immediately with chromium using live contact. Again, an in previous instances, a further improvement was obtained by heating.

Another procedure, which was subsequently investigated, produced even better results. In this method, Procedure *d*, Table 6, a freshly polished and cleaned cast iron surface was plated with a thin deposit of iron, then transferred to a chromium plating bath and treated as a steel part. Average peel strength was now 193 N/cm (110 lb/in.) with a further 30 percent improvement obtained by heating.

Influence of Tensile Strength of Substrate on Adhesion

When a deposit of high cohesive strength is deposited on a substrate of lower cohesive strength and good adhesion is obtained, it would be ex-

TABLE 6—*Peel strength of electrodeposited chromium on steel and cast iron.*

Substrate	Preplate Procedure	Post Treatment	Peel Strength N/cm (lb/in.) Range	Average Mean
1010	vapor blast; anodic etch, 20s[a]	...	123 to 192	158 (90)
		260°C for 1 h	315 to 402	362 (207)
13M cast iron	(*a*) vapor blast; anodic etch, 3 to 5 s[a]	...	60 to 100	79 (45)
		260°C for 1 h	87 to 121	105 (60)
		260°C for 2 h	107 to 123	116 (66)
		260°C for 3 h	123 to 127	123 (70)
	(*b*) vapor blast; anodic etch, >5s[a]	...	little or no adhesion	...
	(*c*) vapor blast; acid etch, 10 to 15s in 2½% H_2SO_4	...	99 to 122	114 (65)
		260°C for 1 h	140 to 159	150 (86)
		260°C for 2 h	178 to 183	182 (104)
		260°C for 3 h	no further change	...
	(*d*) 220 grit polish; cathodic clean; iron strike, 5 A/dm,^{2} 10 min; anodic etch, 20s[a]	...	182 to 218	193 (110)
		260°C for 1 h	241 to 288	258 (147)
		260°C for 2 h	no further change	...

[a] Anodic etch in chromium plating bath, 15 A/dm^2.

pected that rupture would occur in the substrate during peeling and that the physical properties of the substrate would influence the peel strength. This is true. Plated plastics with good bond strength, when peeled, have been shown to rupture not at the interface but in a boundary layer within the plastic substrate [*11*]. Since the physical strength of plastics is low, the peel strength must be necessarily low.

Plated metallic substrates with good bond strength when peeled, likewise, do not rupture at the interface but in a boundary layer within the substrate. For example, with nickel-plated aluminum alloy 3003, as much as 1.8 μm of the aluminum substrate was removed during peeling. Since the physical strength of a metallic substrate is significantly greater than that of a plastic substrate, the peel strength should also be relatively greater.

Peel strength ranges which were obtained with various nickel-plated substrates are presented in Fig. 4 as a function of the tensile strength of the substrate. The extent of increased peel strength obtained by heating

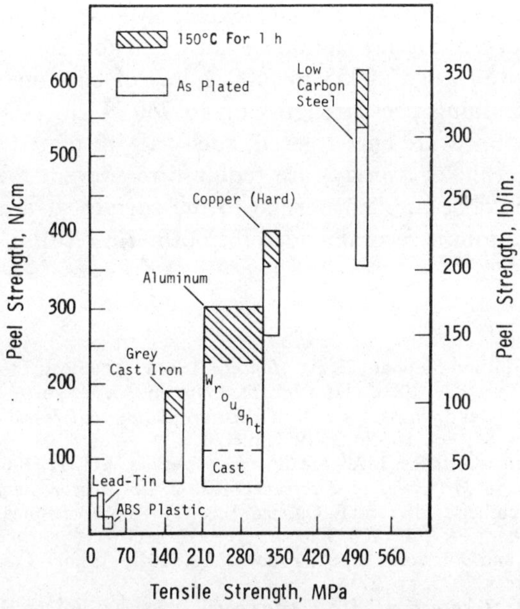

FIG. 4—*Influence of tensile strength of substrate on peel strength of electrodeposited nickel.*

after plating is shown by shaded areas. With the exception of cast aluminum, where the rupture mechanism may differ from the wrought alloys, the maximum peel strengths obtained with the various plated substrates are shown to be directly proportional to the tensile strength of the substrate. A ready prediction can then be made for the ultimate peel strengths possible with other substrates based on their tensile strength.

Summary and Conclusions

A modified Jacquet peel test is a useful method to determine objectively and quantitatively the relative degree of adhesion of plated coatings on metallic substrates. The test is helpful not only in evaluating various procedures for plating but in optimizing procedures or solutions to produce better adhesion. Although a destructive test, it is recommended for monitoring production plating quality either by sacrificing parts or by running suitable test panels through the process.

Preparation of specimens for peel testing can be initiated after normal procedures for plating have been completed. Most electroplated surfaces, if they can be overplated with an adherent coating of nickel, can be tested for peel strength. Although the test is designed primarily for use with flat specimens, it can also be applied to cylindrical-shaped parts, such as pistons, when they are plated with a uniform thickness of the overlay and if they are rotated during peeling to maintain a 90 deg peel angle.

Primary test factors influencing peel results are the thickness of the overlay, the peel angle, and—to a lesser extent—the width of the peel strip. The thickness of the overlay should be sufficient to resist tearing. A thickness of 140 to 180 μm (0.0055 to 0.0071 in.) of sulfamate nickel is adequate for determining peel strengths up to 280 N/cm (160 lb/in.). When higher peel strengths are encountered, such as with plated steel substrates, proportionately thicker overlays are required to prevent tearing. The force for peeling should be applied normal to the surface at a constant rate. A peel width of 25 mm is recommended for optimum results.

References

[*1*] Ferguson, A. L. and Stephan, E. F., *Monthly Review,* American Electroplaters' Society, Vol. 32, 1945, pp. 894, 1006, 1116; Vol. 33, 1946, pp. 45, 166, 620.

[*2*] Solov'eva, Z. A. and Vagramyan, A. T., *Electroplating and Metal Finishing,* Vol. 15, No. 3, 1962, p. 84; Vol. 15, No. 4, 1962, p. 120.

[*3*] Davies, D. and Whittaker, J. A., *Metallurgical Reviews,* Vol. 12, 1967, pp. 15–26.

[*4*] Mittal, K. L. in *Properties of Electrodeposits, Their Measurement and Significance,* R. Sard, H. Leidheiser, Jr., and F. Ogburn, Eds., The Electrochemical Society, Princeton, N.J., 1975, Chapter 17, pp. 273-306.

[*5*] Dini, J. W. and Johnson, H. R., *Proceedings,* 18th Annual Conference, Society of Vacuum Coaters, 1975, pp. 27–41.

[*6*] Jacquet, P. H., *Transactions,* The Electrochemical Society, Vol. 66, 1934, p. 393.

[*7*] Brenner, A. and Morgan, V. D., *Proceedings,* 37th Annual American Electroplaters' Society, 1950, pp. 51–65.

[*8*] Linford, H. B. and Venkateswarlu, A., *Plating,* Vol. 45, 1958, p. 728.

[*9*] Wittrock, H. J. and Swanson L., *Plating,* Vol. 49, 1962, p. 880.

[*10*] Such, T. E. and Wyszynski, A. E., *Plating,* Vol. 52, 1965, p. 1027.

[*11*] Saubestre, E. B., Durney, L. D., Hajdu, J., and Bastenbeck, E., *Plating,* Vol. 52, 1965, p. 982.

[*12*] McNamara, J. B. and Sexton, J., Paper No. 995B presented to the Society of Automotive Engineers at the Detroit Meeting, Jan. 11–15, 1965.

[13] Wittrock, H. J., *Proceedings*, 48th Annual American Electroplaters' Society, 1961, pp. 52-59.
[14] Mohler, J. B. and Shaefer, R. M., *Monthly Review*, American Electroplaters' Society, Vol. 34, 1947, pp. 1361-1364.
[15] Shreider, A. V. in *Theory and Practice of Chromium Electroplating*, A. T. Vagramyan and N. T. Kudryavtsev, Eds., Israel Program for Scientific Translations, Ltd., Jerusalem, 1965, pp 55-60.

DISCUSSION

Carl Dahlquist[1] *(written discussion)*—Do the ribbons undergo ductile yield during the peeling? If so, it would be difficult to estimate the interfacial stress from application of peel mechanics.

O. J. Klingenmaier and S. M. Dobrash (authors' closure)—Measurements were not made *in situ* to determine elasticity of the strip during peeling, and permanent elongation was not normally encountered with the nickel strips. However, in instances of extremely high peel strength, >350 N/cm (200 lb/in.), stretch marks about 1 mm long were produced on the strips during peeling.

H. E. Ashton[2] *(written discussion)*—Have you investigated the angle of peeling, because most adhesive pull tests are carried out at 180 deg and the adhesion group of ASTM D-1 did not find any difference between 90 deg and 180 deg for the tape test, ASTM Measuring Adhesion by Tape Test (D 3359-74), with organic coatings.

O. J. Klingenmaier and S. M. Dobrash (authors' closure)—Using a peel angle of 180 deg was not practical. It was necessary to peel a strip over half its length at a lesser angle to position the jaws of the peeling device for a 180 deg pull. Peel results showed much greater variation at 180 deg and were as much as 40 deg greater than that obtained at an angle of 90 deg. Apparently, higher peel strengths obtained at an angle of 180 deg may be caused by the force required to bend the nickel overlay through an additional 90 deg. In

[1] 3M Company, St. Paul, Minn. 55101.
[2] National Research Council of Canada, Division of Building Research, Ottawa, Ont. Canada.

the tape test for organic coatings, significantly less force is required to bend the overlay since it is thinner and more flexible than the nickel deposit used here.

J. J. Bikerman[3] *(written discussion)*—The dependence of the peel strength on the width of the ribbon, which, by the way, I observed nearly 20 years ago, should be important for those who try to determine the peel strength by pulling at a bent wire. The diameter of the wire certainly is in the range in which this strength is affected by the width; thus, comparison of results obtained with different wires may afford misleading information.

O. J. Klingemaier and S. M. Dobrash (*authors' closure*)—Yes, we would expect the diameter of the wire to influence the results.

H. E. Hintermann[4] *(written discussion)*—The various treatments and cleaning procedures applied to the substrate surface before coating influence the peel strength. Among them, has glow discharge cleaning in inert or reactive gaseous atmospheres been considered (plasma etch)? If so, what were the results?

O. J. Klingemaier and S. M. Dobrash (*authors' closure*)—Glow discharge cleaning in inert atmosphere followed by vapor or ion deposition of a suitable metal has been considered for preparing difficult-to-plate metals for electrodeposition but has not yet been investigated.

[3] Department of Chemical Engineering, Case Western Reserve University, Cleveland, Ohio 44106.

[4] Laboratoire Suisse de Recherches Hologeres, Ch - 2000 Neuchatel, Switzerland.

Summary

Summary

This volume containing 25 papers brings together the state of the art pertaining to adhesion measurement techniques for a variety of thin films, thick films, and coatings.

Mittal, in the introductory paper, has discussed at length some of the problems and controversies involved in adhesion measurement. He has suggested the use of the term "practical adhesion" to denote forces or the work required to disrupt an adhering system (adherend-adherate combination) irrespective of the locus of failure, unless the failure is clearly in the bulk of one of the adhering phases. Failure in bulk constitutes a case of cohesive failure, and is a measure of the cohesive strength of that particular phase. References are made to the recent comprehensive reviews regarding adhesion measurement of thin films, thick films, and bulk coatings; and developments pursuant to these reviews are covered. Some recommendations for future developments in the area of adhesion measurement are highlighted.

The importance of establishing the locus of failure in an adhering system is quite manifest in understanding the failure mechanism as well as to prescribe a proper remedy. Good has described the level of possibility of true interfacial separation based on two theories [weak boundary layer (WBL) theory and theory based on fracture mechanics] and has critiqued both these theories. Good points out that a case sometimes exists which mimics WBL failure in systems where no such material is present. Failure propagates close to, but not at, the phase boundary because of mechanical reasons.

Bikerman has promulgated for a long time that true interfacial separation rarely, if ever, occurs. What is taken as an apparent separation at the interface is a separation in the WBL. In the present paper, he still adheres to his original ideas and cites some recent results to bolster his arguments. In any case, the controversy apropos of the possibility of interfacial separation still persists, and it does not appear that it will be resolved, at least in the near future. Also, Bikerman has criticized some of the so-called adhesion tests for latex paints on the grounds that these simply measure the cohesive strength of the paints.

The paper by Baun describes experimental methods to determine the locus of failure in an adhesive joint as well as in adhering systems. The techniques discussed include ion scattering spectrometry (ISS), secondary ion mass spectrometry (SIMS), scanning electron microscopy (SEM), Auger electron spectroscopy (AES), and electron spectroscopy for chemical analysis (ESCA); their advantages and limitations are discussed.

Mattox has reviewed the various factors which affect thin-film adhesion, and the importance of interfacial regions (in metallic adhering systems) in the measured adhesion of thin films is discussed. He points up that an adhesion testing program must be designed to subject the film-substrate couple to the stresses which it will encounter in production and in service.

Bascom, Becher, Bitner, and Murday discuss the use of fracture mechanics concepts in the testing of film adhesion, and constant-compliance and applied-moment double-cantilever beam tests for adhesive fracture energy have been adapted for the measurement of adhesion of thick-film metallization on alumina substrates. The thick-film tests involve beams soldered to metallization strips and measure the strain energy release rate, ζ_c. The fracture results are compared with peel strengths from solder-wire peel tests of the same metallizations. The constant-compliance and the peel test data exhibit a bimodal distribution. The advantages of the fracture mechanics approach are discussed.

Murr has determined the variation of adhesive energy of metal-ceramic [nickel on thorium dioxide (ThO_2), nickel/chromium (80/20) on ThO_2, and iron/nickel/chromium on aluminum oxide (Al_2O_3)] systems with temperature. What Murr refers to as "adhesive energy" is more commonly known as thermodynamic work of adhesion, W_a. Such determination of W_a should be helpful in predicting the practical adhesion between coatings of these metals on ceramic substrates.

The next seven papers deal with the adhesion of thin films. Crane and Hamermesh have studied the adhesion of thin (1400 Å) plasma polymerized films of styrene and acrylonitrile deposited on a number of plastics. Adhesion of these films was measured by applying a ½ by 2 in. piece of Scotch No. 810 tape to the film and then removing the tape in a 90-deg peel.

Krongelb describes a new technique for measuring practical adhesion of a thin-film structure without requiring any mechanical attachment to the film, and results are presented for evaporated copper film on thermally oxidized silicon. The technique involves application of a known tensile stress to the film; the stress is developed by the interaction of an external magnetic field with an electric current through a suitably patterned specimen of the film to be evaluated. The maximum stress which can be produced is limited by the available magnetic field, the fabrication of the test procedure, and the heating effect of the current through this structure. It is shown that useful stresses of at least 24 000 psi are practical by the electromagnetic method.

Another technique, known as "laser spallation," which does not require any mechanical attachment is described by Vossen. The technique involves impinging a high-energy pulsed laser beam on the back side of the substrate; if the substrate is transparent, then a thick absorbing layer is first deposited on its back side. By increasing the incident laser power, one can find a threshold at which the film is torn loose from the substrate. The film is patterned into dots of about the same diameter as the incident laser beam. If suitable corrections are applied, then it is claimed that the threshold energy can be related directly to the bond strength between the film and the substrate. Spallation thresholds are given for various metal films on fused silica substrates.

The paper by Ahn, Mittal, and MacQueen discusses in the first part the evolution of the scratch technique for measuring practical adhesion of thin films and the difficulties involved in the interpretation of the results. These authors describe the use of a very sensitive surface profilemeter, and a SEM coupled with energy dispersive X-ray spectroscopy to examine scratch topography, modes of film deformation, and the mechanism of film material removal during their studies of scratch hardness and adhesion of multilayer structures using a single point loaded scratch tester. It is found that the scratch behavior is not as simple and general as was originally suggested by Weaver and co-workers.

The next paper, by Oroshnik and Croll, also deals with the scratch technique. They have introduced the concept of Threshold Adhesion Failure (TAF) and the operational definition of TAF as adopted by the authors reads: Threshold Adhesion Failure occurs if, within the boundaries of a scratch and over its 1-cm path, removal of the film from its substrate can be detected by transmitted light with a microscope (magnification $\times$ 40) at even one spot, no matter how small. The TAF criterion differs from the complete removal criterion as was orginally discussed by Weaver and coworkers. The scratch test as used in the TAF mode has been shown to be very sensitive, capable of good reproductibility, and TAF loads are relative values that give some indication of the quality of film adherence. Data are presented for aluminum films on fused quartz.

Faure describes a dynamic (ultrasonic vibration) method for measuring the adhesion of microscopic metallic grains which constitute granular thin films. Data are presented for the adhesion of granular silver films deposited on carbon. Also the adhesion of thick silver films on carbon is discussed.

Kuwahara, Hirota, and Umemoto present their results on the adhesion of thin aluminum and silver films on mild steel and glass, as measured by two techniques. One is to pull down and the other is to twist off a rod whose bottom is cemented to the film with epoxy. The critical force or the torque required to strip off the film from the substrate is taken as a measure of adhesion.

The next six papers deal with the adhesion of thick-film conductors. In the first part of his paper, Hitch briefly reviews the adhesion measurement of thick-film materials and describes the particular requirements of thick-film technology for test methods to measure the adhesion strength of these materials. A number of test methods are treated with reference to their usefulness and practicability. The second part of the paper is devoted to the review of two thick-film adhesion tests and data obtained using these at RCA laboratories. These tests are the thermocompression bonded peel test and the soldered-wire peel test. According to the author, the first test has been used successfully for adhesion strength measurement on gold- and silver-based conductor films.

Morey has evaluated two techniques for adhesion measurement of thick films on alumina. One method is to solder an interconnection onto a thick film pad and measure the strength using a force gage. Another method is to use a wire peel test. The factors affecting adhesion values as obtained by these techniques are discussed.

Ewell has compared the nailhead lead tension and the ribbon lead shear tests for adhesion measurement of thick-film, solderable end terminations on chip resistors, and multilayer monolithic ceramic capacitors. The criteria for test method evaluation are discussed. The nailhead method is compatible with vendor data, and specimens can be easily and quickly prepared.

Leven describes the use of the pull test for measuring the adhesion of conductor-substrate combinations using soldered copper pins to conductor pads. Also, the effects of four environmental stresses on the loss of adhesion are discussed, and it is concluded that the temperature cycling stress appears to be the most practical stress as a screening procedure for solder leaching, which leads to adhesion degradation.

Ingham discusses the considerations involved in testing the bond strength of flame-sprayed coatings and has presented the results (on two different flame sprayed coatings) of round-robin tests for the ASTM Method C 633-69.

Schroter describes a pull-peel at 90-deg test for measuring adhesion of conductors and discusses the factors which affect conductor adhesion.

The next five papers deal with a variety of coatings. Dini and Johnson have described the ring shear, conical-head tension, and I-beam tension tests for quantitative measurement of the adhesion of a number of electrodeposited coating-substrate combinations. The pros and cons of each test are discussed, and adhesion data for various deposit-substrate combinations are included.

Deckert has used resistance to fluoride-containing etchant undercutting as a measure of relative adhesion of photoresist/SiO_2 composites. An empirical equation is given which approximately describes the shape of the undercut oxide edges, thus allowing a numerical measure of the relative adhesion. A variation of this test can be used to detect nonuniform adhesion across a wafer surface.

Masuoka and Nakao describe a general formula for tensile bond strength for the butt joint of internal fracture in terms of a simplified model constructed from measurable parameters. The equations developed indicate that the bond strength depends on both the yield strength and the aspect ratio of the adhesive.

Egorenkov and Belyi discuss the temperature dependence of adhesion, as measured by the peel test, in metal-polymer joints employing temperature gradient in the plane of contact. The strength of adhesion in the case of crystalline polymers is lower than that of amorphous ones, and decreases with increasing test temperature.

The final paper by Klingenmaier and Dobrash describes the use of the Jacquet peel test for determining the adhesion of plated coatings on metallic substrates. The factors affecting peel values are discussed, and a relationship between the tensile strength of the substrate and the possible maximum peel strength is established. Use of the peel test is demonstrated on a production part.

As a group, these papers provide a useful source of information and a guide for further development anent adhesion measurement techniques. Although there are many unanswered questions and moot issues in the realm of adhesion measurement, the brisk activity and interest in this area is quite patent from this collection of papers. The contents of this volume should be of interest to scientists, engineers, technologists, and production and manufacturing personnel whose work involves knowledge of adhesion of thin films, thick films, and coatings. Also, this volume should be utile to both veteran (as a reference source) and novice (as a starting point) in this important area of adhesion measurement.

K. L. Mittal

East Fishkill Facility, IBM Corporation, Hopewell Junction, N.Y. 12533; symposium chairman and editor.

Index

B

C

D

E

F